Energy Storage and Conversion Materials

Inorganic Materials Series

Series editors:
Duncan W. Bruce, *University of York, UK*
Dermot O'Hare, *University of Oxford, UK*
Richard I. Walton, *University of Warwick, UK*

Titles in the Series:

1: Pre-combustion Carbon Dioxide Capture Materials
2: Post-combustion Carbon Dioxide Capture Materials
3: Solar Energy Capture Materials
4: Bioinspired Inorganic Materials: Structure and Function
5: Energy Storage and Conversion Materials

How to obtain future titles on publication:
A standing order plan is available for this series. A standing order will bring delivery of each new volume immediately on publication.

For further information please contact:
Book Sales Department, Royal Society of Chemistry, Thomas Graham House, Science Park, Milton Road, Cambridge, CB4 0WF, UK
Telephone: +44 (0)1223 420066, Fax: +44 (0)1223 420247,
Email: booksales@rsc.org
Visit our website at www.rsc.org/books

Energy Storage and Conversion Materials

Stephen Skinner
Imperial College London, UK
Email: s.skinner@imperial.ac.uk

ROYAL SOCIETY OF CHEMISTRY

Inorganic Materials Series No. 5

Print ISBN: 978-1-78801-090-0
PDF ISBN: 978-1-78801-295-9
EPUB ISBN: 978-1-83916-005-9
Print ISSN: 2472-3819
Electronic ISSN: 2472-3827

A catalogue record for this book is available from the British Library

The Royal Society of Chemistry is a charity, registered in England and Wales, Number 207890, and a company incorporated in England by Royal Charter (Registered No. RC000524), registered office: Burlington House, Piccadilly, London W1J 0BA, UK, Telephone: +44 (0) 20 7437 8656.

For further information see our web site at www.rsc.org

Printed in the United Kingdom by CPI Group (UK) Ltd, Croydon, CR0 4YY, UK

Series Preface

Materials chemistry has grown hugely in the past few decades and continues to evolve rapidly. This is especially the case since the subject now finds itself central to many of the major challenges that face global society, such as those concerning energy and the environment; but applications of materials are also found in many everyday situations, with ongoing developments in technology around electronics and the miniaturisation of devices. Although the study of materials is a highly interdisciplinary topic, the role of innovative, fundamental chemistry continues to lie at the heart of the discovery of new materials, the understanding of their atomic-scale structure and the relationship of structure to a useful property.

With the *Inorganic Materials Series*, our aim is to provide a series of themed volumes, each reflecting the diversity of the topic as it continues to develop over the coming years, with many chosen to relate closely to emerging practical applications. The use of the title *Inorganic Materials* is by no means limiting, but instead emphasises the study of the chemistry of all elements in the Periodic Table and how they impact the development of new materials. The chapters in each book will have a pedagogical flavour with a target audience from first-year postgraduate student upwards, but with wide subject coverage ranging from continuous inorganic solids, to molecular and soft matter systems. We will use expert, active researchers as editors and contributors to provide an up-to-date perspective of their field. At the same time our aim is to provide an international perspective, so to reflect the diversity and interdisciplinarity of the now very broad area of the subject.

Duncan W. Bruce, York, UK
Dermot O'Hare, Oxford, UK
Richard I. Walton, Warwick, UK

Inorganic Materials Series No. 5
Energy Storage and Conversion Materials
Edited by Stephen Skinner

Published by the Royal Society of Chemistry, www.rsc.org

Series Preface

Preface

Increasingly, there is a drive to identify alternatives to fossil fuel combustion processes that will address worldwide growth in energy demand, and so mitigate the effects of climate change. This has led to burgeoning interest in technologies that convert and store energy cleanly. The applications of these technologies are myriad, from large grid scale storage for energy networks through to mobile and portable devices for transport applications. Underpinning each of these technological advances are the development and understanding of fundamental materials chemistry: of the individual bulk components, of the interfaces between components and increasingly in how the materials function in a device. This underpinning knowledge is coupled with the wider energy landscape that considers the distribution of resources, the need for energy generation and storage, its transport and ultimately its cost. There is currently no clear single solution that will address the continuing need to use energy, in mobile devices, in remote locations and in population centres, for both heating and cooling, personal travel, public transport and international freight, as examples. It is clear that a diverse mixture of technologies will provide solutions, and the aim of this book is to provide the reader with an introduction to some of the latest advances in these key technologies.

In this book the focus is "Energy Conversion and Storage Materials" and it is broadly structured to introduce energy conversion technologies before moving on to discuss energy storage. The initial chapters are concerned with the production of energy through a solid oxide fuel cell, converting a gaseous fuel to electricity with high efficiency and at lower temperatures than previously considered feasible. This is followed by a

Inorganic Materials Series No. 5
Energy Storage and Conversion Materials
Edited by Stephen Skinner

Published by the Royal Society of Chemistry, www.rsc.org

chapter on co-electrolysis in which high-temperature ceramic devices are used to convert mixtures of CO_2 and H_2O into useful products such as synthesis gas. Of course, generating energy is only one of the many challenges surrounding the growth in energy demand. The other is the safe and effective storage of that energy, with the most common route being through batteries. The advent of all-solid-state batteries offers a new alternative to flammable liquid electrolytes and Chapter 3 discusses the development of solid state lithium electrolytes. The final two chapters address new approaches to energy storage with thermochemical storage through redox processes and thermoelectric oxides considered. Each of these technologies offer significant advantages over competing conventional power conversion technologies and are likely to have a significant impact either individually, or in hybrid devices.

Stephen Skinner

Contents

Inorganic Materials Series No. 5
Energy Storage and Conversion Materials
Edited by Stephen Skinner

Published by the Royal Society of Chemistry, www.rsc.org

CHAPTER 1

Bilayer Electrolytes for Low Temperature and Intermediate Temperature Solid Oxide Fuel Cells – A Review

ALIREZA PESARAN, ABHISHEK JAISWAL AND ERIC D. WACHSMAN*

Maryland Energy Innovation Institute, University of Maryland, College Park, MD 20742, USA
*E-mail: ewach@umd.edu

1.1 Introduction

Fuel cell technology holds the potential to change the way power is generated, transmitted and utilised in our increasingly energy dependent lifestyles. State-of-the-art solid oxide fuel cells (SOFCs) operate at high temperatures (750–950 °C). Lower operating temperatures would significantly improve the economics of power generation using SOFCs. The aim of this chapter is to evaluate component materials for bilayer electrolyte-based SOFCs that could work efficiently at temperatures below 650 °C.

A fuel cell is an electrochemical device that converts the chemical energy of a fuel and an oxidant into electrical energy. The conversion is direct without the need for intermediate conversion into heat and mechanical energy,

Inorganic Materials Series No. 5
Energy Storage and Conversion Materials
Edited by Stephen Skinner

Published by the Royal Society of Chemistry, www.rsc.org

as in the case of conventional turbine/generator systems and is not limited by the Carnot cycle with up to 50% achievable electrical efficiency.[1] As fuel cells do not involve any combustion process, there is no formation of pollutants such as NO_x, SO_x, hydrocarbons and particulates. Also, fuel cells have limited moving parts (blower) and hence do not generate any noise pollution and require minimal maintenance. Due to these advantages over other energy generators, fuel cells are desired in stationary as well as in mobile applications. Among all of the different types of fuel cells, proton exchange membrane fuel cells (PEMFCs), molten carbonate fuel cells (MCFCs) and SOFCs are considered to be the most advanced and closest to wide-scale commercialisation. However, significant technological and engineering challenges remain. When running on hydrocarbon fuels in addition to external reforming, PEMFCs also require CO removal from the fuel feed as they are susceptible to CO poisoning, which results in lower conversion efficiencies. On the other hand, SOFCs run at high temperatures and ideally can internally reform any hydrocarbon fuel with high efficiencies without the need of expensive catalysts. In addition, the high-quality waste heat from the SOFCs can be utilised in cogeneration to further improve the overall efficiency – up to 85% – of the system.[2]

State-of-the art SOFCs operate at temperatures between 750 and 950 °C using yttria-stabilised zirconia (YSZ) as the electrolyte material, La(Sr)MnO_3-YSZ composite as the cathode material and Ni-YSZ ceramic metal (cermet) composite as the anode material. However, the high operating temperatures of SOFCs put considerable limitations on the choice of materials for the various components and also on the lifetime of the cell. Over the years, research in industry and academia has resulted in the development of materials and fabrication techniques such that the cell components are able to perform well under extreme conditions, to withstand thermal mismatch, to be microstructurally stable and to counter interactions between adjoining components. However, this has led to high material and fabrication costs, making SOFCs less competitive with existing power generation technologies. Decreasing the operating temperatures will considerably improve the economics of power generation using SOFCs by enabling the use of cheap ferritic stainless steel alloys as the interconnect material instead of expensive alloys or ceramics, cheaper balance of plant and insulation along with increased lifetime.[3] Lower operating temperatures will also result in faster start-up that is critical in certain applications. All these factors have led to a global drive towards reducing the operating temperature of SOFCs from 750 to 950 °C to low to intermediate temperatures of 500–750 °C. However, efficient operation at lower temperatures will require new electrolyte materials with higher conductivity and new electrode materials with better catalytic activity at lower temperatures. For the purpose of further discussion, the operating temperature of SOFCs is defined as low temperature (500–650 °C), intermediate temperature (650–800 °C) and high temperature (800–950 °C).

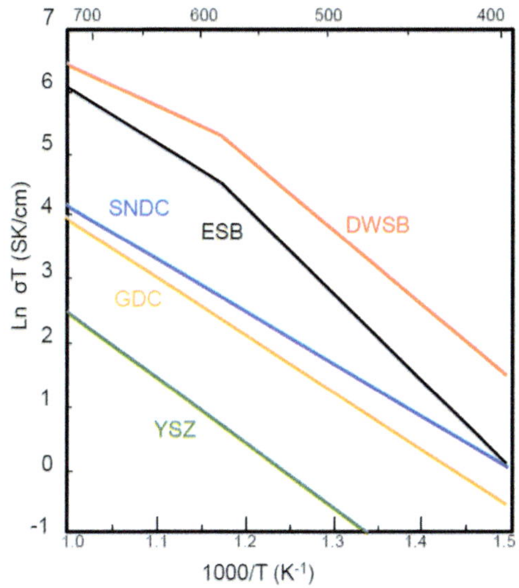

Figure 1.1 Conductivity of oxide-ion conducting solid electrolytes. Reproduced from ref. 3 with permission from AAAS, Copyright 2011.

1.2 Material Review

1.2.1 Electrolyte

Oxide-ion conducting solid electrolytes are the backbone of SOFCs, allowing selective transport of oxide ions for electrochemical oxidation of fuel to generate electrical power. The electrolyte materials for SOFCs are oxides having a fluorite or perovskite structure because they are larger, loose-packed structures that have the ability to accept a wide range of dopants. Oxide-ion conductivity of common solid-oxide electrolytes is shown in Figure 1.1.[3] The ionic conduction is possibly due to the presence of oxide-ion deficiency in the solid electrolyte, which typically is introduced by doping the host electrolyte with lower valent cations. The fluorite type unit cell is shown in Figure 1.2. CeO_2 and ThO_2 exhibit a cubic fluorite structure from room temperature up to their melting points. The open fluorite structure offers the possibility of achieving an unusually wide range of solid solutions with alkaline-earth and rare-earth oxides such as CaO and Gd_2O_3. Generally, when the cation size of the host and the guest are almost equal, the solid solution is easily formed. In the case of ZrO_2 and Bi_2O_3, the high-temperature cubic fluorite structure can be stabilised at lower temperatures by forming a solid solution and hence, the terms stabilised zirconia and stabilised bismuth oxide are used. The electrolyte in the fuel cell is exposed to both the oxidant and the fuel atmospheres, and hence, it should be thermodynamically stable under oxidising and reducing conditions. To

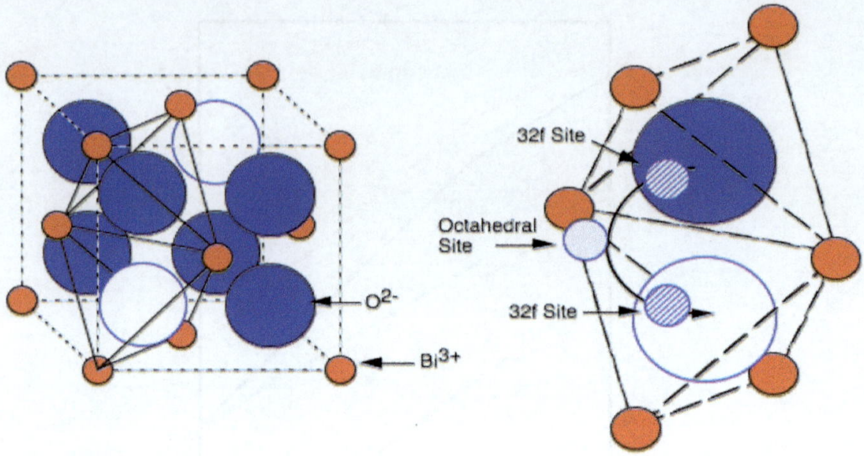

Figure 1.2 Fluorite structure of δ-Bi_2O_3 and an oxide-ion conduction pathway. Reproduced from ref. 91 with permission from Elsevier, Copyright 2001.

avoid the generation of electronic conductivity, the electrolyte should have a large band gap and the dopants introduced into the lattice should not exhibit multiple oxidation states. As the oxygen oxide ion conduction is a thermally activated process, the performance of the electrolyte improves with temperature. However, there are limits in terms of operating temperatures due to material compatibility issues with other components and operational viability over extended periods of time. Reduced operating temperatures will allow the choice of a wider range of materials and significantly improve the economics of power generation using SOFCs. However, operation of SOFCs at intermediate to low temperatures requires better performance electrolyte and electrode materials.

Two approaches have been actively pursued towards reducing the operating temperatures of SOFCs. The first is finding a better electrolyte (with high ionic conductivity and high ionic transference number) and better electrode (with high catalytic activity and high mixed conductivity) materials. At the same temperature, doped ceria and stabilised bismuth oxide shows oxide-ion conductivity that is significantly higher than that of YSZ.[4] However, doped ceria is thermodynamically unstable in the anode atmosphere of the fuel cell due to reduction from Ce^{4+} to Ce^{3+}. The polaronic charge transfer of the mixed-valent Ce^{4+}/Ce^{3+} results in electronic conduction inside the electrolyte, which under fuel cell conditions results in an internal short circuit and lower available power than theoretical.[5] On the other hand, bismuth oxide-based electrolytes show the highest oxide-ion conductivity among all solid electrolytes and are of considerable interest for application in SOFCs, ion transport membranes and oxygen sensors.[6] The high temperature δ-Bi_2O_3 has a cubic fluorite structure with 25% inherently vacant oxygen sites. The high concentration

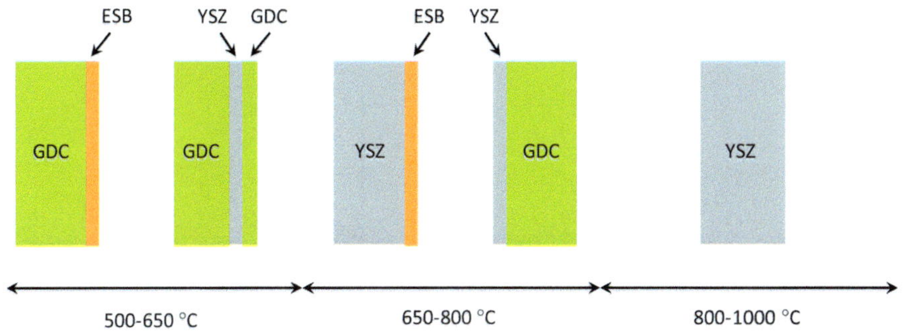

Figure 1.3 Bilayer and trilayer electrolytes for low temperature (LT) and intermediate temperature (IT) SOFCs.

of disordered oxide ion vacancies in δ-Bi_2O_3 along with the high polarisability of Bi^{3+} cation results in oxide-ion conductivity that is one to two orders of magnitude higher than that of YSZ at comparable temperatures. Unfortunately, bismuth oxide is not thermodynamically stable under the reducing atmosphere of SOFCs and decomposes to metallic bismuth.[7] Different combinations of bilayer and trilayer electrolytes for low temperature and intermediate temperature SOFCs have been proposed as shown in Figure 1.3. Yahiro *et al.*[8] presented the concept of a stabilised zirconia/doped ceria bilayer electrolyte for SOFCs where the zirconia layer avoids the reduction of ceria and blocks the partial electronic conduction within the ceria layer. Wachsman *et al.*[9] proposed a doped ceria/stabilised bismuth oxide bilayer electrolyte for low temperature operation. The primary advantage of this bilayer was that instead of using a less conductive YSZ electrolyte to protect the ceria-based electrolyte, a high conductive material (ESB) was used to block the electronic leakage current of ceria (Figure 1.4). At the same time, the ceria layer increases the oxygen partial pressure at the interface providing stability to the bismuth oxide layer. An important consideration in the design of the bilayer electrolyte is the choice of electrolyte layer facing the reducing anode atmosphere. For further discussion, the naming convention used for the anode/electrolyte-1/electrolyte-2/cathode cell was "electrolyte-1/electrolyte-2" in which electrolyte-1 faces the reducing atmosphere.

The second approach towards reducing the operating temperatures of SOFCs is by reducing the thickness of the electrolyte and attaining higher power densities at lower temperatures with electrode-supported electrolyte unit cells. De Souza *et al.*[10,11] demonstrated an anode-supported SOFC by co-firing the anode and electrolyte and subsequent insertion of the cathode and showed a significant improvement in performance. For the processing of thick films on porous supports, the colloidal route has advantages over other film deposition techniques such as vapour deposition (CVD/EVD) and chemical routes (sol–gel) in its simplicity, cost effectiveness, flexibility (thickness

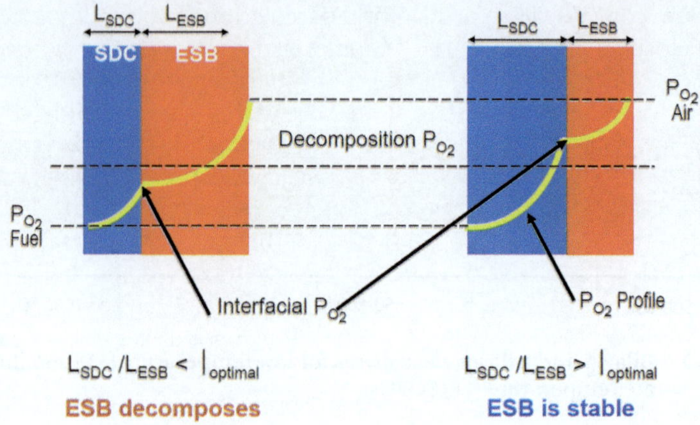

Figure 1.4 Schematic of the effect of the relative thickness of the SDC/ESB bilayer on interfacial oxygen partial pressure (p_{O2}) and ESB stability. Reproduced from ref. 9 with permission from the Electrochemical Society, Copyright 1997.

ranging from 10 to 100 μm can be deposited) and scalability.[12] Following the advent of high-tech lithography and deposition devices, sub-micron and free-standing electrolyte micro SOFCs have also recently gained significant interest.[13–19]

1.2.2 Electrodes

Electrochemical reactions at the cathode and anode are thermally activated and operation at low temperatures result in reduced cell performance. The oxygen reduction reaction at the cathode has a larger activation energy than fuel oxidation at the anode and thus new cathode materials are required. Candidate materials for cathodes should be mixed-ionic electronic conductors (MIEC) with high catalytic activity for oxygen molecule dissociation and oxygen reduction. The material should be compatible with other cell components with respect to chemical reaction and thermal expansion coefficient (TEC). Oxides with a perovskite structure (ABO_3) have been the material of choice for the cathodes; where A is a rare-earth element and B is a transition metal (Mn, Co, Fe). On doping the A site with alkaline-earth cations (Sr^{2+}, Ca^{2+}), the charge compensation occurs by valence change of the transition metal cations and under certain conditions by oxide vacancy formation resulting in an MIEC.[20] La(Sr)MnO_3 (LSM)-YSZ composites have been the material of choice for YSZ-based SOFCs. The oxygen reduction happens at the triple phase boundaries (TPBs) between the electrocatalyst, the electrolyte and the gas phase. Furthermore, the TPBs are active only if electrons, oxide ions and oxygen gas can transport to or away from the TPBs. Hence, the performance of the composite cathode depends critically on the relative ratio, particle size, and spatial distribution of the two solid phases, so as to achieve high concentration of TPBs and percolation for both phases. In addition, the electrode microstructure should also be porous to provide for gas diffusion. It is often found that the high polarisation losses are primarily associated with inadequate ionic transport within the electrode structure. Hence, at lower temperatures LSM (a poor ionic conductor) is replaced by $La(Sr)Co(Fe)O_{3-\delta}$ (LSCF) or $Sr(Sm)CoO_3$ (SSC), which have considerable ionic conductivities.[21,22] YSZ is replaced by doped ceria, with higher oxide ion conductivity, as the electrolyte phase to develop better performance LSCF-GDC cathodes.[23,24]

LSCF has been extensively studied as a candidate cathode material for SOFCs based on ceria electrolytes.[25,26] LSCF is a good electronic conductor (*e.g.* $La_{0.6}Sr_{0.4}Co_{0.2}Fe_{0.8}O_3$ – 300 S cm^{-1} at 750 °C) and also shows fast oxygen surface exchange with higher oxide-ion conductivity than that of LSM (*e.g.* 10^{-3} S cm^{-1} at 750 °C).[24] LSCF also appears to be chemically stable with ceria electrolytes as the pyrochlore compound $La_2Ce_2O_7$ does not exist.[27]

Among this family, the Fe-rich compositions are more attractive than Co-rich compositions as they have a lower thermal expansion coefficient and hence, are better matched with ceria electrolytes.[25] $La_{0.6}Sr_{0.4}Co_{0.2}Fe_{0.8}O_3$-$Ce_{1-x}Gd_xO_{2-\delta}$ composite cathodes have been studied by Dusastre *et al.*[23] and Murray *et al.*[24] and were found to be very effective in improving the performance over single-phase $La_{0.6}Sr_{0.4}Co_{0.2}Fe_{0.8}O_3$ cathodes. Optimum composition was found to be between 30–50 wt% $Ce_{0.9}Gd_{0.1}O_{2-\delta}$ with area specific resistance (ASR) values of 0.33–0.6 Ω cm^2 at 600 °C. Recently, Çelikbilek *et al.*[28] studied the effect of sintering temperature on the microstructure and performance of LSCF-GDC composite cathodes. By using electrostatic spray deposition and synthesis of a nano-structure and lowering the sintering temperature to 800 °C they achieved an ASR of 0.13 Ω cm^2 at 600 °C.[28] A $Ba_{0.5}Sr_{0.5}Co_{0.8}Fe_{0.2}O_3$ (BSCF) based cathode was reported to exhibit low ASR due to its high catalytic activity and high oxygen diffusivity but BSCF shows fast degradation rates due to its low stability. In comparison, $Sm_{0.6}Sr_{0.4}CoO_3$ (SSC) showed reasonably high stability and catalytic activity. Ag-$(Bi_2O_3)_{0.7}$ $_5(Er_{25}O_3)_{0.2}$ and Ag-BIMEVOX cermet cathodes have been also reported for low temperature applications.[29,30] Recently, pyrochlores based on bismuth ruthenate, lead ruthenate and yttrium ruthenate have been studied for application as cathodes in SOFCs.[31–33] Pyrochlore ruthenates are electrically conductive and ruthenium oxide is known to be catalytically active towards oxygen reduction and it can be expected that solid solutions containing ruthenium oxide will be beneficial as cathodes.

The anode material should be chemically compatible with the electrolyte material under reducing atmosphere and elevated temperatures. TPB, porosity, particle size and grain size distribution impact the performance of the anode significantly and must be tuned carefully.[34] Ni-YSZ-based cermet composites have been adopted as an anode material by most of the SOFC groups for high temperature application. Using Ni as the metal component satisfies the major requirements for the electrochemical oxidation of H_2 and CO fuels. The addition of a ceramic component makes the TEC of the composite comparable to that of the electrolyte.[21] However, with reduced operating temperatures, the surface activity of nickel for electrochemical oxidation, as well as for coking and sulfur tolerance, reduces. In this respect, Ni-ceria cermet provides an advantage due to high activity compared to Ni-zirconia for anode reactions. Doped ceria is a MIEC under the reducing conditions at the anode, and the reaction zone is not restricted to the electrocatalyst/electrolyte interface, which results in enhanced performance. Chemical composition and microstructure play important roles in the performance of the anodes as shown by Ohara *et al.*[35] They studied the performance of Ni-samaria doped ceria cermet anodes as a function of Ni content and found that a cermet with a Ni content of around 50 vol% showed the lowest anodic polarisation (~30 mV at 300 mA cm^{-2}, 800 °C).

1.3 Historical Evolution of Bilayer Electrolyte SOFCs

The earliest demonstration of a bilayer electrolyte SOFC dates back to 1988 when Yahiro *et al.*[8] showed the potential of using an electron blocking layer (zirconia) at the anode side of a ceria-based electrolyte to enhance the open circuit voltage (OCV) of the cell. Although the primary focus of their work was to evaluate the conductivity of various ceria-based electrolytes and the effect of cathode on the cell performance, they found that coating the ceria electrolyte on the anode side with a thin layer of zirconia deposited by RF-sputtering significantly improved OCV as shown in Figure 1.5, resulting in 33% higher power density. Following that, many researchers implemented the same approach to empower the concept of replacing the less conductive zirconia-based electrolyte with the more conductive ceria-based electrolytes and lowering the operating temperature of the cell. The initial structural design of SOFCs was constrained to electrolyte-supported cells. Such

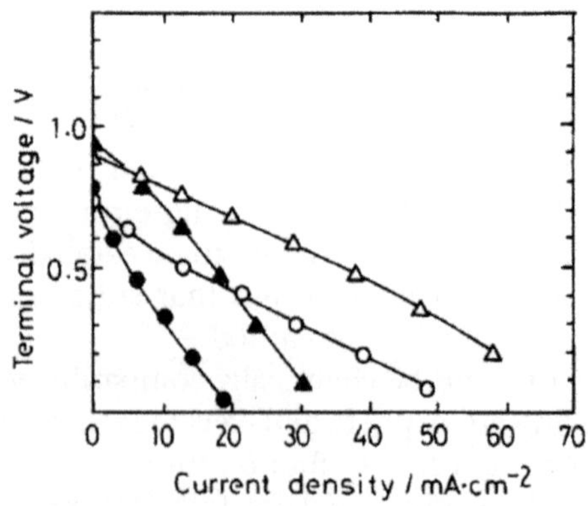

Figure 1.5 *I–V* characteristics of a monolayer GDC cell at 600 °C (closed circles) and 700 °C (open circles) and a YSZ/GDC bilayer electrolyte cell at 600 °C (closed triangles) and 700 °C (open triangles). Reproduced from ref. 8 with permission from The Electrochemical Society, Copyright 1988.

cells suffered from large ohmic losses across the electrolyte layer and thus did not show reasonable power density output. With the advent of sophisticated thick and thin film fabrication methods such as tape casting and pulsed laser deposition (PLD), respectively, the thickness of the electrolyte was significantly reduced, which resulted in much smaller ohmic losses and higher power density output. In this work electrolyte-supported SOFCs will be reviewed briefly. However, electrode-supported SOFCs will be discussed in more detail.

1.3.1 Electrolyte-supported SOFCs

1.3.1.1 Zirconia/Ceria Bilayer IT-SOFCs

The results of Yahiro *et al.*[8] spurred several research activities based on zirconia/ceria bilayer electrolytes for application in intermediate temperature solid oxide fuel cells (IT-SOFCs). There were two key considerations for the development work: firstly, in the design of the zirconia/ceria bilayer, whether the ceria layer should be on the anode side or on the cathode side and the minimum thicknesses of the zirconia layer to block electronic conduction in the ceria layer. Ceria on the cathode side allows the possibility of using cobalt containing cathodes (LSC, LSCF, SSC, BSCF), with higher performance at lower temperatures compared to LSM, which tend to react with zirconia. Secondly, fabrication of the bilayer to match the shrinkage rates of zirconia and ceria layers and to avoid formation of resistive secondary phases at the interface during sintering.

Lim and Virkar[36] measured the chemical potential of oxygen in YSZ/GDC bilayer electrolyte SOFCs and found that the position of the YSZ layer, whether on the anode side or the cathode side, affected the chemical potential profile and the cell performance. The thickness of the GDC and YSZ electrolyte was 45 and 5 μm, respectively, and the electric potential inside the GDC electrolyte was measured using Pt-embedded electrodes. LSM-YSZ and LSC-GDC were used as the cathode for the bilayer cell with GDC at the anode side and the cathode side, respectively. The embedded probe measurements indicated that with GDC at the cathode, electric potential inside GDC was a function of position and load; oxygen partial pressure at the GDC/cathode interface showed a small change (Figure 1.6). However, with GDC at the anode, electric potential inside GDC was independent of position and load and a sharp change in oxygen partial pressure was observed at the YSZ/cathode interface. In terms of performance, the cell with GDC at the anode showed >3 times higher maximum power density (MPD) than the cell with GDC at the cathode (0.52 *vs.* 0.14 W cm^{-2} at 650 °C). However, a majority of the work that followed focused on zirconia/ceria bilayer electrolytes with ceria on the cathode side as it allowed the use of cobalt containing cathodes, which improve performance at lower temperatures. In another study, Kwon *et al.*[37] suggested that a YSZ/GDC thickness ratio of about 2×10^{-7} is sufficient to block the electronic leakage current in the

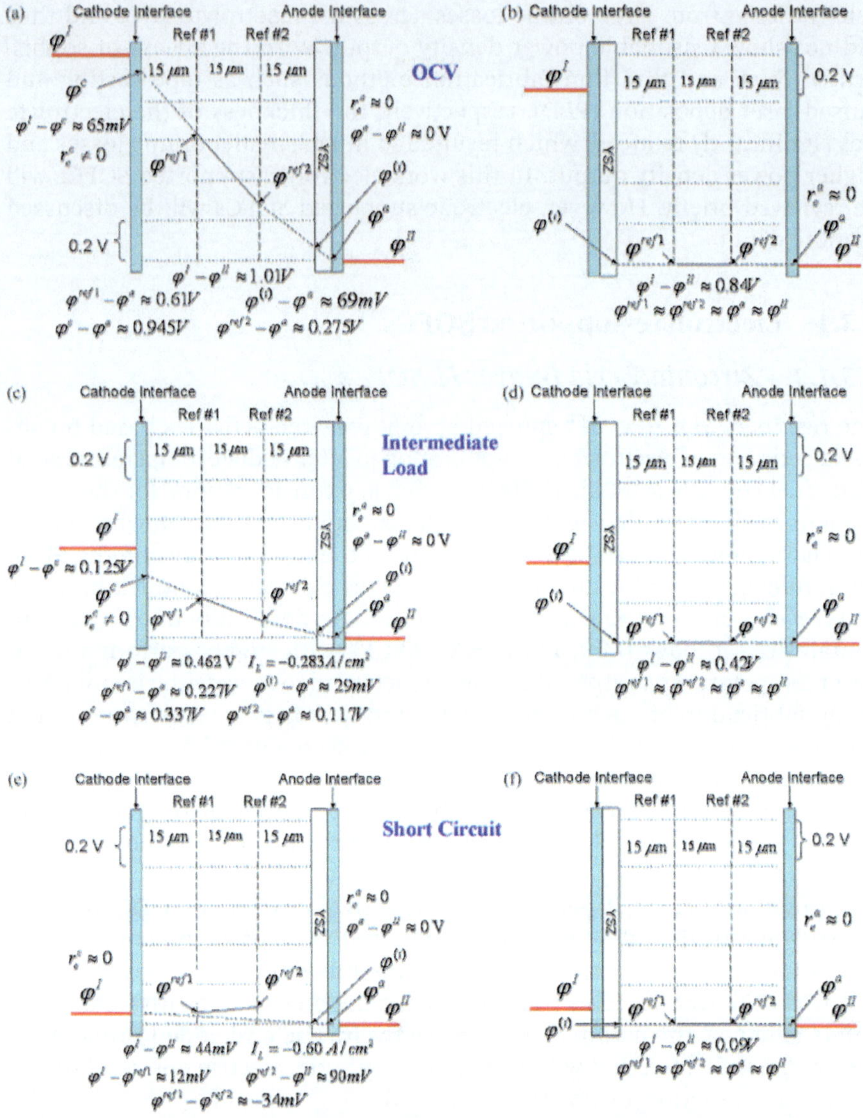

Figure 1.6 Plots of measured electrochemical potential (φ) in a Ni-YSZ/YSZ/GDC/LSC-GDC cell under open circuit (a), under intermediate load (c), and under short circuit conditions (e). Plots of measured electrochemical potential (φ) in a Ni-GDC/GDC/YSZ/LSM-YSZ cell under open circuit (b), under intermediate load (d), and under short circuit conditions (f). Reproduced from ref. 36 with permission from Elsevier, Copyright 2009.

GDC electrolyte. However, fabrication limitation prevents the development of bilayers with such an ultra-thin YSZ layer. Cell configuration, processing and the performance of zirconia/ceria bilayer-based IT-SOFCs are summarised in Table 1.1.

Table 1.1 Cell configuration, processing and performance of zirconia/ceria bilayer-based IT and HT-SOFCs.[a]

Bilayer electrolyte	Anode/Cathode	Processing[b]	OCV, MPD 550 °C	OCV, MPD 650 °C	OCV, MPD 800 °C	Reference
Electrolyte supported						
YSZ (2 μm)/YDC (1.6 mm)	Pt/Pt	SP/SLC/SLC/SP			0.85 V, 0.12 W cm^{-2}	38
ScSZ (8 μm)/GDC (285 μm)	Ni-GDC/LSCF-GDC	SLC/SLC/EX/SLC		1.06 V, 0.16 W cm^{-2}	0.98 V, 0.42 W cm^{-2}	40
Electrode supported						
YSZ (3 μm)/GDC (7 μm)	Ni-YSZ/LSCF-GDC	PP/SPC/SPC/SP			1.05 V, 0.68 W cm^{-2}	53
YSZ (8 μm)/SDC (12 μm)	Ni-YSZ/BSCF	PP/SP/SP/SP		1.10 V	1.06 V	54
ScSZ (2 μm)/SDC (4 μm)	Ni-SSZ/SSC	TC-SP/SP/SP/SP	1.04 V, 0.13 W cm^{-2}	1.03 V, 0.55 W cm^{-2}	0.99 V, 1.80 W cm^{-2}	55
ScSZ (6–7 μm)/SDC (6–7 μm)	Ni-SDC/SSC	TC-SP/PLD/PLD/SP	1.05 V, 0.51 W cm^{-2}	1.04 V, 1.38 W cm^{-2}		56
YSZ (4 μm)/GDC (1.5 μm)	Ni-YSZ/LSCF-GDC	TC/MS/MS/SP			1.12 V, 1.25 W cm^{-2}	57
YSZ (0.7 μm)/SDC (5 μm)	Ni-YSZ/LSCF	TC/PLD/PLD/SP		0.95 V, 1.37 W cm^{-2}		58
YSZ (2 μm)/SDC (6 μm)	Ni-YSZ/SSC-SDC	PP-SLC/PLD/PLD/SP	1.03 V, 0.42 W cm^{-2}	1.1 V, 0.90 W cm^{-2}		59
YSZ (0–0.2 μm)/GDC (1 μm)	Ni-YSZ/LSC	PLD-TC/PLD/PLD/PLD		1.08 V, 1.13 W cm^{-2} (600 °C)		13
YSZ (0.1 μm)/GDC (0.1 μm)	Ni-YSZ/LSCF-GDC	PP-SP/SLC/CSD/CSD	1.01 V, 0.71 W cm^{-2}	1.00 V, 1.35 W cm^{-2}		14
ScSZ (2 μm)/SDC (20 μm)	Ni-SDC/SSC-SDC	SLC/PLD/PLD/SP	1.00 V, 0.14 W cm^{-2}			63
GDC (4 μm)/ScCeSZ (7 μm)/GDC (4 μm)	Ni-GDC/LSCF	SP/TC/TC/TC/SP	1.20 V	1.15 V, 0.19 W cm^{-2}		42

[a]PP – Powder pressing, SPC – Spray coating, SLC – Slurry-based coating, SP – Screen printing (or similar), TC – Tape casting, DCS – DC sputtering, EX – Extrusion, PLD – Pulsed laser deposition, MS – Magnetron sputtering, CSD – Chemical solution deposition.
[b]Anode/Electrolyte-1/Electrolyte-2/Cathode.

An electrolyte-supported YSZ/YDC bilayer electrolyte cell was developed by Kim *et al.*[38] using a thick 1.6 mm YDC pellet made by the sintering of co-precipitated powders at 1600 °C. The YSZ layer was deposited by spin coating and then was sintered at 1400 °C. Compared to a monolayer YDC cell, the YSZ/YDC bilayer cell increased the OCV slightly by 0.05 V between 700 and 1050 °C. However, the MPD of the YSZ/YDC bilayer cell at 800 °C (0.12 W cm^{-2}) was comparable to that of a monolayer YSZ cell at 1000 °C (0.14 W cm^{-2}) due to the higher conductivity of YDC. Based on the model developed by Virkar,[39] p_{O2} at the YSZ/YDC interface was estimated to be two orders of magnitude higher than the stability limit of ceria at 800 °C.

Hsieh *et al.*[40] fabricated and tested electrolyte-supported micro-tubular SOFCs based on a ScSZ/GDC bilayer electrolyte. GDC micro-tubes were extruded and pre-sintered at 1100 °C. A NiO-GDC anode (60 : 40) and LSCF-GDC cathode (80 : 20) were deposited on the inner surface and outer surface of the micro-tubes and sintered at 1400 °C and 1100 °C, respectively. For the bilayer cell, ScSZ was deposited prior to the anode deposition and sintered at 1400 °C. The thickness of the sintered ScSZ and GDC was 8 and 285 μm, respectively. The ScSZ layer in the bilayer cell significantly improved the OCV from 0.92 to 1.06 V at 650 °C by blocking the electronic conduction in GDC, but decreased MPD from 0.20 to 0.16 W cm^{-2} due to high interfacial resistance across ScSZ/GDC and anode/ScSZ interfaces. At 800 °C, the bilayer cell showed a better performance compared to the monolayer GDC cell and the MPD of the cell increased from 0.36 W cm^{-2} to 0.42 W cm^{-2}.

Electrolyte-supported YSZ/GDC bilayer electrolyte SOFCs with an ultra-thin (28–150 nm) YSZ film was fabricated by Jee *et al.*[41] using atomic layer deposition (ALD). A thick 0.43 mm GDC electrolyte support was fabricated by pressing and sintering, and ultra-thin YSZ film was deposited by ALD at 230 °C. The NiO-GDC anode and LSCF-GDC cathode were sintered at 1200 °C and 1400 °C, respectively. High sintering temperatures resulted in significant inter-diffusion of cerium into the YSZ layer as observed by XPS. Furthermore, morphological changes in the YSZ thin film were observed – YSZ grains coalesced at the GDC grain boundaries, uncovering the underlying GDC layer. In an electrochemical test with H_2 as fuel, OCV was found to be stable only for YSZ/GDC bilayer SOFCs with a 150 nm YSZ protection layer, suggesting that an ultra-thin YSZ protective layer may not be practically feasible.

An electrolyte-supported GDC/YSZ/GDC trilayer electrolyte-based SOFC was developed by Timurkutluk *et al.* to suppress the leakage current between the anode and cathode.[42] A thin layer of doped zirconia (YSZ, ScSZ, ScCeSZ) was laminated between two thick GDC layers made by tape casting. A NiO-GDC (60 : 40) anode and LSCF-GDC (50 : 50) were then screen printed. A doped zirconia layer improved the OCV of the trilayer cells at 550–650 °C but the MPD was found to be lower than that of the monolayer GDC cell due to higher resistance. At 650 °C, the MPD of the monolayer GDC cell was ~0.25 W cm^{-2}, while the trilayer cell with ScCeSZ showed ~25% lower MPD (0.19 W cm^{-2}).

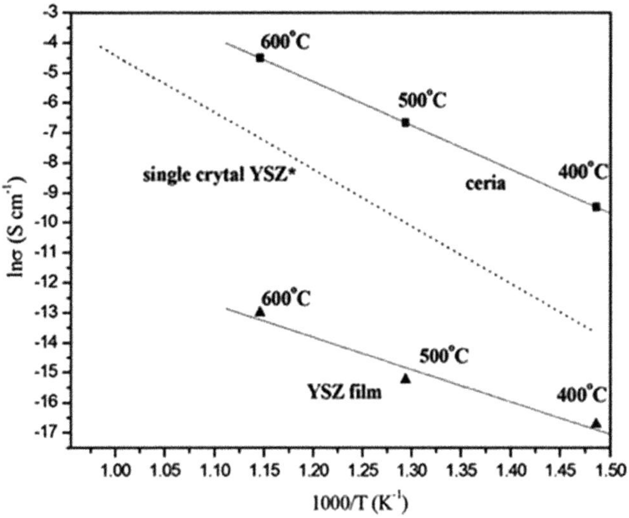

Figure 1.7 Conductivity of SDC, YSZ/SDC bilayer (YSZ film) and single crystal YSZ samples. Reproduced from ref. 43 with permission from Elsevier, Copyright 2011.

In another study, Wu *et al.*[43] measured the conductivity of a YSZ/SDC bilayer consisting of a thin 250 nm YSZ film deposited on a SDC substrate by electron-beam evaporation at 200 °C. YSZ/SDC bilayer samples were sintered at 1000 °C and showed lower conductivity than single crystal YSZ, as presented in Figure 1.7. TEM-EDS analysis indicated the formation of a solid solution layer at the YSZ/SDC interface due to interdiffusion of cerium and zirconium resulting in lower ionic conductivity (Figure 1.8).

1.3.1.2 Ceria/Bismuth Oxide Bilayer IT-SOFCs

Ceria/bismuth oxide bilayer electrolyte-based SOFCs were first proposed by Wachsman *et al.* to combine two low-temperature high-conductivity electrolytes, which cannot operate efficiently as a single electrolyte in a SOFC environment.[9] With ceria on the anode side and bismuth oxide on the cathode side, a functional bilayer electrolyte-based SOFC was demonstrated. It was suggested that the partial electronic conduction in the ceria layer could be blocked by the bismuth oxide layer, while the ceria layer increased the oxygen partial pressure at the interface above the decomposition potential of bismuth oxide. Practical devices require the development of cathodes compatible with stabilised bismuth oxide as platinum and the conventional LSCF cathode for IT-SOFCs show significant reactivity with bismuth oxide. This resulted in the development of new alternative cathodes such as bismuth ruthenate-ESB[44,45] and LSM-ESB[46] composites for low temperature operation. The initial concept demonstration of ceria/bismuth oxide bilayer cells was done with electrolyte-supported cells using noble metal (gold,

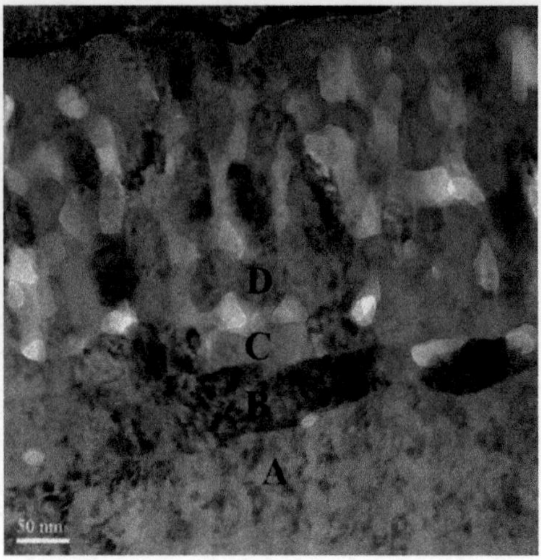

Figure 1.8 Bright field transmission electron microscopy image of a YSZ/SDC sample for EDS elemental analysis. Reproduced from ref. 43 with permission from Elsevier, Copyright 2011.

silver) electrodes, which was then followed with the development of anode-supported cells with thick/thin film bilayer electrolytes showing high performance at low temperatures. Low temperature operation of SOFCs allows the possibility to use stable nanostructured electrodes (without undue sintering during operation) with higher surface-area-to-volume ratio and higher catalytic activity. By shifting the effective particle diameter of the catalytic phase from the micro (10^{-6}) to the nano (10^{-9}) regime, TPB density [$(10^{-6}/10^{-9})^3 = 10^9$] is dramatically increased and can be used to compensate for the exponentially decreasing area-specific electrode reaction rates (activation polarisation) with decreasing temperature.[47] Some of these advances have been employed to improve the performance of bilayer electrolyte SOFCs. Recent work on ceria/bismuth oxide bilayers has focused on utilising new bismuth oxide compositions such as double-doped cubic bismuth oxides (DWSB, DGSB) with high oxide ion conductivity. Cell configuration, processing and performance of ceria/bismuth oxide bilayer-based LT-SOFCs are summarised in Table 1.2.

Wachsman *et al.*[9,48] used different stabilised bismuth oxides (5–60 μm[48]), including $(Y_2O_3)_{0.2}(Bi_2O_3)_{0.8}$ and $(Er_2O_3)_{0.2}(Bi_2O_3)_{0.8}$, to block the electronic transport across thick self-supporting pellets (~0.8–1 mm) of doped ceria – $(CeO_2)_{0.9}(MO_{1.5})_{0.1}$ where M is Y, Sm and Gd. The samples were coated with Au and Pt paste on the cathode and anode side, respectively, and tested with humidified H_2 and O_2 as fuel and oxidant, respectively. The results of the experiments indicated that all bilayer cells have higher OCV compared to the cell with the single layer ceria cell of the same composition. The higher OCV of the bilayer cells was suggested to be due to the electron blocking capability of the bismuth oxide layer, which was stable due to

Table 1.2 Cell configuration, processing and performance of ceria/bismuth oxide bilayer-based LT and IT-SOFCs.[a]

Bilayer electrolyte	Anode/Cathode	Processing[b]	OCV, MPD 450 °C	OCV, MPD 550 °C	OCV, MPD 650 °C	Reference
Electrolyte supported						
SDC (800 μm)/ESB (5–60 μm)	Pt/Au	SP/PP/SP/SP		0.92 V	0.87 V	9
SDC (1 mm)/ESB (30 μm)	Pt/Au	SP/PP/SLC/SP		0.98 V	0.96 V	50
Electrode supported						
GDC (10 μm)/ESB (4 μm)	Ni-GDC/BRO-ESB	TC/SPC/PLD/SP			0.77 V, 1.95 W cm^{-2}	81
GDC (10 μm)/ESB (5 μm)	Ni-GDC/BRO-ESB	TC/SPC/SLC/SP		0.85 V	0.78 V, 1.47 W cm^{-2}	82
GDC (25 μm)/ESB (2 μm)	Ni-GDC/LSM-ESB	TC/SPC/SLC/SP	0.87 V, 0.09 W cm^{-2}	0.80 V, 0.34 W cm^{-2}	0.73 V, 1.01 W cm^{-2}	83
SDC (26 μm)/YSB (6 μm)	Ni-SDC/LSM-YSB	PP/PP/DCS/SP		0.86 V, 0.15 W cm^{-2}	0.85 V, 0.38 W cm^{-2}	84
GDC (20 μm)/YDB (10 μm)	Ni-GDC/LSM-YDB	PP/SLC/SP/SP		1.00 V, 0.28 W cm^{-2}	0.89 V, 0.53 W cm^{-2}	85
SNDC (20 μm)/ESB (20 μm)	Ni-SNDC/LBSM-ESB	PP/PP/SLC/SP	0.97 V, 0.13 W cm^{-2}	0.90 V, 0.52 W cm^{-2}	0.80 V, 0.98 W cm^{-2}	86
SNDC (22 μm)/ESB (20 μm)	Ni-SNDC/PBM-ESB	PP/PP/SLC/SP	0.97 V, 0.10 W cm^{-2}	0.91 V, 0.48 W cm^{-2}	0.80 V, 0.99 W cm^{-2}	90
GDC (20 μm)/DGSB (3–5 μm)	Ni-GDC/LSM-ESB	TC/TC/SP/SP			0.80 V, 1.31 W cm^{-2}	97
GDC (23–70 μm)/xSB (0–25 μm)	Ni-GDC/LSM-ESB	TC/TC/SP/SP		0.84 V, 0.31 W cm^{-2}	0.85 V, 1.06 W cm^{-2}	96

[a]PP – Powder pressing, SPC – Spray/Spin coating, SLC – Slurry-based coating, SP – Screen printing (or similar), TC – Tape casting, DCS – DC sputtering.
[b]Anode/Electrolyte-1/Electrolyte-2/Cathode.

high interfacial p_{O2} provided by the ceria layer. SDC cells (both monolayer and bilayer electrolyte) showed higher OCV compared to YDC cells at all temperatures, due to the higher transference number (t_i) of SDC compared to YDC. For the SDC/ESB bilayer cell, increasing the density of the ESB layer increased the OCV and at 500 °C, the OCV was higher by 90–160 mV depending on the temperature compared to the monolayer SDC cell.[48] The effect of ESB thickness on the performance of the bilayer electrolyte cells was studied using GDC electrolyte and it was found that thick ESB resulted in lower OCV compared to thin ESB, possibly due to the lower interfacial p_{O2} in the case of a thick ESB cell (Figure 1.9). The SDC/ESB cell was operated under an open circuit condition for 1400 h and was found to be stable. Moreover, it was found that the addition of ESB not only increased the OCV but also decreased the cell area specific resistance (ASR), resulting in 33% higher power density for the SDC/ESB bilayer cell.[48] This result favourably compared a ceria/bismuth oxide bilayer to a zirconia/ceria bilayer in which resistive secondary phases form leading to higher resistance and lower performance.[43]

Following the initial promising results with ceria/bismuth oxide bilayer cells, Park *et al.*[49] fabricated and studied the electrical conductivity of a SDC/ESB bilayer electrolyte. SDC powder was produced by conventional solid-state reaction and pressed into ~2 mm thick pellets. The ESB layer was made by two methods: (1) PLD for thin film, and (2) dip coating for thick film. For a PLD target, ESB powder was made by the solid-state route, while for dip coating, ESB powder was fabricated by a citrate process. The ESB powder made by the citrate process required lower calcination temperature (650 °C) compared to the solid-state route (800 °C) and enabled

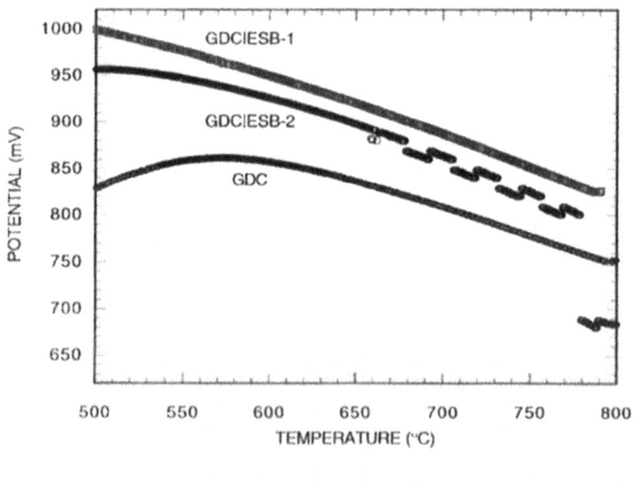

Figure 1.9 OCV as a function of temperature for GDC and GDC/ESB electrolytes. The thickness of the ESB layer for GDC/ESB-2 is six times that of GDC/ESB-1. Reproduced from ref. 9 with permission from the Electrochemical Society, Copyright 1997.

densification of the dip-coated film. The thickness of the ESB layer was controlled by the deposition time and solid loading of the slurry for PLD and dip-coating method, respectively. The XRD pattern of the SDC/ESB bilayer showed only peaks from the SDC and ESB phases and no other secondary phase was identified. An absence of reaction between SDC and ESB was also confirmed through EDX scans across the bilayer interface. Conductivity measurements showed that the SDC/ESB bilayer electrolyte has a slightly higher conductivity compared to the SDC monolayer electrolyte, which was likely due to the formation of a more conductive solid solution phase at the SDC-ESB interface. Further analysis as shown in Figure 1.10 indicates that the grain boundary conductivity of the SDC/ESB bilayer electrolyte was higher than the SDC monolayer electrolyte, which was attributed to the scavenging effect of bismuth oxide for silica-rich impurities at the grain boundaries in SDC resulting in a lower activation energy for oxygen transport through the bilayer electrolyte.

In another study, Park and Wachsman[50] looked at the effect of the relative thickness of bilayer electrolyte layers and anode oxygen partial pressure (p_{O2}) on the OCV and t_i of SDC/ESB bilayer SOFCs at 600–800 °C. The thin ESB layer (1.5 μm) was made by PLD and the thick ESB (9–30 μm) layer was made by dip coating in ESB suspension. At high anode p_{O2} levels, the t_i of the SDC monolayer electrolyte and ESB/SDC bilayer electrolyte was almost identical. However, as the anode p_{O2} was decreased, the bilayer electrolyte showed a higher t_i compared to the SDC monolayer electrolyte and the increase in t_i

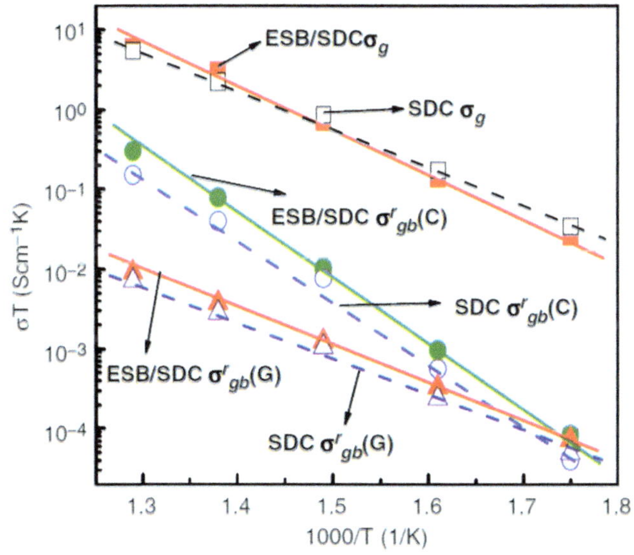

Figure 1.10 Grain conductivities of ESB/SDC and SDC and grain boundary conductivities of ESB/SDC and SDC by the geometric method (G) and capacitance method (C). Reproduced from ref. 49 with permission from John Wiley and Sons, © The American Ceramic Society.

with the bilayer electrolyte was more pronounced at higher temperatures. The effect of relative thickness of the two layers was also examined and it was found that t_i and OCV increased with an increase in the relative thickness of ESB/SDC, showing a sigmoidal (s-shaped) relationship. The t_i of the bilayer electrolyte reached ~0.9 at ESB/SDC relative thickness of 0.03. The performance of the Au and Ag-ESB cathode was also studied and it was found that the Ag-ESB cathode has a lower overpotential resulting in a higher OCV. In another study,[51] Park *et al.* looked at the application of SDC/ESB bilayer electrolytes for ceramic oxygen generators.

1.3.2 Electrode-supported Bilayer Cells

1.3.2.1 Zirconia/Ceria Bilayer IT-SOFCs

1.3.2.1.1 Bilayer cells With Zirconia at the Anode Side. Results with electrolyte-supported zirconia/ceria bilayer electrolyte cells were not promising in terms of power density, which led researchers to evaluate anode-supported configurations using both thick and thin film processing for the fabrication of zirconia and ceria layers. Some studies with cathode-supported and metal-supported configurations have also been reported but the performance was found to be lower compared to anode-supported cells.

To understand the effect of a zirconia/ceria bilayer on the ionic conductivity of the electrolyte, Brahim *et al.*[52] measured the conductivity of the YSZ/GDC bilayer deposited by DC magnetron sputtering on a LSM cathode. Five bilayer samples with different thickness ratios of YSZ and GDC layers were studied. The results of impedance spectroscopy studies showed that the bilayer electrolyte resistance increased with an increase in thickness of both the YSZ and GDC layers and, as expected, the effect of the YSZ thickness on bilayer resistance was more pronounced. Overall, the effect of the thickness ratio was found to be more important than the electrolyte thickness and the bilayer sample with a YSZ/GDC thickness ratio of 1 : 2.2 showed lower conductivity than all the other samples with ratio of 1 : 5 or lower but with higher thicknesses.

Liu *et al.* fabricated Ni-YSZ anode-supported YSZ/GDC bilayer SOFCs with a LSCF-GDC composite cathode. A Ni-YSZ cermet anode was made by powder pressing and pre-sintering at 900 °C. Spray coating was used to deposit YSZ and GDC layers of 3 and 7 μm thickness, respectively. The OCV of the bilayer cell was 1.05 V at 800 °C, which was significantly higher than that of the 10 μm monolayer GDC cell (0.59 V), as shown in Figure 1.11; suggesting that the YSZ layer significantly suppressed the electronic leakage in GDC. However, the MPD of the bilayer cell was low (0.68 W cm^{-2} at 800 °C), which was attributed to gas diffusion limitation inside the Ni-YSZ anode.[53]

Wang *et al.* fabricated and tested YSZ/GDC bilayer electrolyte SOFCs using a BSCF ($Ba_{0.5}Sr_{0.5}Co_{0.8}Fe_{0.2}O_3$) cathode.[54] A NiO-YSZ (50 : 50) anode substrate was prepared by pressing and pre-sintering at 1000 °C. AYSZ layer was screen printed and sintered at 1400 °C, followed by screen

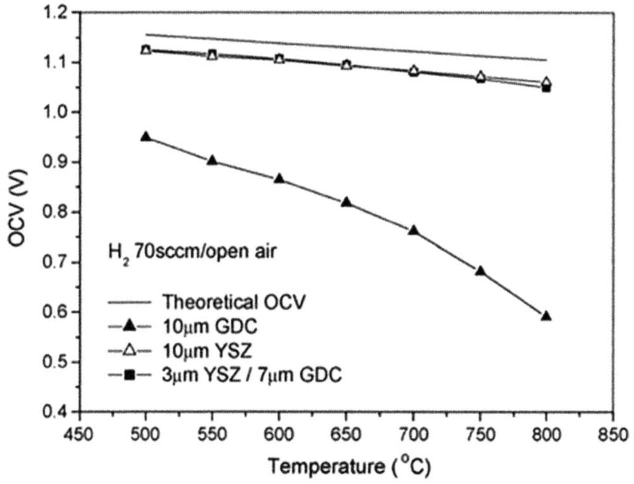

Figure 1.11 Theoretical and measured OCV of a monolayer GDC (10 μm) cell, monolayer YSZ (10 μm) cell and bilayer YSZ (3 μm)/GDC (7 μm) cell. Reproduced from ref. 53 with permission from Elsevier, Copyright 2006.

printing and sintering at 1400 °C of the SDC layer. The thickness of the YSZ and SDC layers was 2–8 and 12 μm, respectively. A reaction between the YSZ and SDC layers after sintering at 1400 °C was confirmed by a shift in XRD peak positions and EDX showed a significant diffusion of cerium and samarium into the YSZ layer. Moreover, conductivity of the YSZ/SDC pellet was found to be seven times lower than that of the SDC pellet. At 750 °C, the OCV of a bilayer cell with 2 μm thick YSZ was 0.88 V, while cells with 5–8 μm thick YSZ showed ~1.05 V, suggesting that the YSZ layer below 5 μm may not be able to completely block SDC leakage current at higher temperatures. The MPD of the bilayer cell with 5 μm YSZ was 0.82 W cm^{-2} at 750 °C, but after only two thermal cycles non-ohmic ASR increased resulting in 35% lower MPD.

Zhang *et al.*[55] used tape casting to fabricate thin ScSZ/SDC bilayer electrolyte SOFCs. NiO-YSZ support was fabricated by tape casting followed by screen printing of a NiO-SSZ (56 : 44) active anode layer. The thickness of the ScSZ and SDC electrolyte layers after co-firing at 1400 °C was 2 and 4 μm thickness, respectively. As shown in Figure 1.12, the OCV of the bilayer cell was 1.03 V at 650 °C indicating that the ScSZ layer successfully blocked the electronic leakage current in the SDC layer. The MPD of the ScSZ/SDC bilayer cell was 0.54 and 1.8 W cm^{-2} at 650 and 800 °C, respectively. Further analysis suggested that a 1 μm thick interdiffusion layer at the ScSZ/SDC interface accounted for more than 90% of the electrolyte ohmic ASR, as the conductivity of the interfacial layer was 2–3 times lower than ScSZ and 1–2 orders of magnitude lower than SDC. Long-term stability studies showed that MPD decreased with time largely due to an increase in non-ohmic ASR

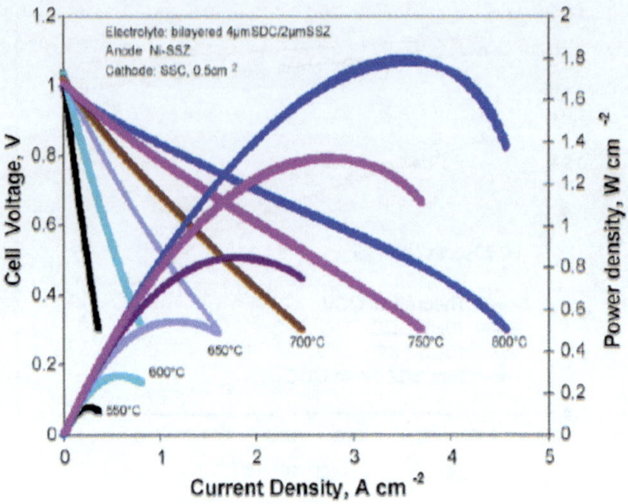

Figure 1.12 *I–V* characteristics of a ScSZ/SDC bilayer electrolyte unit cell using a SSC cathode. Reproduced from ref. 55 with permission from Elsevier, Copyright 2008.

at an approximately constant rate, following an initial 6–7 h decrease. It was suggested that cathode performance degraded over time due to a coarsening of the cathode microstructure and diffusion of strontium away from the SSC cathode.

Thin film ScSZ/SDC bilayer electrolyte-based SOFCs were developed by Yang *et al.*[56] using PLD. PLD process parameters for the two layers were optimised to achieve a cubic phase for ScSZ and SDC at 400 and 200 °C, respectively, and no reaction was observed because of the low processing temperatures. With an increase in substrate temperature, the density of the ScSZ layer was found to decrease while the density of the SDC layer was found to increase due to the different band gaps *vis-à-vis* the laser photons. For the fabrication of full cells, a ScSZ film of 1 μm thickness was first deposited at 200–400 °C on NiO-SDC substrate, followed by deposition of SDC film of 6–7 μm thickness at 400–600 °C. The OCV of the ScSZ/SDC bilayer cell with a SSC-SDC (75 : 25) cathode was 1.04 V and the MPD was 0.5 and 1.6 W cm^{-2} at 550 and 700 °C, respectively.

Solovyev *et al.*[57] used bipolar reactive magnetron sputtering to fabricate YSZ/GDC bilayer electrolyte SOFCs with a NiO-YSZ anode and LSCF-GDC cathode. Monolayer YSZ and GDC cells were also fabricated and tested. It was shown that a bipolar reactive magnetron has almost a three times higher deposition rate than DC sputtering. The deposited YSZ and GDC layers showed a cubic structure without the presence of any secondary phase. Monolayer YSZ cell studies indicated that below 2 μm YSZ thin film did not provide sufficient gas tightness and OCV decreased significantly. Among the different monolayer YSZ cells, a 4 μm thick YSZ cell showed the highest MPD of 0.25 and 0.58 W cm^{-2} at 650 and 800 °C, respectively. In comparison, a 2.8 μm thick monolayer GDC cell showed a higher MPD of 0.54 mW cm^{-2} at 650 °C. On the other hand, a bilayer cell with YSZ and GDC thickness of 4 and 1.5 μm, respectively, showed an OCV of 1.13 V and MPD of 1.25 W cm^{-2} at 800 °C. Unfortunately, the performance of the bilayer samples below 750 °C was not reported in this study.

Lu *et al.*[58] tested YSZ/SDC bilayer electrolyte SOFCs made by PLD. A NiO-YSZ anode substrate was prepared by tape casting and lamination. A YSZ and SDC thin film of 0.5–1 and 5 μm, respectively, were deposited at 1000 °C. A $La_{0.6}Sr_{0.4}Co_{0.2}Fe_{0.8}O_3$ layer was screen printed and then coated with a $La_{0.8}Sr_{0.2}CoO_3$ layer to act as a cathode functional layer and cathode current collector, respectively. At 650 °C, the OCV and MPD of the YSZ/SDC bilayer cell was 0.95 and 0.99 V and 0.95 and 1.37 W cm^{-2} in air and oxygen at the cathode, respectively. The power density improvement in oxygen was due to the significantly lower non-ohmic ASR of the cell. The stability of the YSZ/SDC cell was tested at 1 A cm^{-2} and 650 °C for 400 h and only a slight decrease in cell voltage was observed.

The effect of the anode functional layer (AFL) on YSZ/SDC bilayer electrolyte-based SOFCs was studied by Qian *et al.*[59] Both electrolyte films were deposited on a pressed NiO-YSZ (60 : 40) anode by PLD at 600 °C, followed by a post-anneal at 800 °C. To achieve smoother, denser electrolyte films, NiO-YSZ (50 : 50) AFL of 12 μm thickness was deposited on the anode substrate by drop-coating of AFL slurry. The thickness of the YSZ and SDC films was 2 and 6 μm, respectively. Using a SSC-SDC (70 : 30) cathode, the OCV of the YSZ/SDC cell without AFL was 0.85 V, which was higher than that of the monolayer SDC cell (0.78 V); the MPD of the cell without AFL was 0.75 W cm^{-2}. In comparison, the YSZ/SDC bilayer cell with AFL showed higher OCV and MPD of 1.01 V and 0.9 W cm^{-2} at 650 °C, respectively. The increase in MPD was attributed to the lower polarisation resistance of the cell because of AFL addition. The stability of the YSZ/SDC bilayer cell was tested at 600 °C for 50 h and the OCV slightly dropped by 0.98%.

Myung *et al.*[13] fabricated and tested ultra-thin YSZ/GDC bilayer electrolyte SOFCs by PLD. NiO-YSZ AFL (2 μm thick) was deposited by PLD to modify the surface roughness of the NiO-YSZ anode substrate; a LSC cathode was also deposited by PLD. The thickness of the YSZ film varied between 0 and 0.2 μm, while the thickness of the GDC film was 1.0 μm. YSZ grains in the thin film demonstrated a homoepitaxial relationship with YSZ grains in the anode and a heteroepitaxial relationship with GDC grains. The YSZ/GDC bilayer electrolyte was found to increase OCV from 0.6 V to over 1 V at 600 °C (Figure 1.13). Furthermore, the MPD of bilayer cells was significantly higher (>1 W cm^{-2} at 600 °C) compared to monolayer GDC cells (0.38 W cm^{-2} at 600 °C) as shown in Figure 1.14. It was noted that roughness of the anode substrate played an important role in the reproducibility of OCV among cells with the same thickness.

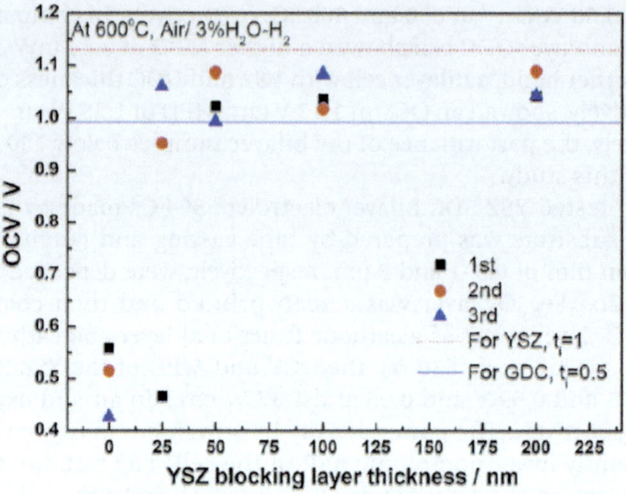

Figure 1.13 Measured OCV value of bilayer YSZ/GDC (1 μm) cells at 600 °C as a function of YSZ thickness. The solid line indicates the calculated OCV of GDC with t_i = 0.5 and the dotted line indicates the OCV of YSZ. Reproduced from ref. 13 with permission from Elsevier, Copyright 2012.

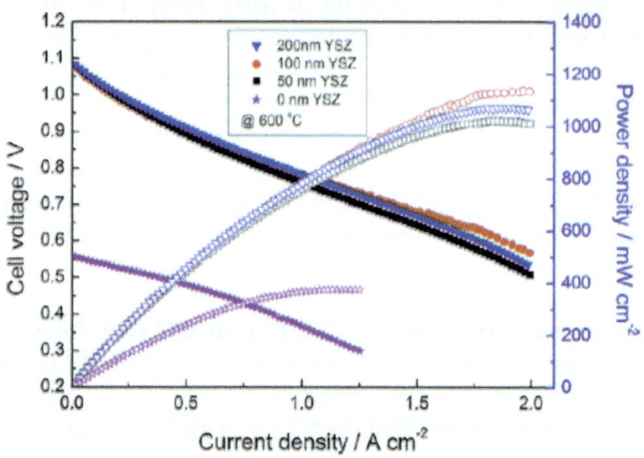

Figure 1.14 *I–V* characteristics of YSZ/SDC bilayer electrolyte unit cells with various YSZ layer thicknesses at 600 °C. Reproduced from ref. 13 with permission from Elsevier, Copyright 2012.

Ultra-thin YSZ/GDC bilayer electrolyte-based SOFCs using a chemical solution deposition (CSD) technique were fabricated by Oh *et al.*[14] A nanoparticle-based solution with 5% volume loading was spin-coated on NiO-YSZ substrate (56 : 44). The CSD required that the film thickness be greater than both the grain size of the film and pore size of the substrate but allowed the fabrication of thin films of YSZ (0.1 μm) and GDC (0.1 μm) at substantially lower sintering temperatures. It was noted that the addition of slow-sintering nanoparticles to the chemical solution was effective in suppressing the generation of macro-defects and directional development of the film microstructure, which typically is observed in CSD due to constrained sintering. Full cells were made with a LSCF-GDC cathode. The OCV of the YSZ/GDC bilayer cells was > 1 V at 500–650 °C and the MPD was 1.3 W cm^{-2} at 650 °C. The authors claimed to have developed the thinnest bilayer SOFCs and suggested that CSD can effectively replace vacuum deposition of thin film SOFCs.

Cathode-supported micro-tubular design was used to evaluate performance of a ScSZ/GDC bilayer electrolyte by Yamaguchi *et al.* in multiple studies, but unfortunately the performance was comparatively lower than that of anode-supported designs.[60–62] LSM cathode support was fabricated by extrusion and a ScSZ/GDC bilayer electrolyte was fabricated by dip coating in GDC and ScSZ slurries and co-firing at 1300 °C. A NiO-GDC anode layer was also made by dip coating. The final sintered bilayer electrolyte cell consisted of 15 μm ScSZ and 20 μm GDC thick layers. The OCV of the bilayer cell was 1.1 V at 500 °C and the MPD was 0.07 and 0.20 W cm^{-2} at 600 and 750 °C, respectively. The electrode ASR accounted for the low power density and a LSM-GDC interlayer was added at the GDC/LSM interface[61] that reduced the electrode ASR from 20 to 5 Ω cm^2 at 600 °C and the MPD of the cell was increased from 0.23 to 0.38 W cm^{-2} at 750 °C. The tubular design allowed rapid heating and cooling cycles without performance degradation after five cycles. In another study, the anode composition was optimised and NiO/GDC at 50 : 50 ratio was found to show six times higher MPD than that at 70 : 30 ratio,[62] due to lower mass transport limitations.

Metal-supported ScSZ/SDC bilayer electrolyte SOFCs with a SSC-SDC cathode were developed by Hui *et al.*[63] Stainless steel (SS430 grade) was used as the metallic support and a NiO-SDC anode was deposited by spin-coating. ScSZ (2 μm) and SDC (20 μm) layers were deposited by PLD and sintered at a comparatively low temperature of 850 °C to minimise oxidation of the metallic support. The OCV of the ScSZ/SDC bilayer cell was found to be 1.00–1.03 V at 400–600 °C. The MPD was low (0.16 W cm^{-2} at 600 °C) but the performance at 500 °C was stable for over 11 days. It was found that most of the voltage loss was due to the ohmic resistance of the cell.

1.3.2.1.2 Bilayer Cells with Zirconia at the Cathode Side. A ceria/zirconia bilayer electrolyte-based micro-SOFC with ceria on the anode side was developed by Lee *et al.*[64] Stainless steel (STS-434L) was used as a substrate to deposit a nano-porous Ni layer by screen printing and sintering in dry hydrogen at 575 °C. A NiO-GDC (60 : 40) anode, GDC, YSZ and $La_{0.7}Sr_{0.3}CoO_3$ layers were deposited by PLD. The OCV and MPD of the cell was 0.91 V and 0.03 W cm^{-2} at 450 °C and remained stable for over 100 h of testing. It was noted that the ohmic resistance of the cell remained constant, indicating

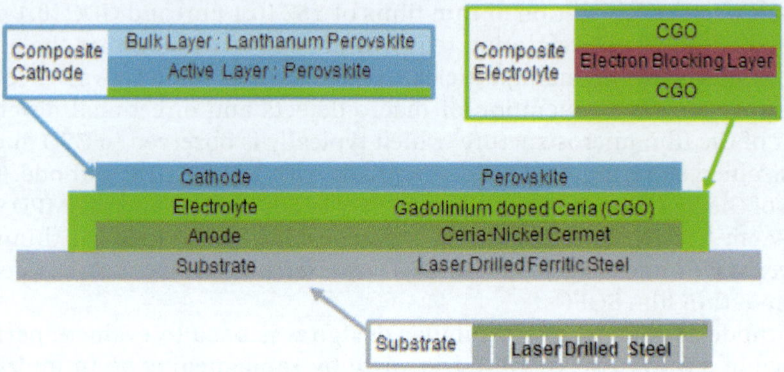

Figure 1.15 Schematic representation of Ceres Power 'Steel Cell'. Reproduced from ref. 65 with permission from The Electrochemical Society, Copyright 2005.

that delamination between the cell components did not occur. Non-ohmic resistance of the cell significantly decreased during the first 100 h and then remained stable, which was attributed to slow kinetics of NiO reduction at 450 °C. The authors claimed their work to be the longest lasting operation of a micro-SOFC based on a STS-NiO metallic support.

Ceres Power is working on the commercialisation of metal-supported trilayer electrolyte-based SOFC technology.[65] The company's 'Steel Cell' is based on a thick film GDC layer that provides mechanical integrity and gas tightness, a thin film electron blocking layer (most likely zirconia based) to block leakage current in GDC, and a thin film GDC layer to avoid reaction between the electron blocking layer (which would be necessary if that electron blocking layer is zirconia based as described in the section below) and a perovskite-based cathode (Figure 1.15). The layers including a nickel-ceria anode are deposited on a ferritic stainless-steel substrate using conventional ceramic processing techniques. Additionally, metal support provides improved mechanical robustness compared to the anode-supported planar design. The OCV of the trilayer electrolyte-based cell was improved to 1.17 V by reduction in electrolyte defects and other improvements. Over the years, power per cell was increased from 10 W in 2010 to 20 W in 2015 largely by the optimisation in materials and mechanical design within the stack as shown in Figure 1.16.[65] Lastly, the metal-supported design showed encouraging long-term stability and thermal cycling results.

1.3.2.1.3 Zirconia-based Zirconia/Ceria Bilayer HT-SOFC. The primary purpose of ceria-based bilayer electrolytes paired with zirconia is to protect the GDC layer from reduction and enhance the OCV of the cell. However, zirconia/ceria-based bilayers have been used for other reasons. Within the structure of zirconia-based bilayer cells, the thinner GDC layer is normally inserted as a buffer layer to limit the diffusion of strontium

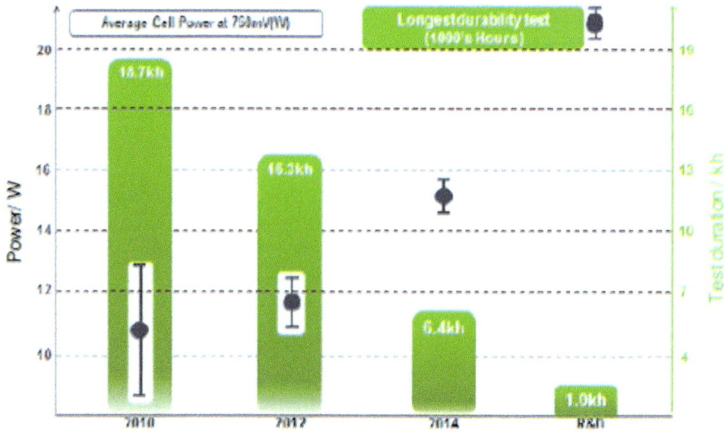

Figure 1.16 Average cell power and longest durability tests of Ceres Power 'Steel Cell' technology over the years. Reproduced from ref. 65 with permission from The Electrochemical Society, Copyright 2005.

and cobalt from the cobalt-containing cathode such as LSC and LSCF into the YSZ, and thus preventing the formation of resistive phases. However, YSZ and GDC are known to have chemical incompatibility. During the conventional high temperature (HT) sintering of the electrolyte at 1450 °C, resistive compounds with two orders of magnitude lower conductivity than YSZ form and surface porosity as a result of the Kirkendhal effect occurs. Hence, the major focus of different works in this category is to develop methods for low temperature fabrication of zirconia/ceria electrolytes such as using the effect of additives for lowering the sintering temperature of the bilayer electrolyte,[66,14,67] using sputtering for deposition of electrolytes,[68,69,15] spray pyrolysis,[70] PLD[71,72] and electron beam physical vapour deposition (EB-PVD).[16]

Due to the relatively low conductivity of YSZ and sluggish oxygen reduction reaction kinetics at lower temperatures, many works targeting 500 °C and lower temperature SOFCs are focused on the fabrication of ultra-thin electrolytes and using noble metal electrodes such as platinum. In such zirconia-based YSZ/GDC bilayer electrolyte SOFCs, a GDC layer is usually implemented to reduce electrode overpotential and achieve high OCV despite the extremely thin (few hundreds nm) electrolytes. It is well known that oxide ion conductivity of the electrolytes adjacent to the cathode and anode impacts the non-ohmic polarisation and a faster oxide ion conductor such as doped ceria compared to zirconia can enhance the performance of the cell, particularly at lower temperature. Although such thin bilayer electrolyte micro-SOFCs showed high performance at lower temperatures (1.3 W cm^{-2} at 450 °C[19]), they are typically fabricated using semi-conductor processing equipment such as plasma-enhanced atomic layer deposition (PEALD),[17] aerosol-assisted chemical vapour deposition (AA-CVD),[18] and a combination of ALD and PLD.[19] They also require noble metals such as platinum as their electrodes and thus are very expensive to fabricate. Following the same rationale, a few works in the literature showed the application of a thin GDC layer at the anode side of a zirconia-based SOFC to reduce the anodic overpotential of the cell.[73–75]

1.3.2.2 Zirconia/Bismuth Oxide Bilayer-based IT-SOFCs

Some researchers have looked at improving the performance of YSZ-based SOFCs for intermediate temperature operation by incorporating a bismuth oxide layer on the cathode side. Joh *et al.*[76] fabricated YSZ/ESB bilayer electrolyte SOFCs using a LSM-ESB cathode. A NiO-YSZ anode (65 : 35) was made by tape casting and pre-sintered at 900 °C. Both YSZ and ESB electrolytes were deposited *via* colloidal drop coating and sintered at 1400 and 800 °C, respectively. The LSM-ESB cathode (50 : 50) was screen printed and sintered at 800 °C. No reaction between the YSZ and ESB layers was observed after heat treatment at 800 °C and after testing of unit cells by EDS analysis. The OCV of the bilayer cell with a 7 μm YSZ layer and 5 μm ESB layer was >1.10 V at all temperatures due to the unity transference number (t_i) of the YSZ electrolyte and at 650 °C the MPD of the bilayer cell was 1.62 W cm^{-2}. The MPD of the bilayer cell was 2.08 W cm^{-2} at 700 °C, which compared favourably to the monolayer 7 μm YSZ cell which showed a MPD of 0.86 W cm^{-2} (Figure 1.17). The significant improvement in the power density of the bilayer cell was attributed to a faster oxygen reduction reaction at the cathode/electrolyte interface in the presence of the ESB layer. The enhanced electrode kinetics was confirmed by the reduction in cathodic overpotential and a higher exchange current density.

In another work, Wang *et al.*[77] looked at YSZ/21NbSB bilayer electrolyte SOFCs made by RF magnetron sputtering. Ni-SDC and silver were used as the anode and cathode, respectively. The YSZ layer was deposited at 300 °C using an Ar/O2 mixture of 30 : 1. For the 21NbSB layer, the composition of the deposited film depends greatly on the gas mixture composition which with an Ar/O2 mixture of 31 : 7 yielded the closest film composition to the target composition. It was found that the bilayer electrolyte deposited at 300 °C was thermally stable and did not crack after a 700 °C anneal. However, the MPD of the cell was rather low at 0.01 W cm^{-2} at 600 °C using a 4 μm YSZ and 1.5 μm NbSB bilayer electrolyte.

1.3.2.3 Ceria/Bismuth Oxide Bilayer LT-SOFCs

As reviewed in the previous section, the initial bilayer electrolyte cells were built using thick electrolyte pellets and noble metal electrodes, and the cell performance was low. To improve the bilayer cell performance, research focused on developing anode-supported cells using a thin-film electrolyte and new cathode materials compatible with bismuth oxide electrolyte has been carried out. Traditional cathode materials for IT-SOFCs such as perovskite LSCF show extreme reactivity with bismuth oxide, and among

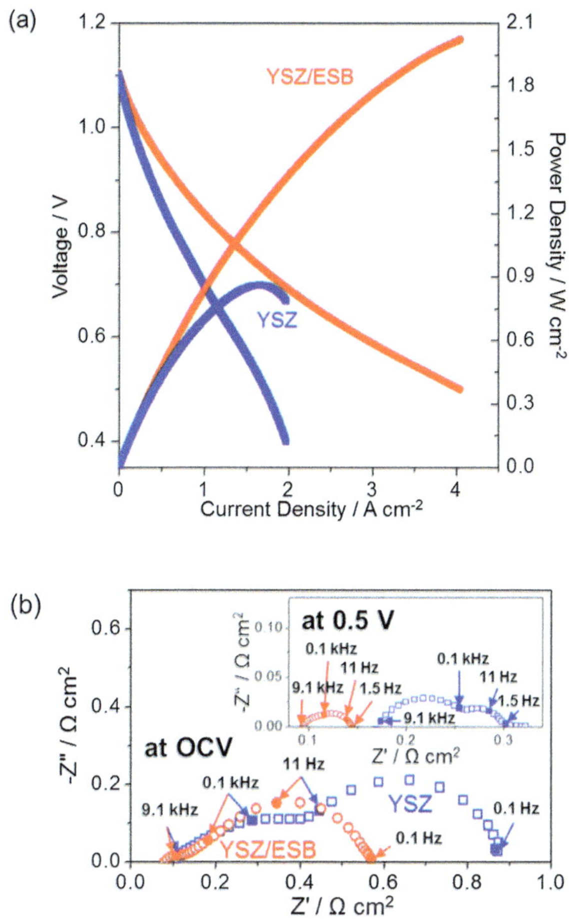

Figure 1.17 (a) *I–V* characteristics and (b) impedance spectra of a monolayer YSZ cell and YSZ/ESB bilayer electrolyte cell at 700 °C using humidified H_2 and air at the anode and cathode sides, respectively. Reproduced from ref. 76 with permission from American Chemical Society, Copyright 2017.

noble metals, Pt is known to form secondary phases such as $Bi_2Pt_2O_7$ with bismuth oxide.[78] One of the cathodes discovered was a bismuth ruthenate (BRO)-ESB composite by Jaiswal.[44] It was shown that the BRO and ESB phases are compatible with each other by XRD and DSC studies. Another system under investigation is the LSM-ESB composite;[79] however, the compatibility results are inconclusive with respect to the reaction between LSM and ESB and the consequent effect on the electrode performance and stability.

Ahn *et al.*[80,81] fabricated anode-supported cells with a thick film bilayer electrolyte by different fabrication routes and studied cell performance using different cathodes. Ni-GDC/GDC half-cells were made by either co-pressing of the anode and GDC powders or by spray coating of the GDC

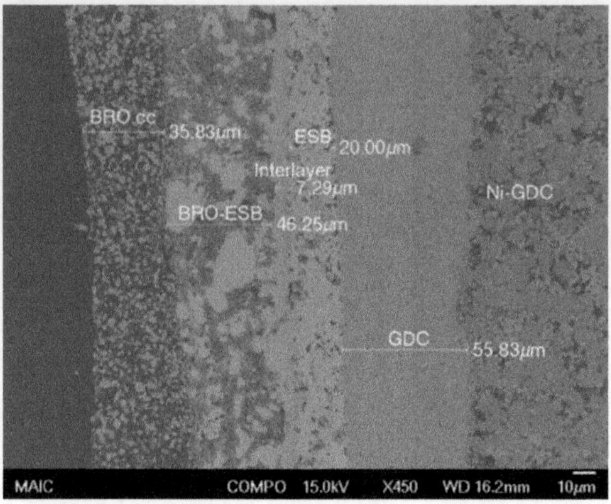

Figure 1.18 Backscattered SEM image showing a cross-section of Ni-GDC/GDC (50 μm)/ESB (20 μm)/BRO-ESB/BROcc unit cell. 'cc' refers to the current collector. Reproduced from ref. 80 with permission from the Electrochemical Society, Copyright 2010.

layer on a tape-casted anode. The co-pressing method resulted in a relatively thick (~50 μm) GDC layer, while the spray coating method resulted in a 10–20 μm GDC layer. For the cells made by the co-pressing method, an ESB layer was made by screen printing using solid state powders. For the cells made by spray coating, an ESB layer was made by either screen printing using co-precipitated powders or by hot/cold PLD. In the last step, a BRO-ESB cathode was brush-painted to complete the cell. A SEM cross-section image of a GDC/ESB bilayer cell is shown in Figure 1.18. The performance of the bilayer cells was compared with a monolayer GDC cell using a standard LSCF-GDC cathode. For the co-pressed cells, the screen-printed ESB layer did not sinter completely into a dense layer due to the large size of the solid state powders and thus was found to not effectively increase the OCV of the bilayer cell. On the other hand, the BRO-ESB cathode reduced the ASR of the bilayer cell resulting in a MPD of 0.59 W cm^{-2} at 650 °C, which was higher than that of the monolayer GDC cell with a LSCF-GDC cathode (0.34 W cm^{-2}). For the cell made by spray coating, the OCV was significantly lower due to the lower thickness (10–20 μm) of the GDC layer, but the screen-printed ESB layer was found to be effective in increasing OCP from 0.67 to 0.75 V, resulting in an increase in MPD of the cell from 0.41 to 0.61 W cm^{-2} at 650 °C. Moreover, the ESB layer made by hot PLD exhibited lower ohmic and non-ohmic ASR and higher OCV (0.77 V *vs.* 0.72 V at 650 °C) compared to that of the monolayer GDC cell, resulting in a 93% increase in MPD from 1.03 W cm^{-2} for the monolayer cell to 1.95 W cm^{-2} for the bilayer cell with a ~4 μm thick ESB layer made by hot PLD (Figure 1.19). The improvement in cell performance was ascribed to the combined effect of higher OCV using an ESB layer and 43% lower ASR

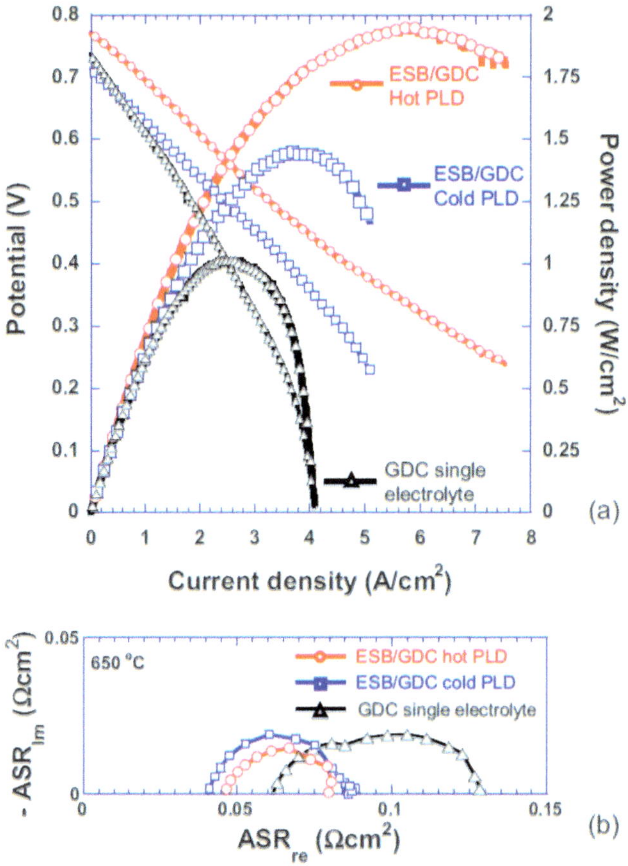

Figure 1.19 (a) *I–V* characteristics and (b) impedance spectra of a monolayer GDC cell and GDC/ESB bilayer electrolyte cell made by hot/cold PLD. Reproduced from ref. 80 with permission from the Electrochemical Society, Copyright 2010.

using a BRO-ESB composite cathode. This performance, ~2 W cm^{-2} at 650 °C, set the benchmark for high power density ceria/bismuth oxide bilayer electrolytes and a subsequent publication[3] laid out the potential and issues for this technology.

As a result, numerous groups have investigated this approach and continued the improvement in OCP and power density. However, while the result of ceria/bismuth oxide bilayer electrolyte-based SOFCs was impressive, the OCV was lower than the theoretical estimate, suggesting that further optimisation of the relative thickness and total thickness of the bilayer electrolyte needs to be carried out. As such, there is a trade-off in optimising the total thickness of the bilayer electrolyte between reducing ASR with thinner electrolytes and increasing OCV with thicker electrolytes.[47] In addition, for bilayer electrolytes, the OCV further depends on the relative thickness of the two layers.

Lee *et al.*[82] studied an anode-supported GDC/ESB bilayer cell with a BRO-ESB cathode to understand the effect of relative thickness of the two electrolyte layers on the cell performance. A GDC electrolyte layer was deposited by spin coating of a GDC slurry on a NiO-GDC anode support. An ESB layer was fabricated by drop coating using powders made from solid state and co-precipitation methods. It was found that powders made by co-precipitation were finer and resulted in a dense ESB layer after sintering at 800 °C. Cells with different GDC and ESB thicknesses were fabricated and it was observed that by increasing the thickness of the GDC and ESB layers, the OCV of the cell increased. The MPD of a bilayer cell with 10 μm GDC and a 5 μm thick ESB reached over 1.5 W cm^{-2} at 650 °C. The increase in MPD compared to the monolayer GDC cell (0.87 W cm^{-2}) was due to both higher OCV (0.8 *vs.* 0.75 V) and 47.6% lower ASR.

Lee *et al.*[83] studied the performance of GDC/ESB bilayer cells with a LSM-ESB cathode using a tape-cast anode and anode functional layer (AFL). GDC and ESB layers were deposited with spin coating of a colloidal solution and the sintered thickness was 25 and 2 μm, respectively. The MPD of the cell was 1.02 W cm^{-2} at 650 °C, which was higher than all previously reported LSM cathode-based SOFCs. In addition, cathode and cell performance was stable for over 80 and 200 h of testing, respectively. Further improvement in cell performance was suggested by replacing the GDC/ESB bilayer combination with higher conductivity doped ceria and stabilised bismuth oxide electrolytes such as SNDC and DWSB.[47] Redox Power Systems is utilising these developments in ceria/bismuth oxide bilayer electrolytes to develop commercial SOFCs operating at low temperatures.

Recent reports on ceria/bismuth oxide bilayer electrolyte-based SOFCs by Zhang *et al.*,[84] Lee *et al.*[85] and Hou *et al.*[86,87] have looked at SDC and SNDC-based ceria electrolytes, YSB-based bismuth oxide electrolytes and LSM-based cathodes. For SDC/YSB bilayer SOFCs using a LSM-YSB cathode by Zhang *et al.*,[84] a NiO-SDC (60 : 40) anode and SDC electrolyte were co-pressed and co-fired at 1250 °C. The YSB electrolyte was deposited using DC magnetron sputtering in an argon/oxygen mixture at room temperature and then annealed at 800 °C for 2 h. The thickness of the SDC and YSB layers was 26 and 6 μm, respectively. Using the bilayer electrolyte, the OCV of the bilayer electrolyte cell increased to 0.85 V compared to 0.78 V of the monolayer SDC cell at 650 °C. MPD also increased from 0.29 W cm^{-2} for the monolayer cell to 0.38 W cm^{-2} for the bilayer cell. Ohmic resistance of the bilayer cell did not change with the addition of the YSB layer due to its high conductivity. However, the non-ohmic resistance of the bilayer cell under load was much lower than that of the monolayer cell due to the positive effect of the YSB electrolyte on cathodic polarisation. The overall increase in MPD was attributed to the higher OCV and lower cathodic polarisation of the bilayer electrolyte.

Lee *et al.*[85] reported on their work on GDC/YSB bilayer SOFCs using a LSM-YSB cathode. A NiO-GDC anode support was fabricated by powder pressing and sintering at 1100 °C. A GDC electrolyte was deposited by drop coating and sintering at 1400 °C, while a YSB electrolyte and LSM-YSB (50:50) cathode were screen printed and sintered at 800 °C and 850 °C, respectively. The thicknesses of the GDC and YDB layer were 20 and 10 μm, respectively. For the reference monolayer GDC cell, LSM-GDC was used as the cathode material. The bilayer cell showed a slightly higher ohmic ASR but a significantly lower non-ohmic ASR compared to the monolayer cell with a LSM-GDC cathode. The lower non-ohmic ASR was attributed to the lower resistance of LSM-YDB for charge transfer at the electrode/electrolyte interface. It was also noted that despite a significantly lower ionic conductivity of LSM compared to LSCF, the LSM-YSB cathode showed a similar performance to that of the LSCF cathode at 650 °C, suggesting the beneficial effect of YSB in the composite cathode. The bilayer electrolyte enhanced the OCV from 0.50 to 0.89 V and the MPD improved from 0.21 to 0.53 W cm^{-2} at 650 °C.

Hou *et al.*[86,87] fabricated $Sm_{0.075}Nd_{0.075}Ce_{0.85}O_2$ (SNDC)/ESB and GDC/ESB bilayer electrolyte SOFCs using $La_{0.74}Bi_{0.1}Sr_{0.16}MnO_3$(LBSM)-ESB (50:50) as the cathode. A Ni-ceria (60:40) anode and ceria electrolyte were prepared by co-pressing and sintering at 1400 °C. An ESB electrolyte was deposited by drop coating and sintering at 800 °C. The thickness of the ceria and ESB layers after sintering was 20 μm each. Bismuth doping of LSM was found to be effective in limiting any reaction between LBSM and ESB after high temperature sintering as confirmed by XRD diffraction pattern analysis. Impedance spectroscopy showed that above 550 °C, ohmic losses accounted for the majority of the cell resistance. At 650 °C, the OCV and MPD of the SNDC/ESB bilayer cell reached 0.80 V and 0.98 W cm^{-2}, respectively, as shown in Figure 1.20.[86] At 450 °C, the MPD of the bilayer cell was 0.13 W cm^{-2}, which the authors claimed to be the highest recorded value at 450 °C.[87] In another work, Hou *et al.* compared the electrochemical performance of SNDC/ESB to that of GDC/ESB bilayer electrolyte SOFCs.[87] At 650 °C, the OCV and MPD of the SNDC/ESB and GDC/ESB cells were 0.81 V and 0.93 W cm^{-2} and 0.83 V and 0.90 W cm^{-2}, respectively. The higher MPD of the SNDC/ESB cell was attributed to the lower ohmic ASR of the SNDC/ESB cell, as confirmed by the impedance spectroscopy. In addition, it was noted that the conductivity of the bilayer electrolyte calculated from the overall thickness of the electrolyte and the ohmic ASR of the cell was higher than that of SNDC and GDC. The authors argued that the measured conductivity in an operating cell was higher than the conductivity measured in air due to consumption of oxygen at the anode. However, impedance measurements were performed at an open circuit condition and hence, consumption of oxygen at the anode should be negligible and cannot be expected to affect the conductivity of the bilayer electrolyte.

Other researchers have studied ceria/bismuth oxide bilayer electrolytes using different cathode materials – Pt,[88] Ag-YSB[89] and $Pr_{0.5}Ba_{0.5}MnO_3$ (PBM).[90] Leng and Chan[88] fabricated GDC/YSB bilayer SOFCs using a Pt cathode. A NiO-GDC anode and GDC electrolyte were fabricated by co-pressing and

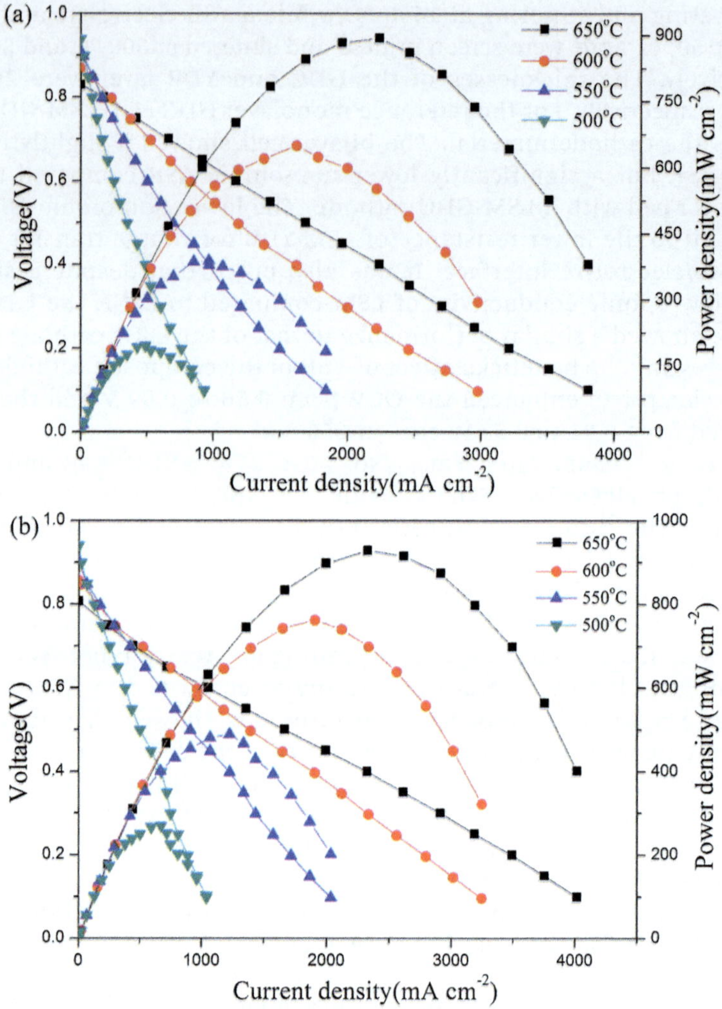

Figure 1.20 *I–V* characteristics of (a) GDC/ESB and (b) SNDC/ESB bilayer electrolyte unit cells using a LBSM-ESB cathode. Reproduced from ref. 87 with permission from Elsevier, Copyright 2015.

sintering at 1350 °C. YSB was applied by screen printing and sintered at 900 °C. A SEM micrograph of the bilayer cell showed a dense, 84 μm thick GDC layer and a porous, 16 μm thick ESB layer. The porous ESB layer was able to marginally enhance the OCV of the bilayer cell from 0.75 to 0.81 V at 700 °C. Although the ohmic resistance of the cell did not change, the non-ohmic polarisation ASR of the bilayer decreased by 36%. The MPD of the bilayer cell was relatively modest but increased from 0.16 to 0.26 W cm^{-2} at 700 °C due to the higher OCV and lower ASR of the bilayer cell. Zhang *et al.*[89] fabricated and tested SDC/YSB bilayer electrolyte SOFCs with an Ag-YSB cathode. A NiO-SDC anode and SDC electrolyte were pressed and co-fired at 1250 °C for 5 h.

A YSB layer was deposited using DC magnetron sputtering and annealed at 800 °C for 2 h. The thickness of the SDC and YSB layer was 26 and 6 µm. An Ag-YSB (60:40) composite cathode was deposited by screen printing. The bilayer SDC/YSB electrolyte improved the OCV from 0.80 to 0.89 V at 500 °C, while the MPD of the cell improved from 0.13 to 0.22 W cm^{-2}. At 600 °C, the MPD of the bilayer cell was 0.57 W cm^{-2}. Impedance studies using symmetrical cells showed that the ASR of the Ag-YSB cathode on the YSB electrolyte was lower than that on the SDC electrolyte and suggested that the increase in MPD of the bilayer cell was due to both higher OCV and lower interfacial polarisation compared to the monolayer SDC electrolyte cell. Hou *et al.*[90] developed a new PBM perovskite and ESB (50:50) based composite cathode for SNDC/ESB bilayer electrolyte SOFCs. PBM and ESB powders were made using a citric acid-nitrate gel combustion method. PBM showed a perovskite structure composed of cubic and hexagonal phases but after firing with ESB at 750 °C for 3 h showed minor secondary phase in the XRD diffraction patterns. The activation energy for the PBM-ESB cathode was estimated to be 1.38 eV. A NiO-SNDC (60:40) anode substrate and SNDC electrolyte were co-pressed and fired at 1400 °C. An ESB electrolyte was deposited by drop coating and sintered at 800 °C. The OCV and MPD of the bilayer cell with a 22 µm thick SNDC layer and a 20 µm thick ESB layer were 0.80 V and 0.99 W cm^{-2} at 650 °C, respectively.

In solid electrolytes, doping of aliovalent cation increases the oxide vacancy concentration but the oxide ion conductivity can be enhanced by increasing the dopant concentration only up to a certain level. Beyond that level, the ionic conductivity decreases due to the association of vacancies or formation of defect complexes in the dilute solution range and the formation of super-lattices or ordering of vacancies in the concentrated solution range.[91,92] The extent of anion sublattice ordering can be reduced by using highly polarisable dopants with ionic radii close to that of the host cation and randomising the site energetics by multi-cation doping.[93] This approach has been used to develop double-doped ceria, such as samarium and neodymium co-doped ceria (SNDC)[94] and dysprosium and tungsten co-stabilised bismuth oxide (DWSB),[93] which show higher oxide ion conductivity compared to ESB and GDC, respectively. In the bismuth oxide system, double doping reduces the amount of dopant required to stabilise the cubic phase to room temperature and hence, increases the ionic conductivity compared to traditional ESB. In addition, the cubic ESB electrolyte suffers from disorder–order transformation on the oxide-ion sublattice at ~500 °C, resulting in decay of oxide ion conductivity with isothermal aging.[91] Recent understanding on the interplay between dopant type/level and phase composition has resulted in the development of a new stabilised bismuth oxide composition, which shows stable ionic conductivity at ~500 °C.[95,96] Ruth *et al.*[97] have studied the performance of bilayer SOFCs using double-doped bismuth oxides, *i.e.* GDC/DWSB and GDC/DGSB. At 650 °C, the conductivity of DGSB and DWSB was 0.94 and 0.53 S cm^{-1}, respectively, which was significantly higher than of ESB.

Ni-GDC and LSM-ESB were used as the anode and cathode material, respectively. The OCV of the bilayer's cells with 20 μm GDC and 3–5 μm bismuth oxide layers at 650 °C was ~0.8 V, while the MPD was 0.6, 0.9 and 1.3 W cm^{-2} for ESB, DWSB and DGSB-based bilayers, respectively.

An interesting work on a layered heterostructure consisting of alternating GDC/ESB layers was reported by Sanna *et al.*[98] with the goal to improve the stability of the ESB phase to room temperature and in reducing environments. The alternating GDC/ESB layers were deposited on a MgO substrate using PLD to result in a MgO/GDC/[ESB/GDC]$_N$ structure, where N was varied from 1 to 30. For a constant film thickness of 60 nm, $N = 20$ showed the highest conductivity in both air and reducing environments. Unlike bulk ESB, the order–disorder transition and related conductivity decay was not observed at 600 °C in air. Moreover, the conductivity of a heterostructure with $N = 20$ was found to increase by two orders of magnitude in 90% H_2/10% N_2 reducing environment at 533 °C. Lastly, the heterostructure was found to be stable in the reducing environment and the electrical conductivity was constant. The conductivity results for the heterostructure were extremely interesting but no full cell results using the heterostructured electrolyte were presented.

1.3.2.4 Other Types of Bilayer Electrolyte SOFCs

As mentioned previously, bilayer electrolytes serve distinct functions in a SOFC structure. In ceria/bismuth oxide or ceria-based ceria/zirconia bilayers, bismuth oxide and zirconia layers block the leakage current through ceria and increase the OCV. However, in other types of bilayer electrolyte SOFCs, the main electrolyte is not reduced and near theoretical OCV can be obtained. The major functionality of the secondary electrolyte is to suppress the reaction between the electrolyte and electrodes or protect the electrolyte from reaction with fuel constituents such as water, CO_2 or SO_2. For instance, LSGM has high oxide ion conductivity and shows near theoretical OCV. However, LSGM and Ni-based cermets react and resistive phases with low conductivity form at the interface. Hence, many researchers have studied the potential of using a thin ceria-based electrolyte such as SDC,[99–103] or LDC[104–107] as a diffusion barrier between the LSGM electrolyte and Ni-cermet anode. Proton conductor-based SOFCs typically use doped $BaCeO_3$ as their electrolyte. Despite its high ionic conductivity $BaCeO_3$ suffers from poor chemical stability under CO_2, acidic gases and steam, which are prevalent under the operating environment of a SOFC. On the other hand, $BaZrO_3$ has a conductivity one order of magnitude lower than that of $BaCeO_3$. However, it possess good chemical stability and hence the potential of protecting the $BaCeO_3$ with a thin layer of $BaZrO_3$ has been studied in the literature.[108–110] Interested readers are referred to other types of bilayers such as ceria/BZCY,[111,112] LBAYO/SDC[113] or biphasic electrolytes[114–116] as a means to improve the OCV of ceria-based SOFCs.

1.4 Final Remarks

As summarised in Tables 1.1 and 1.2, for similar total electrolyte thickness and temperature, ceria/zirconia bilayer IT and HT-SOFCs yield higher OCV values compared to ceria/bismuth oxide IT and LT-SOFCs. However, due to the significantly lower ohmic ASR across the ceria/bismuth oxide bilayer electrolyte layers, these bilayer electrolyte SOFCs have the potential to deliver higher power density at lower temperatures. For example, at 650 °C, ScSZ (6–7 μm)/SDC (6–7 μm) bilayer electrolyte SOFC[56] and GDC (10 μm)/ESB (4 μm) bilayer electrolyte SOFC[81] (total electrolyte thickness of 14 μm) demonstrated OCVs of 1.04 V and 0.77 V, respectively; however, in terms of MPD, they achieved 1.38 and 1.95 W cm^{-2}, respectively.

As such, although many scholarly works in developing IT-SOFCs are primarily focused on depositing a thin stabilised zirconia layer at lower temperatures using alternative processing methods, these bilayers often suffer from higher ohmic losses and hence, despite having higher OCV, they deliver lower power output. Consequently, ceria/bismuth oxide bilayer electrolyte SOFCs are suggested as a more promising approach to reduce the SOFC operating temperature to 650–500 °C. Towards this goal, (1) new doped bismuth oxide compositions with stable oxide ionic conductivity at temperatures below 600 °C must be explored, (2) structure of the cell, *i.e.* bilayer electrolyte thickness ratio and fabrication processes must be engineered, and (3) new cathode materials with low ASR at low temperatures, that are compatible with bismuth oxide, should be developed.

1.5 Conclusion

Lowering the operating temperature of the solid oxide fuel cell below 650 °C reduces the manufacturing cost, decreases the performance degradation rate, increases efficiency and hence can promote the commercialisation of SOFC technology. Two major approaches to reduce the operating temperature are typically pursued: (1) using alternative electrolyte materials with higher ionic conductivity such as doped ceria instead of the traditional YSZ, and (2) few microns or sub-micron electrolyte SOFCs. Sub-micron electrolyte SOFCs show reasonable power density at lower temperatures. However, fabrication requires more expensive semi-conductor processing such as advanced lithography, ALD, or sputtering and hence may not be cost effective for large-scale power generation potential; therefore, they were not reviewed in detail in this work. Doped ceria such as GDC and SDC have ionic conductivity orders of magnitude higher than YSZ and hence have been explored extensively. However, ceria reduces under the fuel conditions of the SOFC, which causes an internal leakage current and power dissipation. Stabilised bismuth oxide compositions such as ESB, YSB and DWSB have been shown to be capable of blocking the leakage current and increasing the OCV of the cell. Although many scholars have investigated the potential of using a thin zirconia layer to block the leakage current, ceria/bismuth oxide bilayer SOFCs showed higher power density values due to the much higher ionic conductivity of stabilised bismuth oxide compared to YSZ. The ceria/bismuth oxide thickness ratio controls the interfacial p_{O2} and the thermodynamic stability of the bismuth oxide layer, and hence is of great importance and must be tuned carefully. Bilayer electrolyte SOFCs can serve other functions (other than blocking the leakage current) such as a diffusion barrier layer between the electrolyte and adjacent electrodes or to enhance the catalytic activity of the cathode as a fast oxide ion conductor.

Abbreviations

BRO – Bismuth ruthenate
BZCY – Barium cerium oxide doped with zirconia and yttria
DGSB – Dysprosium gadolinium-stabilised bismuth oxide
DWSB – Dysprosium tungsten-stabilised bismuth oxide
ESB – Erbium-stabilised bismuth oxide
GDC – Gadolinium doped ceria
LBAYO – Lanthanum barium yttrium oxide
LBSM – Lanthanum bismuth strontium manganese oxide
LSCF – Lanthanum strontium cobalt iron oxide
LSGM – Lanthanum gallate doped with strontium and magnesium
LSM – Lanthanum strontium manganese oxide
LYBO – Lanthanum yttrium-stabilised bismuth oxide
NbSB – Niobium-stabilised bismuth oxide
NGSB – Niobium gadolinium-stabilised bismuth oxide
ScSZ – Scandia-stabilised zirconia
SDC – Samarium doped ceria
SNDC – Samarium neodymium doped ceria
xSB – Double-doped bismuth oxide
YDC – Yttrium doped ceria
YSB – Yttrium-stabilised bismuth oxide
YSZ – Yttria-stabilised zirconia

References

1. Z. Zhan and S. A. Barnett, *Science*, 2005, **308**, 844–847.
2. E. Fontell, T. Kivisaari, N. Christiansen, J.-B. Hansen and J. Pålsson, *J. Power Sources*, 2004, **131**, 49–56.
3. E. D. Wachsman and K. T. Lee, *Science*, 2011, **334**, 935–939.
4. H. Inaba and H. Tagawa, *Solid State Ionics*, 1996, **83**, 1–16.
5. J. B. Goodenough, *Annu. Rev. Mater. Res.*, 2003, **33**, 91–128.
6. T. Takahashi and H. Iwahara, *Mater. Res. Bull.*, 1978, **13**, 1447–1453.
7. T. Takahashi, T. Esaka and H. Iwahara, *J. Appl. Electrochem.*, 1977, **7**, 299–302.

8. H. Yahiro, Y. Baba, K. Eguchi and H. Arai, *J. Electrochem. Soc.*, 1988, **135**, 2077–2080.
9. E. D. Wachsman, P. Jayaweera, N. Jiang, D. M. Lowe and B. G. Pound, *J. Electrochem. Soc.*, 1997, **144**, 233–236.
10. S. De Souza, S. J. Visco and L. C. De Jonghe, *Solid State Ionics*, 1997, **98**, 57–61.
11. S. de Souza, S. J. Visco and L. C. De Jonghe, *J. Electrochem. Soc.*, 1997, **144**, L35–L37.
12. J. Will, A. Mitterdorfer, C. Kleinlogel, D. Perednis and L. Gauckler, *Solid State Ionics*, 2000, **131**, 79–96.
13. D.-H. Myung, J. Hong, K. Yoon, B.-K. Kim, H.-W. Lee, J.-H. Lee and J.-W. Son, *J. Power Sources*, 2012, **206**, 91–96.
14. E. O. Oh, C. M. Whang, Y. R. Lee, S. Y. Park, D. H. Prasad, K. J. Yoon, J. W. Son, J. H. Lee and H. W. Lee, *Adv. Mater.*, 2012, **24**, 3373–3377.
15. H.-S. Noh, J. Hong, H. Kim, K. J. Yoon, B.-K. Kim, H.-W. Lee, J.-H. Lee and J.-W. Son, *J. Electrochem. Soc.*, 2016, **163**, F613–F617.
16. S. Uhlenbruck, N. Jordan, J. M. Serra, H. P. Buchkremer and D. Stöver, *Solid State Ionics*, 2010, **181**, 447–452.
17. W. Yu, G. Y. Cho, S. Hong, Y. Lee, Y. B. Kim, J. An and S. W. Cha, *Nanotechnology*, 2016, **27**, 415402.
18. M. V. F. Schlupp, A. Evans, J. Martynczuk and M. Prestat, *Adv. Energy Mater.*, 2014, **4**, 1301383.
19. J. An, Y.-B. Kim, J. Park, T. M. Gür and F. B. Prinz, *Nano Lett.*, 2013, **13**, 4551–4555.
20. P. J. Gellings and H. Bouwmeester, *Handbook of Solid State Electrochemistry*, CRC press, 1997.
21. C. Xia, W. Rauch, W. Wellborn and M. Liu, *Electrochem. Solid-State Lett.*, 2002, **5**, A217–A220.
22. C. Xia, W. Rauch, F. Chen and M. Liu, *Solid State Ionics*, 2002, **149**, 11–19.
23. V. Dusastre and J. Kilner, *Solid State Ionics*, 1999, **126**, 163–174.
24. E. P. Murray, M. Sever and S. Barnett, *Solid State Ionics*, 2002, **148**, 27–34.
25. L.-W. Tai, M. Nasrallah, H. Anderson, D. Sparlin and S. Sehlin, *Solid State Ionics*, 1995, **76**, 273–283.
26. J.-M. Bae and B. Steele, *Solid State Ionics*, 1998, **106**, 247–253.
27. B. C. H. Steele, *Solid State Ionics*, 1995, **75**, 157–165.
28. Ö. Çelikbilek, E. Siebert, D. Jauffrès, C. L. Martin and E. Djurado, *Electrochim. Acta*, 2017, **246**, 1248–1258.
29. Z. Wu and M. Liu, *J. Am. Ceram. Soc.*, 1998, **81**, 1215–1220.
30. C. Xia, W. Rauch, F. Chen and M. Liu, *Solid State Ionics*, 2002, **149**, 11–19.
31. J.-M. Bae and B. Steele, *J. Electroceram.*, 1999, **3**, 37–46.
32. T. Takeda, R. Kanno, Y. Kawamoto, Y. Takeda and O. Yamamoto, *J. Electrochem. Soc.*, 2000, **147**, 1730–1733.
33. T. Takeda, R. Kanno, K. Tsubosaka and Y. Takeda, *Electrochemistry*, 2002, **70**, 969–971.

34. S. P. Jiang and S. H. Chan, *J. Mater. Sci.*, 2004, **39**, 4405–4439.
35. S. Ohara, R. Maric, X. Zhang, K. Mukai, T. Fukui, H. Yoshida, T. Inagaki and K. Miura, *J. Power Sources*, 2000, **86**, 455–458.
36. H.-T. Lim and A. V. Virkar, *J. Power Sources*, 2009, **192**, 267–278.
37. T.-H. Kwon, T. Lee and H.-I. Yoo, *Solid State Ionics*, 2011, **195**, 25–35.
38. S.-G. Kim, S. P. Yoon, S. W. Nam, S.-H. Hyun and S.-A. Hong, *J. Power Sources*, 2002, **110**, 222–228.
39. A. V. Virkar, *J. Electrochem. Soc.*, 1991, **138**, 1481–1487.
40. W.-S. Hsieh, P. Lin and S.-F. Wang, *Int. J. Hydrogen Energy*, 2014, **39**, 17267–17274.
41. Y. Jee, G. Y. Cho, J. An, H.-R. Kim, J.-W. Son, J.-H. Lee, F. B. Prinz, M. H. Lee and S. W. Cha, *J. Power Sources*, 2014, **253**, 114–122.
42. B. Timurkutluk, C. Timurkutluk, M. D. Mat and Y. Kaplan, *J. Power Sources*, 2011, **196**, 9361–9364.
43. M.-H. Wu, J.-L. Huang, K.-Z. Fung, H.-C. Liu and D.-F. Lii, *Appl. Surf. Sci.*, 2011, **257**, 7871–7875.
44. A. Jaiswal, C.-T. Hu and E. D. Wachsman, *J. Electrochem. Soc.*, 2007, **154**, B1088–B1094.
45. M. Camaratta and E. Wachsman, *J. Electrochem. Soc.*, 2008, **155**, B135–B142.
46. K. T. Lee, A. A. Lidie, H. S. Yoon and E. D. Wachsman, *Angew. Chem., Int. Ed.*, 2014, **53**, 13463–13467.
47. E. D. Wachsman and K. T. Lee, *Science*, 2011, **334**, 935–939.
48. E. D. Wachsman, *Solid State Ionics*, 2002, **152**, 657–662.
49. J. Y. Park, H. Yoon and E. D. Wachsman, *J. Am. Ceram. Soc.*, 2005, **88**, 2402–2408.
50. J.-Y. Park and E. D. Wachsman, *Ionics*, 2006, **12**, 15–20.
51. J.-Y. Park and E. D. Wachsman, *J. Electrochem. Soc.*, 2005, **152**, A1654–A1659.
52. C. Brahim, A. Ringuedé, E. Gourba, M. Cassir, A. Billard and P. Briois, *J. Power Sources*, 2006, **156**, 45–49.
53. Q. L. Liu, K. A. Khor, S. H. Chan and X. J. Chen, *J. Power Sources*, 2006, **162**, 1036–1042.
54. Z. Wang, X. Huang, Z. Lv, Y. Zhang, B. Wei, X. Zhu, Z. Wang and Z. Liu, *Ceram. Int.*, 2015, **41**, 4410–4415.
55. X. Zhang, M. Robertson, C. Decès-Petit, Y. Xie, R. Hui, W. Qu, O. Kesler, R. Maric and D. Ghosh, *J. Power Sources*, 2008, **175**, 800–805.
56. D. Yang, X. Zhang, S. Nikumb, C. Decès-Petit, R. Hui, R. Maric and D. Ghosh, *J. Power Sources*, 2007, **164**, 182–188.
57. A. A. Solovyev, A. V. Shipilova, I. V. Ionov, A. N. Kovalchuk, S. V. Rabotkin and V. O. Oskirko, *J. Electron. Mater.*, 2016, 1–8.
58. Z. Lu, J. Hardy, J. Templeton, J. Stevenson, D. Fisher, N. Wu and A. Ignatiev, *J. Power Sources*, 2012, **210**, 292–296.
59. J. Qian, Z. Zhu, J. Dang, G. Jiang and W. Liu, *Electrochim. Acta*, 2013, **92**, 243–247.

60. T. Yamaguchi, S. Shimizu, T. Suzuki, Y. Fujishiro and M. Awano, *J. Electrochem. Soc.*, 2008, **155**, B1141–B1144.
61. T. Yamaguchi, S. Shimizu, T. Suzuki, Y. Fujishiro and M. Awano, *J. Electrochem. Soc.*, 2008, **155**, B423–B426.
62. T. Yamaguchi, S. Shimizu, T. Suzuki, Y. Fujishiro and M. Awano, *ECS Trans.*, 2009, **25**, 939–943.
63. S. Hui, D. Yang, Z. Wang, S. Yick, C. Decès-Petit, W. Qu, A. Tuck, R. Maric and D. Ghosh, *J. Power Sources*, 2007, **167**, 336–339.
64. Y. Lee, Y. M. Park and G. M. Choi, *J. Power Sources*, 2014, **249**, 79–83.
65. R. T. Leah, A. Bone, M. Lankin, A. Selcuk, M. Rahman, A. Clare, L. Rees, S. Phillip, S. Mukerjee and M. Selby, *ECS Trans.*, 2015, **68**, 95–107.
66. Z. Gao, D. Kennouche and S. A. Barnett, *J. Power Sources*, 2014, **260**, 259–263.
67. E.-O. Oh, C.-M. Whang, Y.-R. Lee, S.-Y. Park, D. H. Prasad, K. J. Yoon, B.-K. Kim, J.-W. Son, J.-H. Lee and H.-W. Lee, *Ceram. Int.*, 2014, **40**, 8135–8142.
68. N. Jordan, W. Assenmacher, S. Uhlenbruck, V. A. C. Haanappel, H. P. Buchkremer, D. Stöver and W. Mader, *Solid State Ionics*, 2008, **179**, 919–923.
69. F. C. Fonseca, S. Uhlenbruck, R. Nedéléc and H. P. Buchkremer, *J. Power Sources*, 2010, **195**, 1599–1604.
70. D. Perednis and L. J. Gauckler, *Solid State Ionics*, 2004, **166**, 229–239.
71. J. Qian, J. Hou, Z. Tao and W. Liu, *J. Alloys Compd.*, 2015, **631**, 255–260.
72. T. Mukai, S. Tsukui, K. Yoshida, S. Yamaguchi, R. Hatayama, M. Adachi, H. Ishibashi, Y. Kakehi, K. Satoh and T. Kusaka, *J. Fuel Cell Sci. Technol.*, 2013, **10**, 061006.
73. T. Tsai and S. A. Barnett, *J. Electrochem. Soc.*, 1998, **145**, 1696–1701.
74. S. Ji, Y. H. Lee, T. Park, G. Y. Cho, S. Noh, Y. Lee, M. Kim, S. Ha, J. An and S. W. Cha, *Thin Solid Films*, 2015, **591**, 250–254.
75. H. Uchida, H. Suzuki and M. Watanabe, *J. Electrochem. Soc.*, 1998, **145**, 615–620.
76. D. W. Joh, J. H. Park, D. Kim, E D. Wachsman and K. T. Lee, *ACS Appl. Mater. Interfaces*, 2017, **9**, 8443–8449.
77. S.-F. Wang, Y.-F. Hsu, Y.-C. Huang and W.-C. Wei, *Ceram. Int.*, 2011, **37**, 2095–2100.
78. C. Zhang and K. Huang, *J. Power Sources*, 2017, **342**, 419–426.
79. Y.-L. Huang, C. Pellegrinelli, A. S. Painter and E. D. Wachsman, *ECS Trans.*, 2017, **78**, 573–580.
80. J. S. Ahn, M. A. Camaratta, D. Pergolesi, K. T. Lee, H. Yoon, B. W. Lee, D. W. Jung, E. Traversa and E. D. Wachsman, *J. Electrochem. Soc.*, 2010, **157**, B376–B382.
81. J. S. Ahn, D. Pergolesi, M. A. Camaratta, H. Yoon, B. W. Lee, K. T. Lee, D. W. Jung, E. Traversa and E. D. Wachsman, *Electrochem. Commun.*, 2009, **11**, 1504–1507.

82. K. T. Lee, D. W. Jung, M. A. Camaratta, H. S. Yoon, J. S. Ahn and E. D. Wachsman, *J. Power Sources*, 2012, **205**, 122–128.
83. K. T. Lee, D. W. Jung, H. S. Yoon, A. A. Lidie, M. A. Camaratta and E. D. Wachsman, *J. Power Sources*, 2012, **220**, 324–330.
84. L. Zhang, C. Xia, F. Zhao and F. Chen, *Mater. Res. Bull.*, 2010, **45**, 603–608.
85. J. G. Lee, M. G. Park, H. H. Yoon and Y. G. Shul, *J. Electroceram.*, 2013, **31**, 231–237.
86. J. Hou, L. Bi, J. Qian, Z. Zhu, J. Zhang and W. Liu, *J. Mater. Chem. A*, 2015, **3**, 10219–10224.
87. J. Hou, F. Liu, Z. Gong, Y. Wu and W. Liu, *J. Power Sources*, 2015, **299**, 32–39.
88. Y. J. Leng and S. H. Chan, *Electrochem. Solid-State Lett.*, 2006, **9**, A56–A59.
89. L. Zhang, L. Li, F. Zhao, F. Chen and C. Xia, *Solid State Ionics*, 2011, **192**, 557–560.
90. J. Hou, L. Bi, J. Qian, Z. Gong, Z. Zhu and W. Liu, *J. Power Sources*, 2016, **301**, 306–311.
91. S. Boyapati, E. D. Wachsman and N. Jiang, *Solid State Ionics*, 2001, **140**, 149–160.
92. S. Boyapati, E. D. Wachsman and B. C. Chakoumakos, *Solid State Ionics*, 2001, **138**, 293–304.
93. D. W. Jung, K. L. Duncan and E. D. Wachsman, *Acta Mater.*, 2010, **58**, 355–363.
94. S. Omar, E. D. Wachsman and J. C. Nino, *Appl. Phys. Lett.*, 2007, **91**, 144106.
95. A. G. Jolley and E. D. Wachsman, *ECS Trans.*, 2017, **78**, 355–360.
96. A. Jaiswal, A. Pesaran, S. Omar and E. D. Wachsman, *ECS Trans.*, 2017, **78**, 361–370.
97. A. L. Ruth, K. T. Lee, M. Clites and E. D. Wachsman, *ECS Trans.*, 2014, **64**, 135–141.
98. S. Sanna, V. Esposito, J. W. Andreasen, J. Hjelm, W. Zhang, T. Kasama, S. B. Simonsen, M. Christensen, S. Linderoth and N. Pryds, *Nat. Mater.*, 2015, **14**, 500–504.
99. T. Ishihara, J. Yan, M. Enoki, S. Okada and H. Matsumoto, *J. Fuel Cell Sci. Technol.*, 2008, **5**, 031205.
100. Y.-W. Ju, H. Eto, T. Inagaki, S. Ida and T. Ishihara, *Electrochem. Solid-State Lett.*, 2010, **13**, B139–B141.
101. T. Ishihara, H. Eto and J. Yan, *Int. J. Hydrogen Energy*, 2011, **36**, 1862–1867.
102. J. Qian, Z. Zhu, G. Jiang and W. Liu, *J. Power Sources*, 2014, **246**, 556–561.
103. J. Yan, H. Matsumoto, M. Enoki and T. Ishihara, *Electrochem. Solid-State Lett.*, 2005, **8**, A389–A391.
104. Z. Bi, B. Yi, Z. Wang, Y. Dong, H. Wu, Y. She and M. Cheng, *Electrochem. Solid-State Lett.*, 2004, **7**, A105–A107.
105. Z. Bi, M. Cheng, Y. Dong, H. Wu, Y. She and B. Yi, *Solid State Ionics*, 2005, **176**, 655–661.

106. D. E. Fowler, J. M. Haag, C. Boland, D. M. Bierschenk, S. A. Barnett and K. R. Poeppelmeier, *Chem. Mater.*, 2014, **26**, 3113–3120.
107. Z. Bi, Y. Dong, M. Cheng and B. Yi, *J. Power Sources*, 2006, **161**, 34–39.
108. E. Fabbri, D. Pergolesi, A. D'Epifanio, E. Di Bartolomeo, G. Balestrino, S. Licoccia and E. Traversa, *Energy Environ. Sci.*, 2008, **1**, 355–359.
109. J. Qian, W. Sun, Q. Zhang, G. Jiang and W. Liu, *J. Power Sources*, 2014, **249**, 131–136.
110. J. Qian, W. Sun, Z. Shi, Z. Tao and W. Liu, *Electrochim. Acta*, 2015, **151**, 497–501.
111. L. Zhao, B. He, J. Shen, F. Chen and C. Xia, *Electrochem. Commun.*, 2011, **13**, 450–453.
112. W. Sun, Z. Shi, Z. Wang and W. Liu, *J. Membr. Sci.*, 2015, **476**, 394–398.
113. K.-Z. Fung and T.-Y. Chen, *Solid State Ionics*, 2011, **188**, 64–68.
114. Y. Mishima, H. Mitsuyasu, M. Ohtaki and K. Eguchi, *J. Electrochem. Soc.*, 1998, **145**, 1004–1007.
115. G. Hao, X. Liu, H. Wang, H. Be, L. Pei and W. Su, *Solid State Ionics*, 2012, **225**, 81–84.
116. F. Liu, J. Dang, J. Hou, J. Qian, Z. Zhu, Z. Wang and W. Liu, *J. Alloys Compd.*, 2015, **639**, 252–258.

CHAPTER 2

High Temperature Co-electrolysis – A Route to Syngas

MIGUEL A. LAGUNA-BERCERO* AND ALODIA ORERA

Instituto de Ciencia de Materiales de Aragón, C/Pedro Cerbuna 12, Zaragoza, E50009, Spain
*E-mail: malaguna@unizar.es

2.1 Introduction

Since the industrial revolution of the 19th century, the emissions of greenhouse effect gases (GHG) to the atmosphere have increased dramatically. For example, the CO_2 atmospheric content was more than 25% higher in 2000 than its average value over the past 400 000 years, the majority of those CO_2 emissions being due to the massive use of fossil fuels. In fact, more than 85% of the energy demand is met by the use of coal, oil and natural gas. In the transport sector, energy is provided mainly by hydrocarbons (*i.e.* gasoline, diesel, alcohols, dimethyl ether, *etc.*), oil being the dominant feedstock to produce these fuels. Due to the shortage of fossil fuel reserves as well as the fluctuations in their price, it seems necessary to search for alternatives. Furthermore, in order to mitigate greenhouse gas emissions, carbon capture and storage (CCS) technologies are under continuous development.[1] An overview of these CCS technologies is summarised in Figure 2.1.

In addition to CCS technologies, significant research into the use of low-carbon energy technologies is being conducted, including renewable

Inorganic Materials Series No. 5
Energy Storage and Conversion Materials
Edited by Stephen Skinner

Published by the Royal Society of Chemistry, www.rsc.org

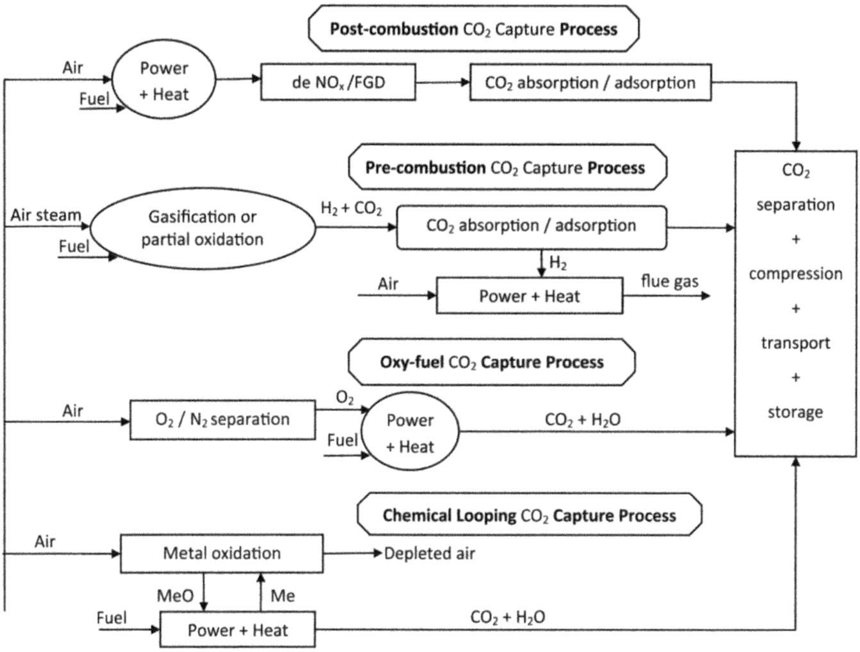

Figure 2.1 Scheme showing the different CO_2 capture technologies. Reproduced from ref. 1, https://doi.org/10.1016/j.rser.2014.07.093, under the terms of the CC BY 3.0 licence, https://creativecommons.org/licenses/by/3.0/.

energies (solar,[2,3] wind,[4,5] hydropower,[6] geothermal[7,8]), nuclear power[9,10] or the use of biofuels.[11,12] It is well known that most renewable energies present an intermittent nature, which is a great challenge in order to control the grid, not to mention the important challenges associated with nuclear power. In any case, fossil fuels coupled with CCS is at the moment the only technology with the potential to significantly reduce greenhouse emissions while allowing energy needs to be met securely and affordably in developed countries, as is, for example, the case in the United Kingdom.[13]

Another alternative, next-generation technology that will contribute to the reduction of emissions is the use of fuel cells, existing in several variations of their basic concept that enable their operation over a wide range of temperatures. In this case, hydrogen would be the main energy carrier for clean generation. However, there are still some issues related to hydrogen storage, in addition to the difficulties associated with the infrastructure needed for its distribution. For example, hydrogen must usually be pressurised or liquefied to overcome its low volumetric energy density. Nowadays, those costly operations make hydrogen uncompetitive with present oil prices. However, there are numerous hydrogen storage options still under technological development, as for example, chemical storage using metal hydride materials,[14] carbon nanotubes[15] or metal–organic frameworks (MOFs).[16] Having these limitations in mind, solid oxide fuel cells

(SOFCs) also present very important advantages over other fuel cell systems, as they can employ different fuels (hydrocarbons, liquid fuels *etc.*) apart from hydrogen.[17–19] There has been great development of SOFC technology in the past 20–30 years, including important efforts devoted to the use of SOFCs in reverse mode as solid oxide electrolyser cells (SOECs). In this case, the typical reactions occurring in a SOFC are reversed by applying a voltage higher than the open circuit voltage (OCV) between the cell electrodes. Water stands as the reactant and is supplied to the cathode side of the cell (the anode in SOFC mode). Oxide ions are transported to the anode (cathode in SOFC mode) through the electrolyte, and hydrogen is produced at the cathode. Thus, it is possible to produce both hydrogen and oxygen by steam electrolysis, according to reactions (2.1) and (2.2). The overall reaction (2.3) consists in the production of hydrogen and oxygen from water. To avoid confusion, the electrodes will be hereinafter referred to as the fuel electrode (cathode in SOEC mode) and the air electrode (anode in SOEC mode).

$$H_2O + 2e^- \rightarrow H_2 + O^{2-} \quad \text{(fuel electrode)} \tag{2.1}$$

$$O^{2-} \rightarrow 1/2\ O_2 + 2e^- \quad \text{(air electrode)} \tag{2.2}$$

$$H_2O \rightarrow H_2 + 1/2\ O_2 \qquad \Delta H_{298\,K} = 286\ \text{kJ mol}^{-1} \tag{2.3}$$

High-temperature steam electrolysis (HTSE) is becoming of great interest because of its most favourable thermodynamic conditions, which lead to a lower electrical energy consumption than low-temperature electrolysis. By increasing the temperature, energy demand is significantly lowered by the vaporisation of water into steam.[20]

The minimum electric energy supply required for the electrolysis process is equal to the change in the Gibbs free energy (ΔG):

$$\Delta G = \Delta H - T\Delta S \tag{2.4}$$

where ΔH is the enthalpy change, T the temperature and ΔS the entropy change. The electrical energy demand, ΔG, decreases with increasing temperature. Operating at higher temperature can, therefore, decrease the cost of the hydrogen produced, especially if the increase in heat energy demand can be fulfilled by an external heat source, such as nuclear power, renewable energy, or waste heat from high-temperature industrial processes. A detailed description of SOEC operation, performances and different associated issues, can be found in several complete reviews.[21–28]

In a SOEC, it is also possible to perform the electrochemical reduction of carbon dioxide into carbon monoxide and oxygen (reaction (2.5)).

$$CO_2 \rightarrow CO + ½\ O_2 \qquad \Delta H_{298\,K} = 283\ \text{kJ mol}^{-1} \tag{2.5}$$

Although both H_2O and CO_2 electrolysis require about the same amount of electrical energy, in practice the latter operating mode usually exhibits poorer performance compared to steam electrolysis.[29–31] Additionally, depending on the CO concentration, carbon could be deposited according to the Boudouard reaction (2.6) and/or CO electro-reduction (2.7) could occur at high CO_2 conversion rates. This carbon deposition is especially concerning when nickel-based fuel electrodes are used, since nickel is known to catalyse the dissociation of carbon-containing gases.

$$2CO \rightarrow CO_2 + C \tag{2.6}$$

$$CO + 2e^{-} \rightarrow C + O^{2-} \tag{2.7}$$

This coke formation is patently undesirable, as carbon will reduce the number of active sites for the electrolysis reaction thus reducing the electrochemical performance of the cell. The formation of carbon can even result in the removal of nickel from the electrode, ultimately causing its fracture.[32,33] In any case, coke deposition is only expected in SOECs when operating with high CO concentrations, as the equilibrium of the Boudouard reaction (2.6) is shifted toward CO at typical CO_2 electrolysis conditions (at temperatures above 800 °C and *p*CO below 90%).[34] It is also well-known that the addition of steam suppresses coke formation during the reforming.[35]

Another alternative scenario for the transitory period from fossil fuels to a clean energy economy could be the use of greener synthetic liquid fuels, as they could easily make use of the current available infrastructure for both storage and transportation. Moreover, liquid fuels present higher energy density than hydrogen. In this respect, SOECs are also capable of converting electrical energy into chemical energy by splitting steam and CO_2 into synthesis gas (syngas), which is a mixture of H_2 and CO in a variable percentage, which can be used as an efficient energy storage method. Although this syngas cannot be burnt directly, it is used as a fuel source or as an intermediate to produce other chemicals, including liquid fuels. The current technologies to produce syngas include steam methane reforming (SMR), partial oxidation of hydrocarbons, gasification of coal or biomass and the co-electrolysis of steam and CO_2.[36] If electricity is obtained from a carbon-free source, syngas can be then used as a feedstock to produce different cleaner synthetic fuels (such as synthetic methane, methanol or dimethyl ether, DME) by a Fischer–Tropsch (FT) process. This technology opens new opportunities in the fuel sector for recycling CO_2 into synthetic hydrocarbon fuels using renewable energies.[37–43] Coupling these renewable energies with co-electrolysis could solve the problems associated with their intermittent nature, since fuels would be produced at peak energy generation and could be used at off-peak periods. Alternatively, power-to-liquid (P2L) processes could also be carried out, as is the case for the production of methanol, DME, *etc.* Ideally

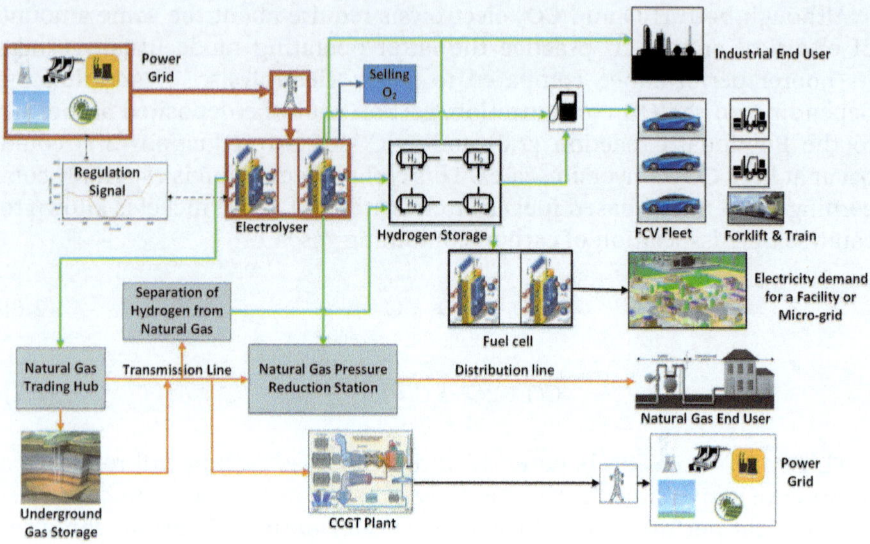

Figure 2.2 Schematic representation of power-to-gas. Reproduced from ref. 47, https://doi.org/10.3390/en10081089, under the terms of the CC BY 4.0 licence, https://creativecommons.org/licenses/by/4.0/.

it could also be used to recycle the CO_2 emitted by different industries such as those in the steel, cement, or waste incinerator sectors. In addition, oxygen by-products can further increase the added value of the co-electrolysis process, if used for oxy-combustion purposes, for example. Finally, the fuel produced could later be used in the same device to generate power in reversible mode (SOFC).

Therefore, power-to-gas (P2G) and P2L technology would involve the use of the excess carbon-free electricity coming from renewable or nuclear sources, together with the use of waste CO_2 from industrial facilities, to produce highly valuable products, thus storing electrical energy in some way.[44–52] A schematic representation of the P2G process is shown in Figure 2.2.[47]

P2G is a key technology for the development of hydrogen technologies and fuel cells, as part of the European Strategic Energy Technology Plan (Europe 2020 and Europe 2050).[53] It must be taken into account that, for example, several car manufacturers have announced that commercialisation of H_2 vehicles in Europe (Toyota, Honda and Hyundai are already marketing fuel cell cars mainly in USA and Japan) is close and thus hydrogen produced from electrolysis could be essential in order to reduce GHG emissions in the transport sector. In addition, the existing natural gas pipeline system could also be used for supplying hydrogen mixed with natural gas. If that is the case, hydrogen/natural gas mixtures can then be delivered to end users with or without separation. As hydrogen and natural gas present different physical and chemical properties, the allowed hydrogen fraction is limited to up to 20%.[44] In conclusion, as P2G could use the current natural gas infrastructure, it would provide a natural transition to the hydrogen economy.

Natural gas has become very important in recent years due to its abundant distribution infrastructure, and its expected low price in the near future. In the short term, natural gas provides an opportunity for the immediate phase-out of coal. For example, natural gas replacement of diesel fuel in large fleet vehicles would contribute to the reduction of air pollution.[54]

Another method for the convenient exploitation of the natural gas pipeline system is the methanation of hydrogen with carbon dioxide, ideally from waste sources, to produce methane, in accordance with the Sabatier reaction (2.8).

$$4H_2 + CO_2 \rightleftharpoons CH_4 + 2H_2O \qquad \Delta H_{298\,K} = -164.9\ \text{kJ mol}^{-1} \qquad (2.8)$$

This synthetic gas would be identical to conventional natural gas and could be stored and transported over the natural gas system and used without restrictions. The present approach is being currently considered by many important companies by using fuel cells operating in reverse mode. In fact, several companies working traditionally in SOFCs are now focusing their efforts on their reverse operation mode (SOECs). This is the case, for example, of Haldor Topsoe. They announced their decision to put their entire SOFC technology programme on hold in 2014 to focus on the development of selected applications of SOECs. This company has recently proved that the methanation technology at a demonstration plant is able to produce 10 m^3 of biogas per hour while ensuring that the gas is of at least the same quality as natural gas.[55] In addition to the methanation work line, Haldor Topsoe has achieved a breakthrough using SOECs to produce CO from CO_2 and electricity (CO_2 electrolysis). With this approach, CO could be produced on-site at the final destination where it will be used, as would be the case of numerous industrial processes such as the production of medical goods, electronics or fine chemicals. In a similar regard, an innovative concept was recently suggested at Northwestern University[56] using methane-rich synthesis gas as an intermediate product for energy storage, as illustrated in Figure 2.3.

An excellent overview of the different sustainable energy scenarios in which solid oxide cells (SOCs) are used was described by Hansen[57] The author concluded that, although there is no doubt about the important progress in terms of materials and fabrication of SOC stacks, it is still very difficult to compete with fossil fuels as their current low prices have not taken into account local and global environmental cost.

It can be concluded that, with feasible technology development and mass production of the different components, CO_2-recycled hydrocarbon fuels can be produced at the scale needed to replace transportation fuels at a price that is competitive with most of the conventional fossil-derived hydrocarbons, especially if oil price is high and the costs of consequent CO_2 sequestration

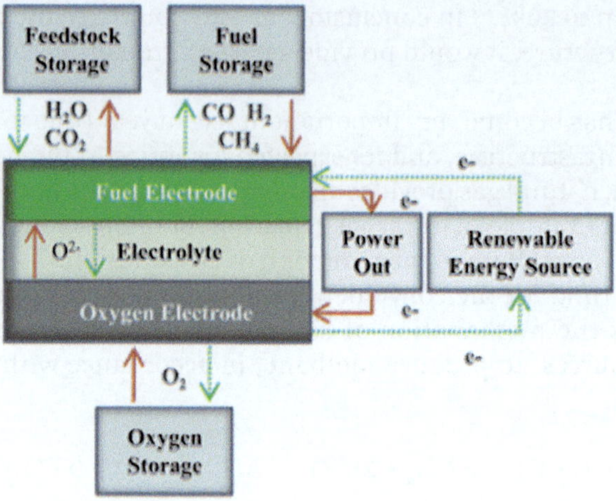

Figure 2.3 Simplified reversible SOC system using hydrogen/methane rich fuel. Reproduced from ref. 56 with permission from the Royal Society of Chemistry.

are taken into account. The potentially greater sustainability of CO_2-recycled fuels over fossil or biomass derived fuels, as well as their independence from the geographic and supply restrictive issues of conventional fuels, could also assure these CO_2-recycled fuels an advantage and a leading position in the market.

2.2 Description of the Co-electrolysis Process

During co-electrolysis of CO_2 and steam, endothermic reductions of both H_2O and CO_2 take place at the same time according to reactions (2.1) and (2.5), and the overall reaction (2.9):

$$H_2O + CO_2 \rightarrow H_2 + CO \tag{2.9}$$

In the electrochemical cell, steam and CO_2 are fed to the fuel electrode, where H_2 and CO are formed. Oxide ions (O^{2-}) are conducted through the solid electrolyte by an applied electric field and are recombined to form gaseous oxygen at the air electrode.

Co-electrolysis is more energy efficient than the two separate processes of steam and CO_2 electrolysis, as the energy consumption for co-electrolysis is slightly lower than that for CO_2.[58]

The reverse water gas shift reaction (RWGS) (2.10), which is a kinetically fast equilibrium reaction at high temperatures, may also occur in parallel during the co-electrolysis process.

$$H_2 + CO_2 \rightarrow H_2O + CO \qquad \Delta H_{298\,K} = +41\ \text{kJ mol}^{-1} \tag{2.10}$$

Another secondary reaction that can occur during syngas production is the Sabatier reaction (2.11):

$$3H_2 + CO \rightarrow H_2O + CH_4 \tag{2.11}$$

There are still many fundamental issues concerning the electrochemical mechanism of the co-electrolysis process to be learned, as it is much more complicated than single electrolysis mechanisms. In fact, there is no agreement in the literature about whether the RWGS reaction contributes to CO production or not.[59,60] For example, it has been reported that at temperatures above 820 °C the equilibrium of the WGS reaction is shifted towards CO and H_2O (RWGS reaction).[61]

Thermodynamics for the co-electrolysis of both H_2O and CO_2 are rather similar to that of steam electrolysis explained above. At high temperature, a significant part of the total energy demand can be provided in the form of heat, as can be observed in Figure 2.4.[62] Then, through the use of the Joule heat that is produced by passing an electrical current through the cell, the electrical consumption of energy is significantly reduced and the efficiency of the system increased.

2.3 Considerations

In general, it has been reported that H_2O electrolysis outperforms pure CO_2 electrolysis, as higher cathodic overpotentials are found for CO_2 splitting.[31,59] Remarkably, the performance of CO_2 and steam co-electrolysis has been found to be very similar to that of steam electrolysis or at least in-between those of H_2O and CO_2 reduction. As a consequence, some uncertainty exists

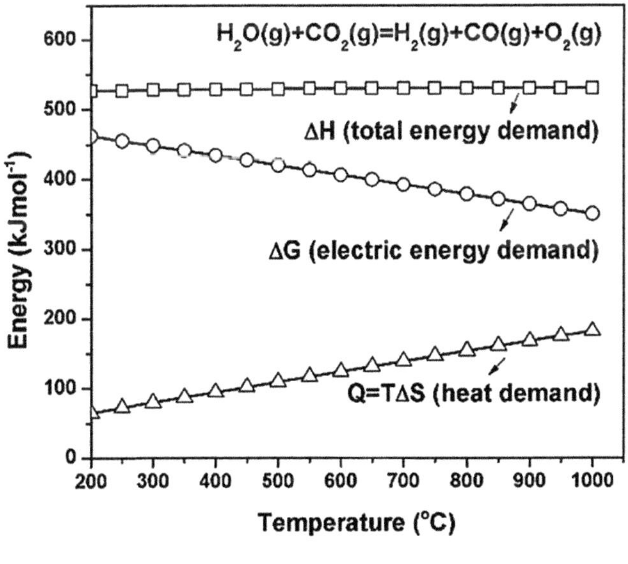

Figure 2.4 Thermodynamic calculation as a function of temperature for the co-electrolysis process. Reproduced from ref. 62 with permission from Elsevier, Copyright 2015.

about the extent of direct CO_2 electrolysis and the magnitude of the chemical reaction of CO_2 with electrochemically produced hydrogen *via* the RWGS reaction.

In this sense, Stoots *et al.*[59] found that the principal electrochemical reaction taking place in co-electrolysis mode was steam reduction, as they obtained similar area-specific resistances (ASRs) for steam and co-electrolysis processes. They reported an ASR of 1.38 Ω cm^2 for co-electrolysis at 800 °C, close to that of 1.36 Ω cm^2 for steam electrolysis and much lower than the 3.84 Ω cm^2 found for HT CO_2 electrolysis. These figures led them to conclude that during CO_2/steam co-electrolysis, CO is primarily formed *via* the RWGS reaction. Similar findings were reported by Zhan *et al.*[63] and also by Chen *et al.*[64] They analysed the gas transport in the porous electrode and the adsorption/desorption processes over the nickel surface, and found that cathodic reactions are probably dominated by the reduction of steam to hydrogen, whereas carbon monoxide is mainly produced *via* the RWGS. However, the results of Ebbesen *et al.*[65] and Zhan *et al.*[66] were contrary to previous findings. Graves and co-workers found that the ASR values of co-electrolysis are between those of CO_2 and H_2O electrolysis, and thus both H_2O and CO_2 are reduced electrochemically in the co-electrolysis mode,[31] which is consistent with modelling studies performed by Ni.[67] A very interesting contribution was made by Li *et al.*,[68] studying the effects of different cathode characteristics (porosity, particle diameter and ionic conductivity) and operating conditions (gas composition and temperature), and then simulating the relative polarisation curves, the ratio of CO_2 participating in each reaction, the mass and charge transfer

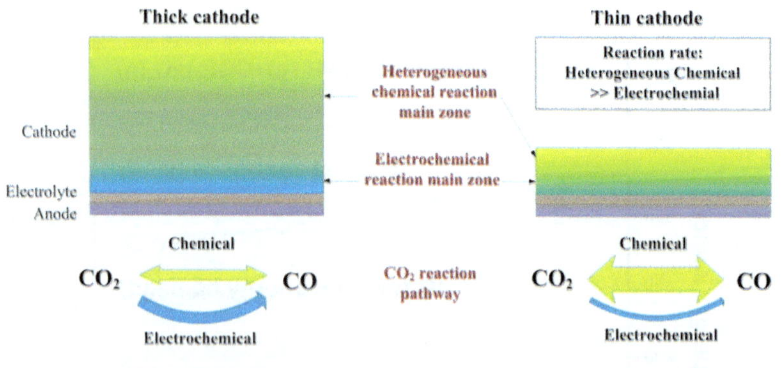

	Governing equation	Boundary of main zones*	Determinant of zone size
Heterogeneous chemical reaction main zone	Mass balance $\nabla\left(-D_k^{\mathrm{eff}}\nabla c_{k,g}\right) = R_{k,g}$	90% of $\Delta c_{k,g}$ at OCV	Mass transfer flux $D_k^{\mathrm{eff}}\nabla c_{k,g}$
Electrochemical reaction main zone	Charge balance $\nabla\left(-\sigma_{\mathrm{ion}}^{\mathrm{eff}}\nabla V_{\mathrm{ion}}\right) = Q_{\mathrm{ion}}$	90% of ΔV_{ion} at 1.4V	Charge transfer flux $\sigma_{\mathrm{ion}}^{\mathrm{eff}}\nabla V_{\mathrm{ion}}$

* Artificial definition in the simulation

Figure 2.5 Heterogeneous chemical/electrochemical reaction main zones, CO_2 reaction pathway and mathematical analysis. Reproduced from ref. 68 with permission from Elsevier, Copyright 2015.

flux and the zones where heterogeneous chemical/electrochemical reactions mainly take place. A scheme showing the results regarding the study of the main reaction zones is shown in Figure 2.5. According to their mathematical analysis, the mass transfer flux determines the main zone of heterogeneous chemical reactions, while the charge transfer flux determines that of the electrochemical reactions. When the zone of the heterogeneous chemistry is enlarged, the polarisation curves of CO_2/H_2O co-electrolysis would shift to those of H_2O electrolysis. Meanwhile, in the case of the enlarged electrochemistry zone, more CO_2 could react through this pathway, and the polarisation curves of CO_2/H_2O co-electrolysis tend to those of CO_2 electrolysis.

Zhang *et al.*[69] also performed a deep analysis in order to understand the electrochemical reactions taking place in the co-electrolysis of steam and carbon dioxide in conventional cells. They studied the impact of CO_2, H_2O, CO and H_2 contents on the polarisation systematically, concluding that ASR data for co-electrolysis are between those of pure steam and CO_2, with those for pure CO_2 electrolysis being about three times those of pure H_2O electrolysis. Thus, with a decrease in the CO_2 content from 50% to 10%, the ASR decreased significantly, implying that an appropriate control of CO_2 concentration is required. Moreover, their results imply that the performance of the co-electrolysis process could be dominated by a CO_2 diffusion process. Besides, the charge transfer processes associated with CO–CO_2 oxidation/reduction are slower than the corresponding H_2–H_2O processes. Therefore, an appropriate electrode microstructure and a better understanding of the CO–CO_2 exchange kinetics are needed to improve gas diffusion.

In order to elucidate the reaction mechanisms, a recent model was proposed by Feng *et al.*[70] They investigated the electro-reduction of CO_2 and electro-oxidation of CO on a ceria-based cell by using *in operando* ambient-pressure X-ray photoelectron spectroscopy (AP-XPS). They found that during CO_2 electrolysis, the steady-state concentration of CO_3^{2-} increases over a 400 μm-wide active region on the ceria electrode surface at positive applied voltages. However, the CO_3^{2-} concentration and surface Ce^{3+} concentration decrease slightly during CO electro-oxidation. The increase of CO_3^{2-} at positive biases suggests the formation of an intermediate product requiring the pre-coordination of CO_2 to the ceria surface to form this intermediate ($CO_{2(g)} + O^{2-}{}_{(surface)} \rightarrow CO_3^{2-}{}_{(surface)}$). This reaction step precedes a rate-limiting electron-transfer process to give CO and oxide ions ($CO_3^{2-}{}_{(surface)} + 2Ce^{3+} \rightarrow CO_{(g)} + 2O^{2-}{}_{(surface)} + 2Ce^{4+}$). The reaction is taking place *via* a stable carbonate intermediate, which is coupled to the surface Ce^{3+} concentration. They concluded that the strong adsorbate–adsorbate interaction explains the coverage-dependent reaction mechanism.

Regarding the important issue of carbon deposition on the electrodes, there are several studies such as those from Sun *et al.*[71] claiming that during co-electrolysis, carbon formation is suppressed due to the CO removal by H_2O *via* the RWGS (eqn (2.7)).[72] However, other investigations showed that carbon is formed during co-electrolysis when operating at high voltages and fuel utilisation. For example, as shown in Figure 2.6,[73,74] carbon fibres have been detected at the electrode/electrolyte interfaces, limiting the number of TPBs (triple phase boundaries). The study of the mechanism for this carbon

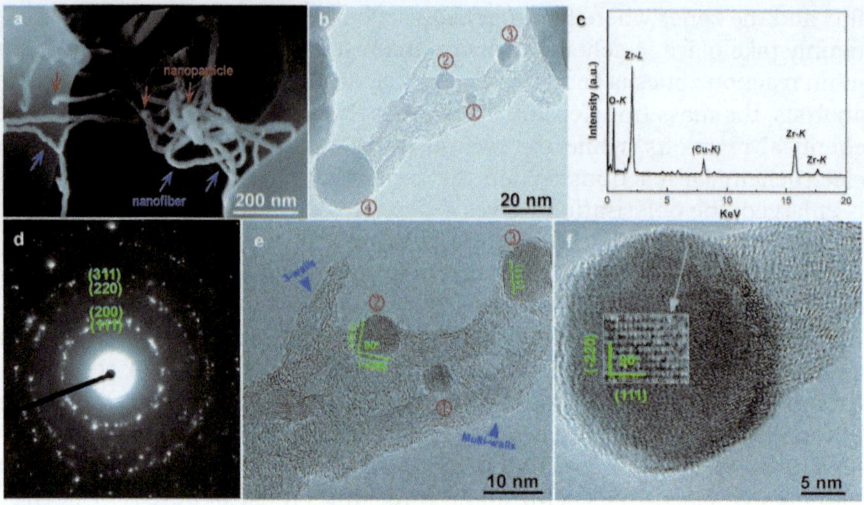

Figure 2.6 (a) SEM image of the carbon nanostructures at the Ni–YSZ cathode/YSZ electrolyte interface; (b) TEM image of the carbon nanotubes and nanoparticles; (c) EDX and (d) electron diffraction of the nanoparticles; (e), (f) HRTEM images of the structures shown in (b). Reproduced from ref. 74 with permission from John Wiley and Sons, © 2014 WILEY-VCH Verlag GmbH & Co. KGaA, Weinheim.

deposition suggested that the growth of carbon nano-structures was mainly due to the presence of zirconia (in contrast to what happens in SOFCs, where nickel acts as a catalyser for the coking). This is explained by zirconia turning into a MIEC (mixed ionic and electronic conductor) during high cathodic polarisation through the reduction of Zr (IV). To conclude, in order to avoid carbon deposition in the cell it is very important to select the optimal operational conditions and/or alternative carbon-resistant electrodes, as discussed for CO_2 electrolysis.

In conclusion, the great influence of both the chemical kinetics and mass transport in the fuel electrode has been established, since they not only determine the dominant reaction pathway, but also the syngas yield and its selectivity.[75] Thus, in order to maximise the production of high-quality syngas and thus store electrical energy effectively and efficiently, there is a requirement for a complete optimisation of different parameters. These include the operating conditions, the incorporation of catalysts for both the steam electrolysis and the RWGS reaction, and the fuel electrode microstructure; all of which affect the chemical kinetics and mass transport.

2.4 History

Archibald Cochrane (1749–1831), the ninth Earl of Dundonald, was probably the first to manufacture tar from coal.[76] He used the obtained product in a lamp for lighting his family estate, but no commercial development followed. This technology was also used in 1820 to illuminate Paris. However, when electricity and fossil fuels were discovered, syngas started to lose economic interest. Later on, during the Second World War (1939–1945), it became more important either as direct fuel or as a raw material to produce liquid fuels for countries with limited access to coal and hydrocarbons stocks. The origin of electrochemical conversion of CO_2 comes from the US space programmes in the 1960s, where removal of CO_2 from breathing air became a fundamental requirement for life support in manned spaceflights. In those missions, oxygen was produced by CO_2 electrolysis using a SOC, this methodology being also proposed for life support and propulsion in submarines. The first studies of CO_2 electrolysis are probably those of Chandler *et al.* in 1961–1962,[77,78] where they proposed an electrolytic method for the regeneration of oxygen from carbon dioxide to design a closed-cycle life-support system. In these first prototypes, the electrodes employed were based on molybdenum oxides and/or carbon, while the solid electrolyte was a mixture of thorium oxide with either yttrium or lanthanum oxides. The operating temperatures ranged from 400 to 900 °C, and the production of carbon monoxide was confirmed by gas chromatographic analyses. Another of those pioneering works was performed at the Lockheed Palo Alto Research Laboratory in 1965–1966.[79] In this work, dissociation of CO_2 and H_2O was performed using a $(ZrO_2)_{0.85-x}(CeO_2)_x(CaO)_{0.15}$ solid electrolyte, where x varied from 0 to 0.45. They employed tubular zirconia–calcia cells and planar zirconia–ceria–calcia cells, with porous platinum electrodes, and a large number of electrolysis experiments were performed using air, oxygen, CO_2–H_2O, and CO–CO_2 mixtures in the temperature range from 840 to 1007 °C, demonstrating the concept. They also found that a large amount of ceria as dopant does not give reproducible values for oxygen current efficiencies, and this behaviour seemed to be related to changes in the oxidation state of the electrolyte.

The idea of CO_2-recycled synthetic hydrocarbon fuels had particular interest during the oil crisis in the 1970s. In 1976, Steinberg patented a method and system for synthesising methanol from the CO_2 obtained from air, using electric power and an alkaline electrolyser.[80] The CO_2 captured from the atmosphere is absorbed by a solution of KOH to form K_2CO_3, which is electrolysed to produce liquid methanol as fuel. The use of nuclear fission or fusion for supplying the electricity and heat was also proposed. Regarding the CO_2 source, other alternatives considered in this work were removing CO_2 from seawater and using exhaust CO_2 from industrial plants. The idea was further explored during the period from 1980 to 2000, when several patents were filled. For example, Lewis *et al.*[81] and also Corbett *et al.*[82] proposed similar methods for the production of synthetic hydrocarbons from air using low-temperature electrolysers. The scheme proposed by Lewis *et al.* is shown in Figure 2.7. Yamauchi *et al.*[83] developed a process involving the use of nuclear heat in a high-temperature, gas-cooled nuclear reactor. This heat was used for decomposing the steam into hydrogen by means of a steam electrolyser, and this hydrogen was used for synthesising

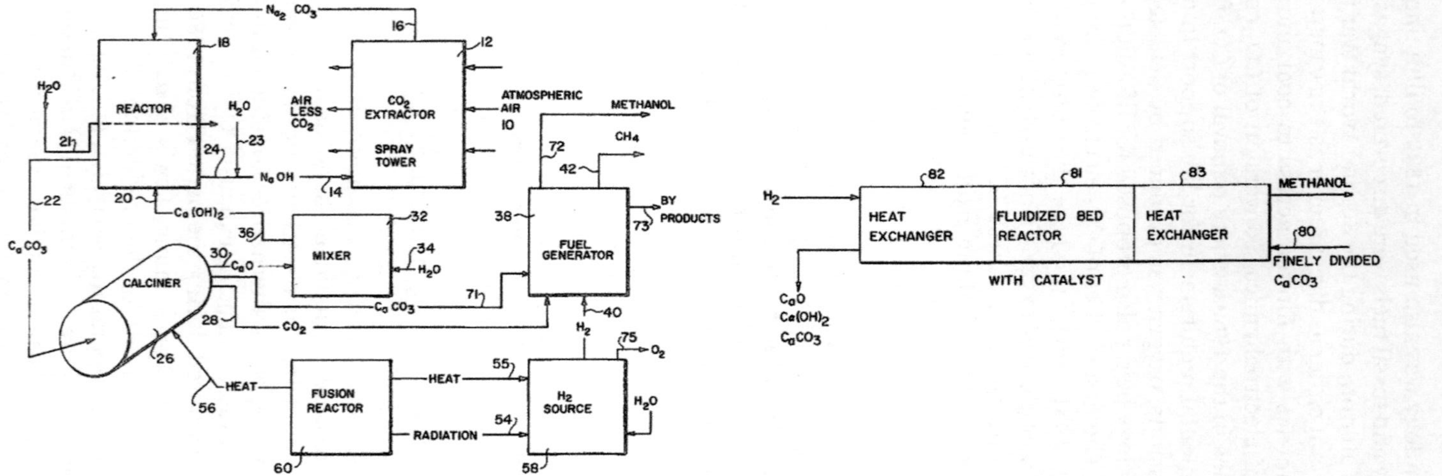

Figure 2.7 (Left) block diagram of the overall processing system proposed by Lewis *et al.*, and (right) block diagram of that part of the process for converting carbonates to combustible fuels, as described in ref. 81.

methanol together with carbon dioxide obtained from a carbon dioxide source. In addition, a power-generating plant is provided in which oxygen-enriched air, generated by the mentioned steam electrolyser, is used as combustion air. Finally, Jensen *et al.*[84] proposed that the red shift in the ultraviolet absorption spectrum of CO_2 with increasing temperature permits the use of sunlight to photolyse CO_2 to CO. They projected preheating CO_2 to near 1800 K, then expose the preheated CO_2 to sunlight producing CO, O_2 and O, and quenching the hot product with room temperature CO_2. The excess thermal energy could be used to produce electricity and to heat additional CO_2 for subsequent process steps. CO could be also used to generate H_2 by the shift reaction or to synthesise methanol.

2.5 Current Large-scale Projects

In Europe, Haldor Topsoe is probably the most active company in the field. Nowadays, the SOEC team at Haldor Topsoe is composed of about 40 people, being most likely one of the largest SOEC groups in the world. They have participated in several projects over the past decade, such as GreenSynFuels,[85] Biogas-SOEC,[86] CO_2 Electrofuels[87] or Green Natural Gas.[88] Already, in 2011, they constructed a three-reactor pilot plant for purifying biogas, which was under operation at Århus University for 1000 h, as part of the project 'Towards the Methane Society' (2011–2012).[89] The idea of the project was the coupling of electrolysis, biogas upgrading and the natural gas grid. For this, an electrolyser would generate hydrogen and heat based on wind power, then, this hydrogen would react with CO_2 (biogas) to form CH_4 and high-grade steam, and finally the natural gas grid would be used for the distribution and storage of the generated methane. In addition, Chen *et al.*[90] performed durability studies using a Topsoe Fuel Cell (TOFC®) 10-cell stack, containing three different types of Ni/yttria-stabilised zirconia (YSZ) electrode supported 12 × 12 cm SOEC. The stack was operated at 800 °C and −0.75 A cm^{-2} with 60% conversion for a period of 1000 h. After the experiment, from the three types of cells, two of them showed long-term degradation, while the other one showed activation during the entire electrolysis period. These results suggest that long-term electrolysis is feasible without notable degradation under those operating conditions. Current plans of Haldor Topsoe A/S include the recent installation of a 40 kW SOEC unit prototype in Foulum for the production of synthesis gas coupled with methanation, as part of a project on converting biomass and wind power into synthetic fuels.[91] The expected consumption of electricity is 13–14 kWh per Nm^3 methane. They have also successfully proven a complete sulfur clean up. Their economic analysis has shown that upgrading by SOEC can compete with traditional upgrade by CO_2 removal. By the use of this technology, it is estimated that about 10% of Danish energy consumption could be met saving 1 MT CO_2 per capita per year.

Sunfire GmbH has been involved since 2014 in the HELMETH project (Integrated High-Temperature Electrolysis and Methanation for Effective

Power to Gas Conversion).[92] Other partners include KIT (Karlsruhe Institute of Technology), Politecnico di Torino, EIRC (European Research Institute of Catalysis), National Technical University of Athens, Ethos Energy Italia and the German Technical and Scientific Association for Gas and Water. Their goal is to achieve efficiencies above 85% in P2G systems using a 15 kW SOEC electrolyser operating at about 860 °C and −0.5 A cm^{-2}, coupled to a CO_2 methanation system using two reactors in series at 300 °C and 30 bar with intermediate water removal.[93] The Sunfire SOEC electrolyser is composed of 3YSZ (3 mol% yttria-stabilised zirconia) electrolyte supported cells with an active area of 128 cm^2 per cell, connected in stacks of 30 cells. Ni-CGO (cerium gadolinium oxide) is used as the fuel electrode and LSCF (lanthanum strontium cobalt ferrite) as the air electrode. For methanation, they selected layered Ni–Al hydrotalcite derived catalysts, aiming at an expected production of up to 5.4 m^3 h^{-1}. In addition, although SOEC technology is in a pre-commercial and fundamental research stage, according to its website, Sunfire is the only company offering hydrogen generation systems, in particular devices of 150 kW with an electrical efficiency of 82%.[94]

In a parallel approach, the performance of a 4-cell planar stack (1 kW) was tested in co-electrolysis mode by Cinti *et al.*[95] The stack was supplied by Forschungszentrum Jülich. They concluded that an increase of CO_2 does not lead to any significant increase in ASR. Finally, the study of the gas outlet composition showed that equilibrium is always obtained and the syngas quality can be easily adjusted to the desired downstream upgrade process based on the inlet H_2O/CO_2 ratio and thermodynamic operating conditions of the stack.

In Europe, additional stack experiments in co-electrolysis mode were performed by CEA (Commissariat à l'Energie Atomique) in France,[96] and also by SOFCPOWER/HTceramix in Italy/Switzerland.[97]

In the USA, the leading company is Ceramatec, cooperating with the Idaho National Laboratory (INL). Ceramatec has over 30 years of experience designing, fabricating, and demonstrating SOFC and SOEC systems at a multi-kW scale. In particular, they have been very active in studying the feasibility of SOEC for large-scale, nuclear-powered, hydrogen production as well as researching high-temperature co-electrolysis of steam and CO_2,[98] having demonstrated the feasibility of using high-temperature SOEC to produce syngas. In 2007 they developed a model for a 300 MW co-electrolysis plant, coupled to a high-temperature, gas-cooled nuclear reactor. Moreover, if the carbon dioxide source is based on biomass, the entire process would be climate neutral.[99] For this study, a 10-cell stack was used, (as shown in Figure 2.8) where the cells were electrolyte-supported with an active area of 64 cm^2. The materials of choice were scandia-stabilised zirconia (ScSZ) for the electrolyte (~150 μm thick), doped manganite as the air electrode and a nickel cermet as the fuel electrode. In their initial experimental studies, they found that ASRs did not vary substantially between steam electrolysis and steam/CO_2 co-electrolysis, but increased dramatically with CO_2 electrolysis.

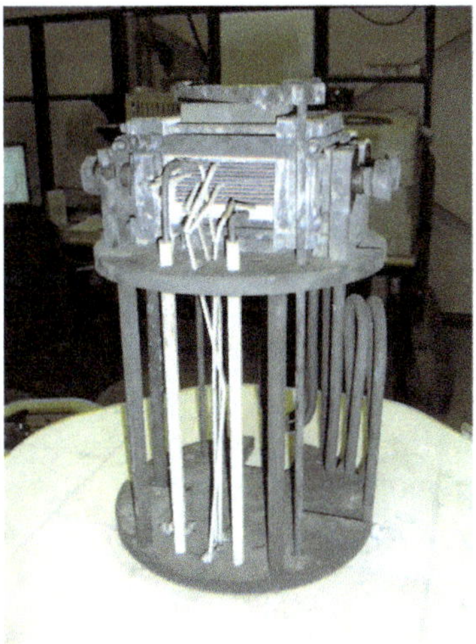

Figure 2.8 A 10-cell stack from Ceramatec used for co-electrolysis experiments. Reproduced from ref. 59 with permission from Elsevier, Copyright 2009.

Furthermore, they also proved the concept of directing co-electrolysis products through a methanation reactor. For the methanation step, a commercial steam reforming catalyst based on nickel supported on magnesium aluminate (R-67 R from Haldor Topsoe) was chosen. The reactor tube was placed within a zinc–aluminium sleeve to homogenise the axial temperature gradient and the testing was performed at 300 °C, obtaining yields of 40–50 vol% in the production of methane.[100] The same group from INL, in collaboration with NASA Ames research Center and NASA Marshall Space Flight Center, have explored the steam/CO_2 co-electrolysis for resource utilisation in space exploration.[101] They performed a mathematical analysis and described the concept of a closed-loop life support system based on individual and combined process models of co-electrolysis, considering Sabatier, Bosch, Boudounard and hydrogenation reactions.

Currently, they are also conducting research into the design of systems for producing synthetic liquid fuels by Fischer–Tropsch synthesis, using natural gas or synthesis gas from co-electrolysis as a feedstock.[102] In particular, the work is now focused on the fabrication of modules for a specific application associated with the use of synthesis gas derived from municipal solid waste. This SOEC system, fabricated in collaboration with the Jet Propulsion Laboratory and Massachusetts Institute of Technology, is expected to be included on the US NASA 2020 Mars launch. This project will test the potential of this system to electrolyse the CO_2 in the Martian atmosphere to provide oxygen for future manned missions and potentially provide fuel for the return trip.

Another active company in the USA is Fuel Cell Energy, who acquired Canadian partner Versa Power at the end of 2012. Fuel Cell Energy, with their *SureSource* power plants (ranging from 1.4 to 3.7 MW) is leading the fuel cell industry with millions of megawatts of *ultra-clean power* installed in three continents. In October 2016, they announced a market-driven energy storage solution through high efficiency electrolysis using SOEC.[103]

In Japan, the Toshiba Corporation opened the Toshiba Group Hydrogen Energy Research & Development Center (HRDC) in Tokyo in 2015. They are developing a SOEC system (flat tubular geometry) as part of a New Energy and Industrial Technology Development Organization project. This SOEC system is to be used in combination with solar photovoltaic power generation systems, fuel cells and other devices to carry out demonstration experiments, toward achieving practical use. The SOEC electrolyser is able to produce 0.5 $Nm^3\ h^{-1}$ of hydrogen that is stored in a tank with a capacity of 12 000 Nm^3. Demonstration of power-to-gas is also being conducted.[104]

2.6 Economic Considerations

Several feasibility studies can be found in the literature, integrating SOC technology with different types of carbon-free energy sources, such as nuclear,[105–108] solar,[109,110] or geothermal power.[111,112] The integration of systems with power plants has been also investigated. The main issue is that, as power plants are about 40% efficient, their use to produce synthetic fuels or hydrogen at about 50% electrical efficiency is not economically viable. The only feasible scenario for a coupled SOC/power plant would be to store surplus energy from the coal-base load in the transition period to a carbon-free energy system.

Possibly, the first study of the economic feasibility of an electrolyser coupled with synthetic fuel production was that of Graves *et al.*[113] Their estimations of the synthetic fuel cost, dependent on various factors including the electrolyser performance, full or intermittent operation and electricity price, are featured in Figure 2.9.

In this particular co-electrolysis-based cycle, if the state-of-the-art technologies at each stage of the cycle are combined to work together efficiently, considering an electricity-to-liquid fuel conversion efficiency of about 70% and a mass production of the components, economic viability seems feasible. Assuming an electricity price of less than 3 U.S. cents kWh^{-1} from a constant power supply (by the use of geothermal, hydroelectric or nuclear sources), the synthetic fuel price could be competitive with gasoline at around U.S. \$ 2 gal^{-1} (\$ 0.53 L^{-1}). When intermittent operation is considered, economical fuel production will require additional technology development on the electrolysis system to reduce the capital cost (*via* achieving durable high current density operation and/or lowering manufacturing cost). Thus, regarding the most advantageous circumstances for initial implementation, it is concluded that sites with excess inexpensive constant-supply renewable power would be ideal, as it is for example, in the case of Iceland.[114]

Fang *et al.*[115] reported a life-cycle analysis of a combined CO_2 capture and conversion membrane reactor model system. The reactor is composed of a high-temperature mixed electronic and carbonate-ion conductor (MECC)

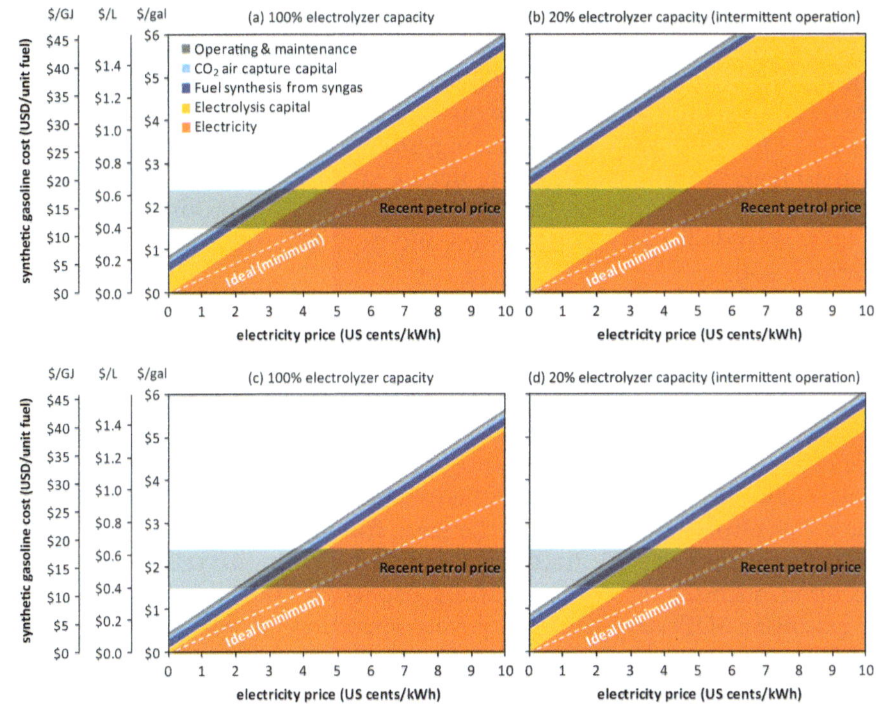

Figure 2.9 Estimations of synthetic fuel cost *versus* electricity price using (a) the assumptions summarised in table 3 of ref. 112 including constant operation, (b) the same assumptions but with highly intermittent operation (a 20% capacity factor), and operating the electrolyser at a higher current density with (c) constant operation, and (d) highly intermittent operation. Reproduced from ref. 113 with permission from Elsevier, Copyright 2011.

membrane for the CO_2 capture, and a solid oxide electrolysis cell (SOEC) for CO_2/H_2O co-reduction. They showed that the parasitic energy using a MECC membrane to capture CO_2 is less than half of the consumption of the conventional MEA (monoethanolamine) plant. Moreover, the energy efficiency of the combined system can reach 82%. A more detailed cost analysis further shows that at low SOEC areas (lower capital cost) the cost of electricity dictates the price of the synthetic fuel produced by the reactor, while both electricity cost and SOEC capital cost play an important role for higher SOEC areas (higher capital cost).

Additional economic analysis can be found in references 116–125.

2.7 Materials

The typical materials employed for high-temperature co-electrolysis are in general those also used for fuel cell, steam electrolysis and CO_2 electrolysis applications.[126–131] It is important to bear in mind that during co-electrolysis operation, both electrode materials are required to offer active sites for an efficient electrochemical reaction in both sides, including the oxygen evolution and the reduction of steam and CO_2. The main difference with a SOFC is that in this case, the fuel electrode needs to be stable under high steam concentration, as well as carbon-tolerant. In addition, H_2 and CO will be generated during operation, these mixtures proving easy to leak due to their small molecule sizes, and the use of advanced sealings will be then required for co-electrolysis operation.[132] Generally used materials for both pure steam and pure CO_2 electrolysis are firstly summarised.

2.7.1 Materials for Steam Electrolysis

Yttria-stabilised zirconia (YSZ) is the most wide-spread electrolyte material for SOEC applications. Using YSZ as the electrolyte, Jensen *et al.*[133] reported the highest value of current density: −3.6 A cm^{-2} at 1.48 V and 950 °C. The electrodes used were Ni–YSZ for the fuel electrode and LSM/YSZ–LSM for the air electrode. In addition to the work from DTU (Technical University of Denmark), several authors have reported the use of YSZ for electrolysis applications.[134–137] One important aspect to take into account when operating in SOEC mode is the limiting operation voltage. It was reported by several authors that high voltages can lead to electronic reduction of the YSZ.[138] Degradation of the cell can then be observed by the presence of voids, cracks and even delamination of the air electrode.[137,139–141] This severe electrolysis-induced degradation can be significantly overcome by reversibly cycling between electrolysis and fuel-cell modes, similar to a rechargeable battery.[142]

An alternative to YSZ electrolytes is the use of ScSZ. For example, electrolyte-supported 10Sc1CeSZ (10 mol% Sc_2O_3 and 1% CeO_2 stabilised zirconia) cells were tested using different air electrodes,[143,144] with measured current densities of up to −0.7 A cm^{-2} at 1.3 V and 850 °C. Furthermore, Ceramatec tested 10-cell stacks (also electrolyte supported) and current densities of up to −0.23 A cm^{-2} at a stack voltage of 13 V were measured at 830 °C.[145]

In addition, $La_{0.9}Sr_{0.1}Ga_{0.8}Mg_{0.2}O_3$ (LSGM) has also been proposed as an electrolyte for SOEC by Ishihara *et al.*, achieving current densities of −0.35 A cm^{-2} at 1.8 V and 600 °C.[146] However, according to studies from Ceramatec, suitable and stable hydrogen/steam electrodes need to be found for the applicability of LSGM, since commonly used nickel-based electrodes can cause severe degradation.[147] An additional alternative electrolyte material is the oxygen excess phase $LaNb_{0.84}W_{0.16}O_{4.08}$, where oxide ion conduction paths include interstitial sites. This composition was reported as a novel SOEC electrolyte, yielding current densities of −0.25 A cm^{-2} at 1.3 V and 950 °C for a 360 μm thick electrolyte-supported cell.[148]

As for the air electrode, an interesting alternative is the use of K_2NiF_4 structure type materials, such as $Ln_2NiO_{4+\delta}$ (Ln = La, Nd, Pr). The capability of these materials to accommodate oxygen excess is believed to favour the catalytic activity of air electrodes in SOEC mode for oxygen evolution. For example, neodymium nickelate was first proposed by Chauveau *et al.*,[149] where they measured current densities of about −0.87 A cm^{-2} at 850 °C and 1.3 V. Since then, other oxygen-excess materials such as $Pr_2NiO_{4+\delta}$, $La_2NiO_{4+\delta}$, or $La_{1.7}Sr_{0.3}Co_{0.5}Ni_{0.5}O_{4.08}$, and also oxygen-deficient phases such as $Nd_{2-x}Sr_xNiO_{4-\delta}$ have also been studied for SOEC applications.[150–155]

Further detailed information regarding high-temperature steam electrolysis materials can be found in references 21–27.

2.7.2 Materials for CO_2 Electrolysis

In the case of CO_2 electrolysis, it is of particular interest to find new electrodes with stable operation under high CO_2 partial pressure. As for steam electrolysis, nickel-cermet electrodes are the state-of-the-art fuel electrode for CO_2 electrolysis. For example, current densities in the range of −0.8 A cm^{-2} at 1.2 V and 850 °C were obtained by Ebbesen *et al.*[156] In principle, when low voltages are applied, no carbon deposition is detected. However, it is known that nickel tends to catalyse the Boudouard reaction (eqn (2.6)), especially at low operating temperatures and high partial pressures of carbon monoxide, which means high conversion rates. Therefore, research into alternative fuel electrode materials has received great attention for electrolysis of CO_2 and co-electrolysis applications.

In this trend, Huang *et al.*[157] reported on the use of copper, silver and platinum for CO_2 reduction, finding that copper behaves better for CO oxidation than for CO_2 reduction. However, direct sintering of Cu-YSZ is not possible as a solid solution is formed above 1150 °C[158] and lower temperatures led to insufficient adherence.[159] Despite this, the use of direct sintered Cu-CGO electrodes has been recently reported by Cheng *et al.*,[160] as CGO requires lower sintering temperatures to form a connected network than YSZ. Their performance was only slightly lower than that of a Ni–YSZ electrode and no additional phases due to solid–state reactions between CuO and CGO were detected. Another possible solution would be the use of impregnation techniques, as will be later discussed in Section 2.8.1.

In principle, every alternative material proposed for SOFCs could be also applied for solid oxide electrolysis. For example, the total or partial substitution of nickel by other metals such as Sn, Ag, Au, Pt, Ru and Pd has been proposed to avoid coke formation during steam CO_2 reforming.[161–164] In any case, with this substitutional approach there is usually a trade-off between coke formation rates and the system efficiency. Of the listed metals, Sn is of particular interest, being effective for coke suppression.[160] In fact, the enhancement of the stability of modified Ni–YSZ fuel electrodes[165] has been explained by the bimetallic nature of the Sn–Ni surface alloy.[166] Furthermore, tin has also been shown to be effective against sulfur poisoning for the studied range of 1–5 ppm H_2S.[167] Fabrication methods included impregnation of SnO particles or directly Sn into the NiO-YSZ electrode, as metallic Sn melts at 231 °C.[168] In a similar approach, a reduction of the carbon formation using Ni–Fe alloys has been reported by Zhu *et al.*[169] Among the precious metals, Ru showed a promising performance to suppress this carbon formation.[170,171]

An interesting alternative is the use of perovskite-based electrodes, as they are more redox-stable. In addition, oxide-based electrodes have a lower tendency to catalyse the formation of C–C bonds, and thus provide certain resistance against coke formation.[162] In this trend, Bidrawn *et al.*[172] reported the use of $La_{0.8}Sr_{0.2}Cr_{0.5}Mn_{0.5}O_3$ (LSCM), $Ce_{0.48}Zr_{0.48}Y_{0.04}O_2$ and Pd infiltration into an YSZ scaffold as the fuel electrode. They obtained current densities of −1.8 A cm^{-2} at 1.5 V for pure CO_2 feed at 800 °C. Similar LSCM-GDC/YSZ scaffolded composite cathodes prepared by a co-infiltration method were also recently reported.[173] The addition of copper or iron into LSCM also showed an increase in the performance.[174,175]

Furthermore, other perovskite electrode materials have been proposed, including doped titanates[176–178] and ferrites.[179] For example, Ye *et al.*[178] suggested a dual strategy of A-site deficiency on the perovskite titanate structure by using redox active dopants such as Mn/Cr linked to oxide vacancies to facilitate CO_2 reduction, and thus promoting adsorption/activation at the cathode surface. They found that the A-site deficient $(La_{0.2}Sr_{0.8})_{0.9}Ti_{0.9}Mn(Cr)_{0.1}O_{3-\delta}$ electrode presented optimal performance even after 100 h of high- temperature operation and three redox cycles, suggesting a promising cathode material for CO_2 electrolysis. As for the ferrites, Addo *et al.*[180] demonstrated that the $La_{0.3}Sr_{0.7}Fe_{0.7}Cr_{0.3}O_{3-\delta}$ (LSFCr) perovskite is stable towards CO_2 electrolysis (SOEC mode) and the oxidation of CO (SOFC mode). They showed that the cell performance is higher during the electrolysis of CO_2 than for the oxidation of CO. In addition, their cell showed very stable activity during CO_2 electrolysis, with a degradation rate of only 0.057 mV h^{-1} after testing under constant current load at −100 mA cm^{-2} in 90% CO_2 : 10% CO at 800 °C for 135 h. For the case of doped strontium titanate, a novel approach of cathode surface engineering has been recently reported for high-performance direct CO_2 electrolysis.[181] In this strategy the key factor is the tailored surface structure of doped strontium titanate, which is created through SrO nanoparticle exsolution on the perovskite surface. They found both by experimental observations and density functional theory (DFT) calculations that these nanostructures enable high-temperature chemical CO_2 adsorption and activation, obtaining performances four or five times higher than those observed for titanates without surface modification. Another alternative of redox-stable fuel electrode is $Sr_2Fe_{1.5}Mo_{0.5}O_{6-\delta}$ (SFM).[182] High electrochemical performance was demonstrated for a single-phase SFM cathode using 100% CO_2 as the feeding gas. For example, a current density of −0.71 A cm^{-2} was obtained using a fuel cell supported on LSGM electrolyte operated at 800 °C and an applied voltage of 1.5 V. The electrolysis performance was further improved by using a SFM-$Sm_{0.2}Ce_{0.8}O_{2-\delta}$ composite cathode, and the current density increased to -1.09 A cm^{-2} under the same operation conditions. In addition, a durability test at 800 °C for 100 h demonstrated a relatively stable performance for CO_2 electrolysis. Furthermore, Opitz *et al.*[183] also recently studied the performance of an acceptor-doped perovskite-type as electrodes for electrochemical CO_2 reduction. In particular, they studied the surface chemistry of a (La, Sr)$FeO_{3-\delta}$ and (La, Sr)$CrO_{3-\delta}$ based perovskite-type at SOEC operating temperatures. They found that under cathodic polarisation an intermediate carbonate develops on the oxide surface. Since the amount of carbonate increases with the concentration of oxygen vacancies, these vacant lattice sites must be the predominant adsorption sites for carbon dioxide. These results suggest that acceptor-doped oxides with high electron concentration and high oxygen vacancy concentration will be of interest for CO_2 reduction.

The effect of the microstructure, especially at the supporting electrode, plays an important role, as confirmed by the study of Singh *et al.*[184] They obtained significant changes in the cell performance by using different electrode microstructures for Ni-CGO synthesised by various routes. Fuel electrodes prepared by the spray pyrolysis method demonstrated smaller ASR (~0.05 Ω cm^2) compared to that of electrodes prepared by the mechanical mixing method (~0.35 Ω cm^2), both measured at 1000 °C under a fuel composition of 45% CO_2 - 10% H_2 - 45% N_2.

Additionally, a layered double perovskite ($PrBaMn_2O_{5+\delta}$) was found to enhance electrocatalytic activity for CO_2 reduction with limited coking. In this compound, the presence of mixed valent transition metal cations such as Mn could provide high electrical conductivity and maintain a large oxygen vacancy content, contributing to fast oxide ion diffusion.[185]

2.7.3 Materials for Co-electrolysis

2.7.3.1 Electrolytes

As in the case of SOFCs, electrolyte materials need to present high ionic conductivity, compatible TEC (thermal expansion coefficient) values with the electrodes and also mechanical and chemical stability with the rest of the cell components, in both reducing and oxidising atmospheres. Besides, regarding the cell fabrication, the electrolyte will need to be as thin as possible in order to reduce ohmic losses. Thus, the state-of-the-art electrolyte material is stabilised zirconia. For example, ScSZ was selected for the cells fabricated at Idaho National laboratories and Ceramatec.[29,186] However, YSZ is probably the most widely used electrolyte, as is the case, for example, of the cells fabricated at Northwestern University,[187] University of Pennsylvania[188] or DTU.[31] In addition, zirconia co-doped with Sc and Yb was tested by Torrell *et al.*[189] using LSCM symmetrical cells (YbScSZ thickness of about 165 μm), but significant degradation was observed after 6.5 h under current load.

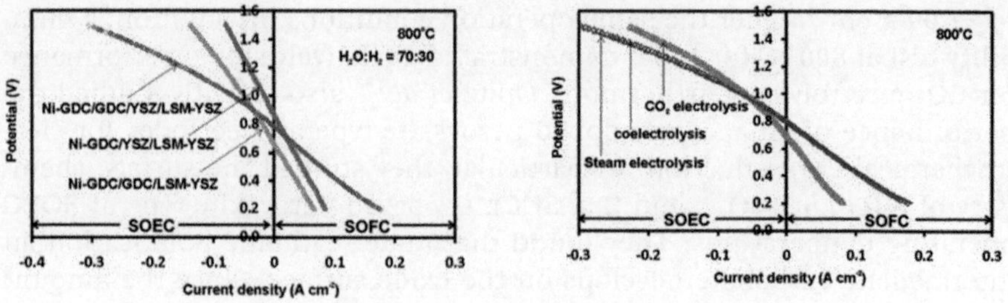

Figure 2.10 (left) *J*–*V* curve of the cells with different electrolytes (YSZ, CGO, and bi-layered CGO/YSZ electrolyte) in SOFC/SOEC modes at 800 °C (H_2O/H_2 = 50/50 to H_2 electrode) and (right) *J*–*V* curve of the bi-layered electrolyte cell at 800 °C under steam electrolysis (H_2O/H_2 = 50/50: 38% H_2O – 38% H_2 – 24%N_2), CO_2 electrolysis mode (CO_2/CO = 50/50: 25% CO_2 – 25% CO – 50% N_2), and co-electrolysis (22% CO_2 – 22% CO – 22% H_2O – 22% H_2 – 12% N_2). Reproduced from ref. 30 with permission from Elsevier, Copyright 2011.

It is important to note that in the particular case of electrolysis and co-electrolysis applications, dense electrolytes are required to present very limited electronic conduction. As a consequence, contrary to the SOFC case, the use of ceria-based electrolytes has been generally avoided. However, the use of bi-layered CGO/YSZ electrolytes, where the YSZ layer is normally added in order to block off any electronic short circuit in the cell and avoid the decrease in OCV from the electronic properties of the ceria, has been reported. Ni-CGO/YSZ/CGO/LSM-YSZ cells have been investigated and compared to those with pure single electrodes,[30] as shown in Figure 2.10. The layered electrolyte cell exhibited significantly higher performance, suggesting that the CGO electrolyte with a YSZ buffer layer (YSZ/CGO) can be preferable for electrolysis and co-electrolysis operation, achieving current densities of about −0.2 A cm^{-2} at 1.3 V and 800 °C.

In the same trend, commercial cells from Ningbo Institute of Industrial Technology were tested by Chen *et al.*[64] The cells were composed of Ni–YSZ/YSZ/CGO/LSCF, where CGO is acting as a buffer layer. They reported current densities in the range of −0.7 A cm^{-2} at 1.3 V and 800 °C using a fuel composition of 40% CO_2 – 40% H_2O – 20% H_2. From their systematic study changing gas composition (fixing H_2 at 20% and varying CO_2 from 20–80% while the remaining corresponds to steam), they found that the RWGS reaction plays an important role in the formation of CO during the co-electrolysis process, and the electrochemical performance of HT steam/CO_2 co-electrolysis is then close to that of steam electrolysis and higher than that of CO_2 electrolysis. They also found that the outlet syngas composition after steam removal varies significantly, with a H_2/CO content ratio of 1.0 : 6.0 as a function of the CO_2 concentration in the inlet gas, and a ratio of 1.9 : 1.0 as a function of the electrolysis current density, confirming the feasibility of controlling the syngas composition by adjusting the operating conditions.

2.7.3.2 Fuel Electrodes

An excellent recent review from Zhang *et al.* summarises the different materials and requirements as fuel electrodes for co-electrolysis of CO_2 and H_2O.[190] Ni–YSZ composite is still the most widely used material for the fuel electrode, as it presents high catalytic activity, good chemical stability, adequate TEC as well as a relatively low cost. Optimisation of the microstructure of nickel cermet electrodes is one of the major issues for electrolysis and co-electrolysis applications, mainly due to nickel coarsening, as will be discussed in the durability section. In addition, long-term operation with Ni–YSZ is feasible only in CO/CO_2 gas mixtures, where the presence of CO maintains a sufficient reducing atmosphere.[155] Under realistic conditions, redox cycling of Ni becomes an important issue leading to electrode degradation or even delamination.[191] Ni–YSZ was, for example, selected for co-electrolysis applications by Ebbesen *et al.*[65] and Zhan *et al.*[66] Zhang *et al.*[190] demonstrated that state-of-the-art Ni–YSZ electrode supported SOFCs can work reversibly to efficiently produce syngas. For a typical feedstock composition of $2H_2O/CO_2$, the electrolysis performance decreased only slightly compared to H_2O electrolysis, and was significantly better than for pure CO_2 electrolysis. For a SOEC operated on 25% H_2 – 25% CO_2 – 50% H_2O at 800 °C and 1.3 V, a current density of −1 A cm^{-2} was measured, yielding syngas production rates of ~7 sccm cm^{-2}. Another detailed co-electrolysis study using standard planar cells of Ni–YSZ/YSZ/LSM–YSZ was performed by Ebbesen *et al.*[192] in order to investigate the reaction mechanisms taking place. These experiments were performed using different gas mixtures to understand the possible effect of the WGS/RWGS reaction. Current densities of about −1 A cm^{-2} at 1.3 V and 850 °C were reported using a fuel composition of 25% CO_2 – 25% H_2O – 25% CO – 25% Ar. They proved that feeding the SOC with mixtures of CO_2–H_2O–CO and CO_2–H_2, the system reaches the equilibrium of WGS/RWGS, and CO may be therefore produced *via* the RWGS reaction during the co-electrolysis. Furthermore, they observed that different reaction mechanisms and segregation of impurities to the TPB explain the different behaviours under H_2O–H_2 and CO_2–CO feeding mixtures.

Li *et al.*[193] also studied planar Ni–YSZ/ScSZ/LSM–ScSZ cells and found that the presence of CH_4 observed for the feeding gas of 28.6% H_2O – 14.3% CO_2 – 57.1% Ar and voltages above 2 V, could be effectively suppressed by impregnating Ru in the porous Ni–YSZ cathode. This catalyst avoids the deposition of carbon in the electrode and thus prevents the reaction $C(s) + 2H_2 \rightarrow CH_4$, which has been proposed as one of the possible pathways for CH_4 production during SOEC H_2O–CO_2 co-electrolysis. Kim-Lohsoontorn *et al.*[194] also studied the effect of addition of Ru at the fuel electrode of Ni/CGO, finding a better performance for co-electrolysis than a cell with a standard fuel electrode.

An alternative electrode for low-temperature thin-film based solid oxide cells (TF-SOC) includes the incorporation of a Pd catalyst in the fuel electrode.[195] This multi-layered electrode was prepared by alternating sputtered Pd layers with the NiO–YSZ layers fabricated by pulsed laser deposition. This way, the catalytic activity of the formed Pd–Ni alloy improves the electrochemical performance at temperatures below 600 °C. The presence of Pd is believed to promote the WGS reaction, supressing methanation due to CO depletion. For example, using a reactant gas composition of 30% H_2O – 30% CO_2 – 10% H_2 – 30% Ar, current densities of 0.9 A cm^{-2} were achieved at 1.4 V and only 600 °C. From a more general point of view, one of the main conclusions of this work is the positive influence of the heterogeneous catalyst insertion, although it is not enough to overcome the limitations of low-temperature electrolysis.

In order to find the most suitable metal as an electrocatalyst for co-electrolysis, Cho *et al.* recently performed DFT calculations,[196] including the study of 11 transition metals (Ni, Cu, Ir, Pt, Pd, Rh, Au, Ag, Co, Ru and Fe), on the basis of an oxygen spill-over mechanism. On the basis of the Brønsted–Evans–Polanyi (BEP) relationship and scaling relation, these authors found that Fe, Ru, Co and Fe-doped Ni are the most suitable catalysts.

Following the trend of the materials used as the electrolyte, Ni-CGO cermets have been proposed as alternative fuel electrode materials in their conventional form or modified by replacing Ni for Au,[197] Mo,[198] Fe,[199] Ru[192] or Pd.[195,200,201] Other materials also proposed for the fuel electrode are $La_{0.75}Sr_{0.25}Cr_{0.5}Mn_{0.5}O_{3-\delta}$ (LSCM)[189,202–205] $La_xSr_{1-x}TiO_{3-\delta}$ (LST),[206,207] $Sr_{0.5}Fe_{1.5}Mn_{0.5}O_{6-\delta}$ (SFMn) or $Sr_2Fe_{1.5}Mo_{0.5}O_{6-\delta}$ (SFMo),[208] $SrTi_{0.9}Nb_{0.1}O_{3-\delta}$ (STN), FeV_2O_4 (FVO),[209] $La_{0.8}Sr_{0.2}FeO_{3-\delta}$ (LSF),[210,211] $La_xSr_{1-x}VO_{3-\delta}$ (LSV),[212–214] $La_{0.6}Sr_{0.4}Fe_{0.5}Cr_{0.5}O_{3-\delta}$ (LSCrF),[215] MoO_2-based[216] or $Cu/Ce_{0.6}Mn_{0.3}Fe_{0.1}O_{2-\delta}$.[217]

One of the main disadvantages of Ni-based fuel electrodes is the need to add H_2 or CO to the feeding gas in order to avoid the partial oxidation and agglomeration of the Ni particles that lead to the degradation of the cell performance. Li *et al.*[208] reported the use of SFMo electrocatalyst nanoparticles deposited on to an YSZ scaffold. The nanostructured SFMo-YSZ electrodes demonstrated excellent performance for CO_2–H_2O electrolysis, obtaining performances comparable to those reported for the Ni–YSZ fuel electrodes, without the need of hydrogen or CO in the feeding gas. However, according to the reported *J–V* curves, at voltages above 1.3 V it seems that YSZ electroreduction is taking place and as a consequence the values at high current densities will not be accurate. In any case, SFMo-based electrodes seem promising, allowing current densities of about −0.45 A cm^{-2} using a gas mixture of 20% H_2O – 80% CO_2 at 800 °C, and exhibiting also good stability in the short-term test (about 25 h).

In some other cases, impregnation techniques are used to improve not only the catalytic activity of the electrode, but also its electronic conductivity. For example, Cu was impregnated on LSCM electrodes to increase its conductivity in reducing conditions. LSCF/LSGM/LSCM–Cu cells were studied under steam and carbon dioxide co-electrolysis[203] compared with cells without Cu impregnation for steam electrolysis under the same

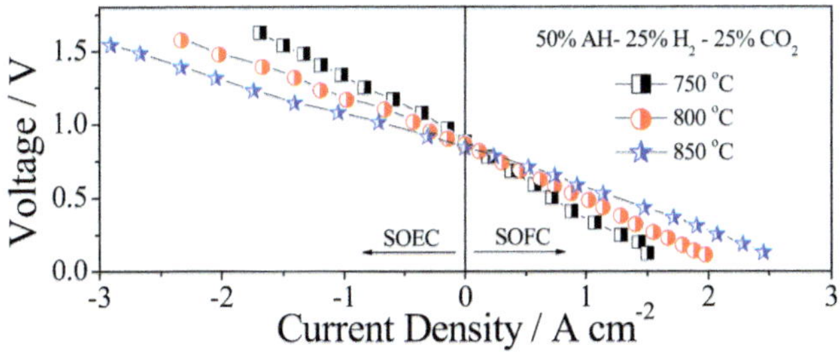

Figure 2.11 *J-V* curves of a LSCF/LSGM/LSCM-Cu cell with 50% H_2O – 25% H_2 – 25% CO_2 at different temperatures. Reproduced from ref. 203 with permission from Elsevier, Copyright 2015.

conditions. From EIS (electrochemical impedance spectroscopy) results, the impregnated cell displays lower ohmic resistance and better electrochemical performances. Under a voltage of 1.65 V at 750 °C, the maximum consumed current density increases from −1.31 A cm^{-2} without CO_2 to −1.82 A cm^{-2} with 37.5% CO_2, these current densities (Figure 2.11) being some of the highest reported in the literature to date at this temperature. Besides, durability tests confirmed their stability for more than 50 h with the electrolysis current density of 0.33 A cm^{-2} under 50% H_2O – 25% H_2 – 25% CO_2 at 750 °C.

A similar methodology was performed by Huang *et al.*,[215] as they modified the LSCrFe–GDC electrode with $Ce_{0.9}M_{0.1}O_{2-\delta}$ (M = Fe, Co, Ni) by infiltration, showing an enhanced performance by the introduction of transition metal dopants. For example, for the CeCo–LSCrFe–GDC cell they obtained a current density of −0.65 A cm^{-2} at 1.5 V and 800 °C, which was about 2.4 times higher than that of the cell without infiltration. They also found by EIS analysis that both CO_2 adsorption and the carbonate intermediate dissociation at TPBs are accelerated by the introduction of the catalysers.

Another interesting approach was recently proposed by Dogu *et al.*[218] using $La_{0.2}Sr_{0.8}TiO_{3\pm\delta}Cl_{\sigma}$. This novel material presented enhanced activity for the co-electrolysis process in comparison with the standard Ni/YSZ catalyst, as a lower amount of coke formation possibly associated with the formation of carbonate species on the surface.

Since there is a need to develop sulfur-tolerant electrodes for co-electrolysis, composites of YSZ/Sr-doped LSV have been tested as fuel electrode with Pd and CeO_2 added catalysts.[212] The electronic conductor reduced phase of LSV was found to be stable under an operating CO/CO_2 (30% H_2O) atmosphere. The presence of both catalysts determined the reactions taking place in the electrode: when both catalysts are present, steam was electrolysed into hydrogen and CO formed from CO_2 through the RWGS reaction, while the absence of the Pd catalyst led to the electrolysis of CO_2. Within this electrode, the electrolysis of dry CO_2 showed very poor long-term stability.

Sulfur-tolerant fuel electrodes (LSV-based) were also tested by Ge *et al.*[214] under reversible SOFC/SOEC mode for over 500 h using different simulated gases such as H_2, syngas, biogas, town gas and coal gas. The electrodes exhibited better performance when operated in the SOEC mode, and in all cases the polarisation resistances of the LSV electrodes were in the range of 0.1 to 0.5 Ω cm^2. Interestingly, LSV-based reversible SOCs showed negligible performance degradation both under carbon-forming and H_2S-containing gases.

A MoO_2-based cathode was recently proposed.[216] While the MoO_2-based cathode cell displays similar electrocatalytic activity as a standard Ni/YSZ-based cell in the co-electrolysis mode, it shows a much improved electrocatalytic activity towards the electrolysis of CO_2. In co-electrolysis mode, current densities of about −0.25 A cm^{-2} at 1.3 V and 750 °C were achieved using a fuel composition of 33.3% CO_2 – 33.3% H_2O – 33.3% H_2.

In any case, the reaction mechanisms of the fuel electrode are still not fully understood. It is normally accepted that CO_2 and H_2O are reduced to $CO + O^{2-}$ and $H_2 + O^{2-}$, respectively. But the detailed dissociation processes are still unknown.[70,219,220] In this sense, an interesting investigation on the electrode kinetics of the H_2/H_2O and CO/CO_2 reactions were recently performed by Graves *et al.*[221] They performed an extensive study varying fuel composition: H_2/H_2O and CO/CO_2, which are the two reactions competing in the co-electrolysis mode. For this purpose, they tested different fuel electrodes including Ni/ScYZ (scandia and yttria-stabilised zirconia), CGO, and CGO nanoparticles coating Nb-doped $SrTiO_3$ backbones (CGO/STN) in both model electrode and porous electrode geometries and with different micro/nanostructures. The ratio of electrode polarisation resistance (R_p) in CO/CO_2 *vs.* H_2/H_2O reactions decreased from 33 to 2 when comparing Ni/ScYZ and porous Ni–ScYZ electrodes. Their studies suggested that a greater concentration of impurities at the TPB in the model electrode lead to this higher ratio. In addition, the Ni/ScYZ interface was more active for oxidation than reduction reactions. Surprisingly, they found that CGO/STN model electrodes show the opposite behaviour, with higher activity for CO/CO_2 and higher activity for reduction. This is probably the first report of a higher electrocatalytic activity for CO/CO_2 than for H_2/H_2O reactions on CGO nanoparticles in the absence of gas diffusion limitations. It is proposed that this behaviour is associated with an enhanced surface reduction of the doped ceria in CO/CO_2, at the same pO_2.

Despite the material choice, as previously mentioned, the microstructure of the electrodes is even more important that for fuel cell operation, especially the microstructure of the fuel electrode. The pore size and distribution will be designed to ensure a good gas transport (injection of steam and CO_2 and evacuation of H_2, CO plus the unreacted gases) also providing an increased number of TPBs.

2.7.3.3 *Air Electrodes*

For the air electrode, a composite of LSM and YSZ is the most widely used material. Although LSM is a poor ionic conductor, it is typically used in combination with YSZ to increase both its ionic conductivity and the compatibility with the electrolyte material. However, higher performances are obtained with alternative air electrodes, such as LSF[194] or LSC ($La_{0.8}Sr_{0.2}CoO_3$).[222,223] For example, Lo Faro *et al.*[223] tested commercial planar cells from ELCOGEN (ASC-400B) obtaining current densities in the range of −0.35 A cm^{-2} at 1.3 V and 700 °C for the inlet composition of 12.8 H_2 – 5.3 H_2O – 5.1 CO_2 – 76.8 He (vol%).

Other materials include pure LSCF or LSCF/CGO composites,[224,225] BSCF (barium strontium cobalt ferrite)[226] or other mixed perovskites such as $La_{0.8}Sr_{0.2}Co_{0.8}Ni_{0.2}O_{3-\delta}$,[227] $La_{0.7}Sr_{0.2}Ni_{0.1}Co_{0.1}Fe_{0.8}O_{3-\delta}$,[228] $La_{0.4}Sr_{0.6}Co_{0.2}Fe_{0.7}Nb_{0.1}O_{3-\delta}$[229] and $La_{0.9}Ca_{0.1}Fe_{0.9}Nb_{0.1}O_{3-\delta}/Zr_{0.8}Sc_{0.2}O_{2-\delta}$ (LCFNb/ScSZ).[230]

It is, in general, well established that the critical processes related to the durability of the cells in co-electrolysis mode (as well as for SOFC and SOEC operation modes) usually occur at the electrode/electrolyte interface. Several examples regarding air electrode delamination were already mentioned in the steam electrolysis section. Most of these aspects are elegantly covered in a recent review article by Irvine *et al.*[231] As an example, Figure 2.12, extracted from this article, illustrates a variety of activation/passivation phenomena in TPB regions of different electrodes (12(a): LSM/YSZ air electrode, 12(b): Ni/YSZ fuel electrode). In this scheme, it is shown that MIEC oxygen-electrodes exhibit the same activation and degradation mechanisms as LSM at high negative and positive potentials. At low overpotentials, MIEC oxygen-electrodes present an active solid/gas two-phase boundary, similar to that of LSM under cathodic polarisation (a) but without Mn^{2+}-enriched area. In the case of fuel electrodes (b), one of the most common passivation/activation processes are related to sulfur poisoning. Another issue could be an undesired oxidation of Ni to NiO, either by exposure to a high pO_2 or by high anodic polarisation. The microstructure of the electrode is then essential to avoid the rupture of the porous YSZ matrix. Finally, a long exposure to high cathodic overpotential can lead to microstructural changes such as the loss of contact between Ni and YSZ. Then, the effects of high anodic or cathodic overpotentials on MIEC fuel-electrodes Ni/YSZ are of great importance and need to be further investigated.

Besides the explanation of different potential-driven electrode phenomena, Irvine *et al.* show that the composition and structure of the electrode–electrolyte interface can be optimised by chemical methods such as impregnation and by control of the stoichiometry to improve properties and stability. Processes such as dissolution/exsolution and surface cation segregation are linked to each other and to the chemistry governing ageing and activation.

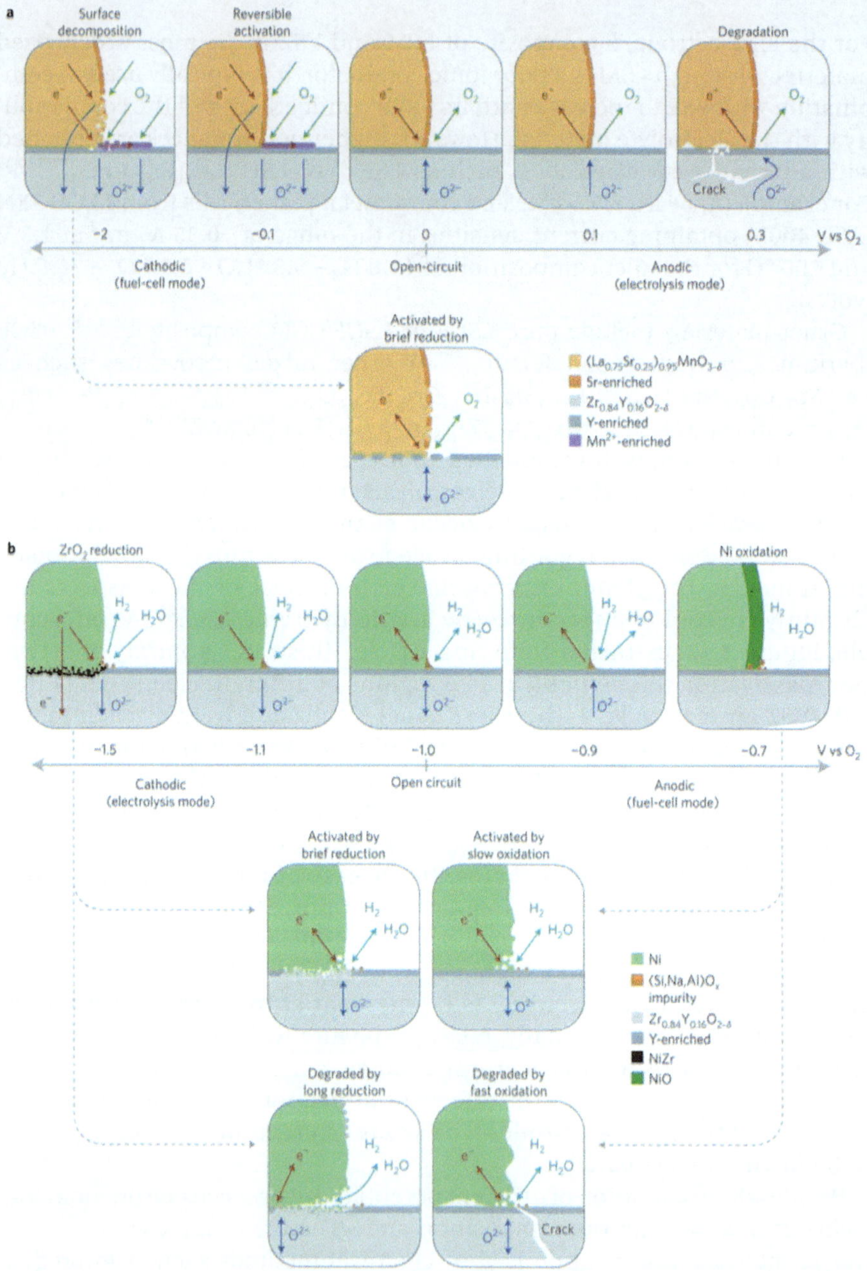

Figure 2.12 Electrode-potential-driven activation and passivation phenomena at TPB of (a) LSM/YSZ/O_2 air electrode and (b) Ni/YSZ/H_2, H_2O fuel electrode. Reproduced from ref. 231 with permission from Springer Nature, Copyright 2016.

Understanding and controlling the evolution of this interface provides the tools to develop the next generations of SOCs; however, there is still much to learn about the details of the different processes involved.

2.7.3.4 *Proton Conductors*

The use of proton conductors for CO_2 electrolysis is scarce, as there are only a few investigations reported in the literature. For example, Xie *et al.*[232] demonstrated the potential of CO_2 electrolysis using an anode-supported proton conductor cell (Ni-BZCYZn/BZCYZn ($BaCe_{0.5}Zr_{0.3}Y_{0.16}Zn_{0.04}O_{3-\delta}$)/Fe-BZCYZn). They measured current densities of −1.5 A cm^{-2} at 614 °C but using voltages as high as 2.6 V. Ruiz-Trejo *et al.*[233] studied different cathode materials using BZCYZn supported electrolytes, obtaining very low current densities (−40 mA cm^{-2} at 1.5 V and 600 °C) using Pt-modified Fe as the cathode. Similar results were reported by Wu *et al.*,[234] where current densities of about −20 mA cm^{-2} were achieved for a $(La_{0.75}Sr_{0.25})_{0.95}Mn_{0.5}Cr_{0.5}O_{3-\delta}$-BZCYZn/BZCYZn/Ni-BZCYZn cell at 1.5 V and 600 °C. More recently, promising results were reported,[235] as current densities of −1.23 A cm^{-2} at 1.5 V were measured at 550 °C, being comparable values to those for oxide conductors operating at 800 °C. The authors also found from EIS analysis that the presence of CO_2 in the cathode is beneficial to accelerate the cathode reaction. In addition, they found from *in situ* Raman spectroscopy that CO_2 is more prone to be adsorbed over BZY *via* the CO_3^{2-} radical, and CO as well as HCO_3^- can be formed over all the samples.

Despite these results, CO_2 electrolysis using proton conductors is still a challenge. Although different reaction routes of CO_2 reduction to CO have been proposed and seem feasible—in which protons are suggested to be key for effective CO_2 conversion in proton conducting SOECs—there is an urgent need to further study the reaction mechanisms of CO_2 conversion when using proton conductors.

2 8 Novel Fabrication Procedures and Designs

2.8.1 Catalyst Infiltration/Impregnation

Infiltration of catalyst nanoparticles into a ceramic porous matrix or into a standard electrode has been successfully proposed for electrode fabrication in SOCs, as summarised in different articles.[236–238] By this technique, the electrode high-temperature sintering step could be omitted, avoiding undesirable reactions with other cell elements. In addition, when the infiltration is performed into a porous backbone of the electrolyte material, the thermal expansion mismatch between the electrolyte and the electrode is decreased, thus reducing the risk of electrode cracking. Another advantage will be an increased TPB length compared with standard cathodes due to the smaller size of the dispersed catalyst particles, since the lower firing temperature inhibits the coarsening of the catalyst during the

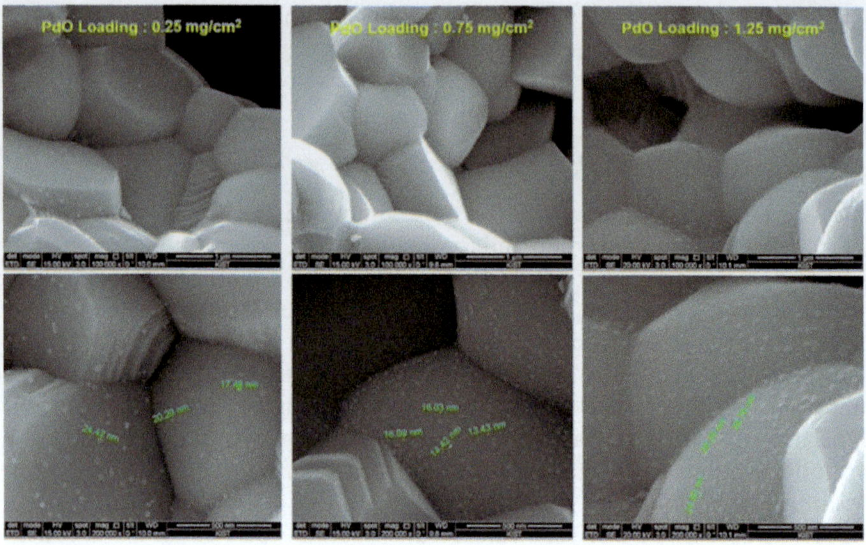

Figure 2.13 Homogeneous distribution and morphology of PdO nanoparticles anchored on the NiO/YSZ fuel-electrode substrate as the PdO loading is raised. Reproduced from ref. 201 with permission from the Electrochemical Society, Copyright 2016.

sintering process. For example, it was demonstrated that the high stability of nanostructured LSM–YSZ electrodes fabricated by infiltration is related to the microstructure stability of infiltrated LSM nanoparticles as a consequence of a LSM lattice shrinkage effect under electrolysis polarisation conditions.[239]

Zheng *et al.*[227] studied the use of $La_{0.8}Sr_{0.2}Co_{0.8}Ni_{0.2}O_{3-\delta}$ (LSCN) impregnated CGO–LSM composite air electrodes for H_2O/CO_2 co-electrolysis. The LSCN impregnated cell reached current densities of −1.6 A cm^{-2} at 1.5 V at 800 °C with a H_2O/CO_2 ratio of 2 : 1 in co-electrolysis mode. In addition, this cell showed a very stable performance without obvious degradation for more than 100 h under current load.

The addition of Pd- or Fe-based nano-catalysts by infiltration into the Ni-based fuel electrode was studied by Jeong *et al.* to improve CO_2 conversion.[240] Medium-term durability studies were conducted for a period of 350 h under a current of −0.3 A cm^{-2}, after which the Pd-infiltrated cell had a degradation of 0.8%, while the Fe-infiltrated cell degraded 4.8%. The lower degradation rate of the Pd-infiltrated cell can be explained by the relatively higher carbon coking tolerance of the Pd catalysts. Additional studies about Pd infiltration were performed by Kim *et al.*,[201] and the microstructures of the fuel electrodes are shown in Figure 2.13.

Ahn *et al.*[241] studied the impregnation of various metals (Fe, Cu, Ni, Co and Pd) into LSCM-YSZ composite electrodes. They found that Pd was the best catalyst, but Ni and Co also showed good catalytic activity in co-electrolysis mode. On the other hand, LSCM-YSZ with Fe and Cu showed similar behaviour to LSCM-YSZ without a metal catalyst, due to the oxidation of Fe in the co-electrolysis environment and the poor catalytic activity of Cu, respectively. A remarkable finding was that the cell with the LSCM-YSZ composite electrode electrolysed CO_2 into CO, followed by the WGS reaction to convert H_2O to H_2. However, the cells infiltrated with Ni, Co or Pd electrolysed steam to H_2, followed by the RWGS reaction to convert CO_2 to CO.

The addition of catalysts such as CeO_2 or Pd drastically improved the performance of the cells using LSCM and LSV fuel electrodes.[213] Similar results for CGO infiltration were found on standard Ni–YSZ fuel electrodes.[242]

Additional improvements in co-electrolysis cells were performed by the addition of Ag into LSCF air electrodes[243] or nickel into cerium-doped STO fuel electrodes.[244]

2.8.2 Exsolution of Metallic Particles

In recent years, tailored exsolution of metallic particles from A-site deficient perovskites has achieved much attention. By this procedure, catalysts are incorporated as cations within the host lattice during the synthesis process under oxidising conditions, and the metallic catalyst is exsolved in the form of nanoparticles during reduction. Compared with conventional techniques, such as physical mixing and infiltration, this process can produce finer and more evenly distributed nanocatalysts, in a time-effective manner.[245] In this sense, Myung *et al.*[246] fabricated $La_{0.43}Ca_{0.37}Ni_{0.06}Ti_{0.94}O_3$ (LCNT)/ScSZ/LSM–ScSZ cells and demonstrated the growth of a finely dispersed array of anchored metal nanoparticles on an oxide electrode by applying a potential of 2 V for a few seconds. The resultant cell reached 2 W cm^{-2} at 900 °C in SOFC mode (97% H_2 – 3% H_2O) and −2.75 A cm^{-2} at 1.3 V in SOEC mode (50% H_2 – 50% H_2O). These results prove that *in operando* fabrication methods can yield promising nanomaterials, as it was confirmed that the cells fabricated by this technique do not degrade during at least 150 h of testing. *In situ* exsolved Fe–Ni alloy nanospheres have also been formed on an oxygen-deficient perovskite by reducing the Sr and Ni from $La_{0.6}Sr_{0.4}Fe_{0.8}Ni_{0.2}O_{3-\delta}$ (LSFN).[247] This system was proposed for direct CO_2 electrolysis, achieving about −1 A cm^{-2} at 1.3 V and 850 °C. Besides, no carbon deposition was detected after 100 h under operation.

Similarly, active nanostructures on titanate surfaces were recently investigated.[248] The exsolved metal nanoparticles coupled with tailored oxygen vacancies through Cr and Mn substitution produce a strongly interactive interface. As an example, SEM (scanning electron microscopy) and high-resolution TEM (transmission electron microscopy) images of a reduced $(La_{0.2}Sr_{0.8})_{0.95}Ti_{0.85}Mn_{0.1}Ni_{0.05}O_{3+\delta}$ ($LSTMN_{A-}$) sample are shown in Figure 2.14. Chemical adsorption/activation of CO_2 as well as CO_2 electrolysis was investigated, obtaining maximum current densities of −0.91 A cm^{-2} at 2.0 V and 800 °C for CO_2 electrolysis.

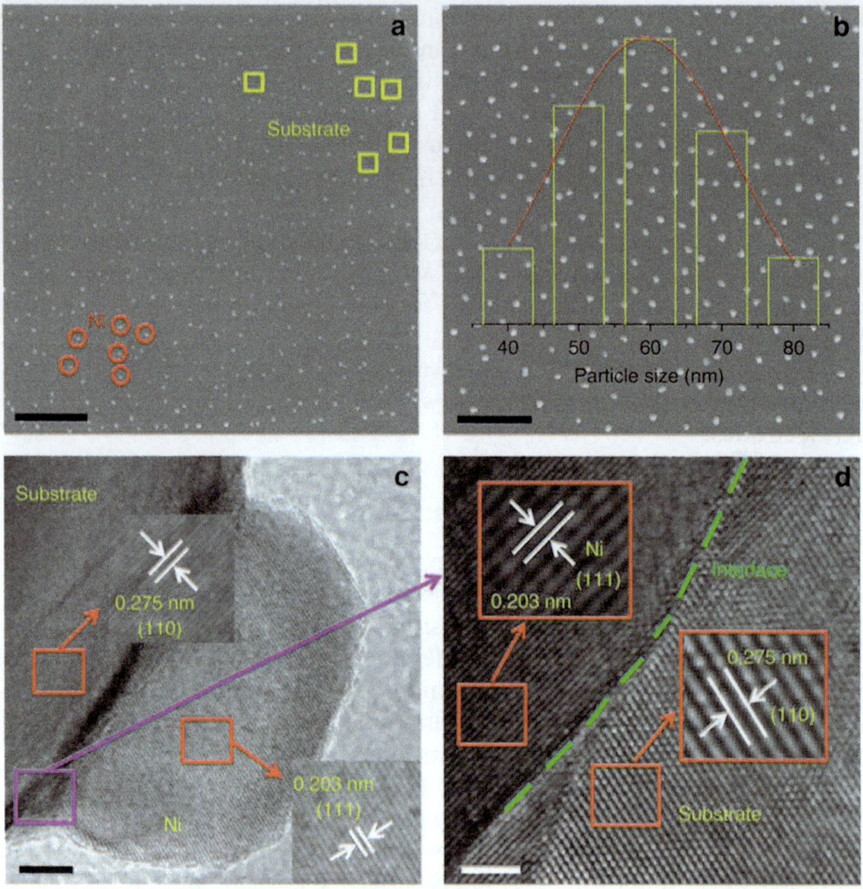

Figure 2.14 Scanning electron microscopy and high-resolution transmission electron microscopy images of a $LSTMN_{A-}$ sample. (a) Scanning electron microscopy images of the reduced $LSTMN_{A-}$, scale bar is 5 µm. (b) Ni nanoparticles size distribution of the reduced $LSTMN_{A-}$ (inset), scale bar 3 µm. (c) High-resolution transmission electron microscopic results for the reduced $LSTMN_{A-}$, scale bar 5 nm. (d) The heterojunction between Ni and substrate along the pink square shown in (c), scale bar 2 nm. Reproduced from ref. 248, https://doi.org/10.1038/ncomms14785, under the terms of the CC BY 4.0 licence, https://creativecommons.org/licenses/by/4.0/.

In situ exsolution of $Ni_{1-x}Cu_x$ alloy from $(La_{0.75}Sr_{0.25})_{0.9}(Cr_{0.5}Mn_{0.5})_{0.9}(Ni_{1-x}Cu_x)_{0.1}O_{3-\delta}$ (LSCM-$Ni_{1-x}Cu_x$) perovskite electrodes was demonstrated to be highly efficient to produce syngas in a solid oxide electrolyser with CO_2 electrolysis in the cathode and CH_4 oxidation in the anode.[249] In addition, the cell presented enhanced coking resistance and catalyst stability. The authors also performed *in situ* infrared characterisations and revealed the interface activation of CO_2 at a transition state between a CO_2 molecule and a carbonate ion. This novel strategy claims optimal performance with no degradation after 300 h of high-temperature operation and 10 redox cycles, suggesting a reliable process for conversion of CH_4 into syngas using CO_2.

2.8.3 Microtubular Configuration

The use of other alternative designs to the conventional planar configuration was tested in the co-electrolysis mode. One of these interesting alternatives is the microtubular design, as it presents higher energetic density, better resistance to thermal stresses, and smaller start-up times compared to conventional planar stacks.[250] Lee *et al.*[251] studied the performance of a tubular solid oxide co-electrolysis cell fabricated using standard materials (Ni/YSZ–YSZ–LSM/YSZ). They reported that CO_2 reduction was predominantly governed by thermochemical reactions, including the RWGS reaction, which influenced the overall electrochemical performance of the H_2O–CO_2 co-electrolysis process near the limiting currents. For example, they reported current densities of about −0.5 A cm^{-2} at 1.3 V and 800 °C using a fuel composition of 20% CO_2 – 40% H_2O – 30% Ar – 10% H_2.

Yu *et al.*[252] studied tubular cells composed of Ni–YSZ based fuel-electrode supports, YSZ or ScSZ electrolytes, and a composite air-electrode of LSM and LSCF. They found maximum current densities for the ScSZ cell of about −0.7 A cm^{-2} at 1.3 V and 850 °C using a fuel composition of 30% CO_2 – 60% H_2O – 10% H_2.

Direct synthesis of CH_4 from a CO_2–H_2O feedstock was also demonstrated using a tubular design with standard materials, obtaining a CH_4 yield of 11.8% by combining the CO_2–H_2O co-electrolysis and methanation in a single tubular unit. A scheme of the concept is shown in Figure 2.15. Current densities of about −0.5 A cm^{-2} at 1.3 V and 800 °C were achieved using a fuel composition of 8.5% CO_2 – 40% H_2O – 51.5% H_2.[253]

Recently, Lei *et al.*[254] performed similar experiments combining co-electrolysis of H_2O–CO_2 and low temperature methanation into a microtubular reactor. In these experiments, the CH_4 yield was increased to 21.1%, along with CO_2 conversion ratio of 87.7%. Furthermore, a detailed study of the effect of the inlet gas composition (H : C ratio) on the CO_2 conversion rate and CH_4 yield allowed an increase to 23.1%.

Additionally, simulation studies using microtubular configuration were performed considering the chemical reforming reactions for both SMR and WGS, electrolysis of CO_2 and H_2O, mass transport and charge transport.[255] The model was validated for different CO_2 and CH_4 conversion ratios and it was demonstrated that WGS reaction and co-electrolysis can be improved by modifying the inlet gas from CO_2 rich to CO rich gas, which is subsequently used to synthesise CH_4. Besides, from mass transport analysis, partial CH_4 back diffusion is observed between the FT section and SOEC section.[256]

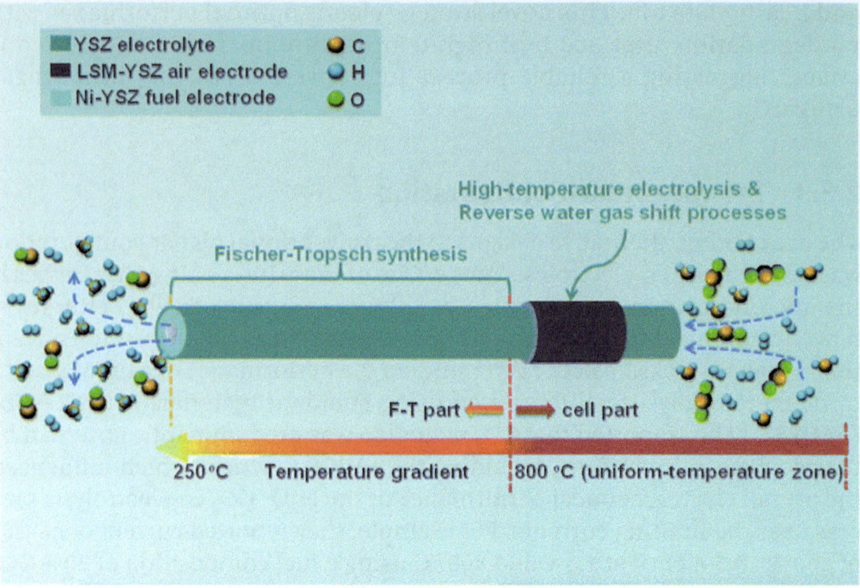

Figure 2.15 Illustration of direct methane synthesis from CO_2–H_2O co-electrolysis in a tubular unit combining a high-temperature SOEC and a reduced temperature Fischer–Tropsch reactor. Reproduced from ref. 253 with permission from the Royal Society of Chemistry.

2.9 Development of the Technology and Applications

2.9.1 Direct Synthesis of Methane

As previously mentioned, it has been demonstrated that it is possible to convert CO_2/H_2O into hydrocarbon in a single step in an oxygen-ion conducting solid oxide electrolyser. In this case, methane is directly synthesised in an efficient electrolyser with configuration of (air electrode) $(La_{0.8}Sr_{0.2})_{0.95}MnO_{3-\delta}$/YSZ/$La_{0.2}Sr_{0.8}TiO_{3+\delta}$ (fuel electrode) by combining co-electrolysis of CO_2/H_2O and *in situ* Fischer–Tropsch-type synthesis (see Figure 2.16).

A high faradaic yield of CO/H_2 and lower methane yield was demonstrated, showing that the limit on conversion efficiency comes from the heterogeneous catalysis process. Their electrochemical results also showed that the electrochemical reduction of the $La_{0.2}Sr_{0.8}TiO_{3+\delta}$ cathode is the main process at low electrical voltages while the co-electrolysis is the main process at high voltages.[207] Xie *et al.* reported on the synthesis of methane from CO_2/H_2O in a single step at the rate of about 0.005 ml min^{-1}. In addition, they reported a Faraday yield of 74.2% on the synthesis of CO/H_2. They also observed a higher conversion of H_2O (25%) than for CO_2 (11.5%), in concordance with other studies previously mentioned, which confirm that steam electrolysis is more efficient than CO_2 electrolysis.

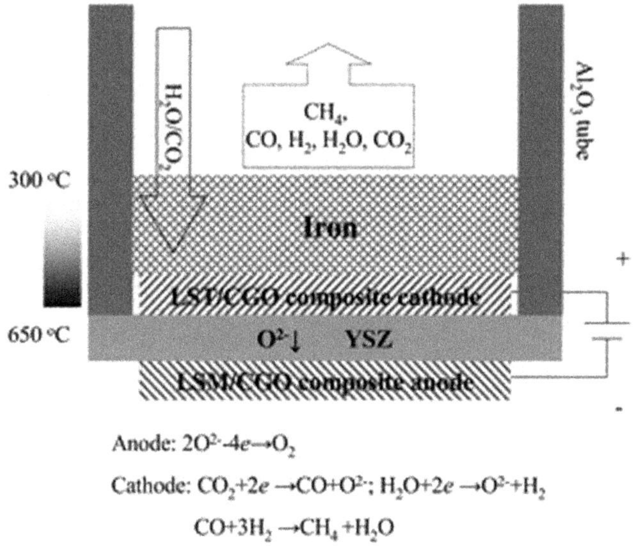

Figure 2.16 Schematic of cell testing conditions and temperature region for a porous iron catalyst. Reproduced from ref. 207 with permission from the Royal Society of Chemistry.

2.9.2 Development of Pressurised Systems

The conversion of CO_2 and steam into syngas in a pressurised solid oxide co-electrolysis cell is considered one of the most promising pathways towards the production of sustainable fuels. Mehran *et al.*[257] designed a high-pressure tubular SOC system using Ni/YSZ-ScSZ/LSCF-CGO cells, as shown in Figure 2.17. The pressurised tubular SOC cell was operated between 1 and 8 bar pressure at 800 °C in both fuel cell and co-electrolysis modes. The maximum power density in the fuel cell mode of the tubular SOC cell increased from 586 mW cm^{-2} to 845 mW cm^{-2} when the pressure was increased up to 8 bar, and this increase was attributed to the higher OCV and reduced polarisation resistance of the electrodes at higher pressures. In the co-electrolysis mode, they obtained a voltage for the pressurised cells of 1.46 V (1 bar) and 1.28 V (8 bar) at −800 mA cm^{-2}. EIS results confirmed that at higher pressure, there is a significant decrease in the polarisation resistance of the electrodes. It is also remarkable that post-test material characterisation by electron microscopy did not show any significant degradation.

Bernadet *et al.*[258] performed experiments using pressurised cells composed of Ni–YSZ/YSZ/CGO–LSCF up to 10 bar. They also concluded that by increasing the pressure of the system, higher current densities could be achieved for the same operation voltage. In addition, the methane formation was only activated at 10 bar under polarisation. They also developed a model to predict the polarisation curves under pressure, as well as the production

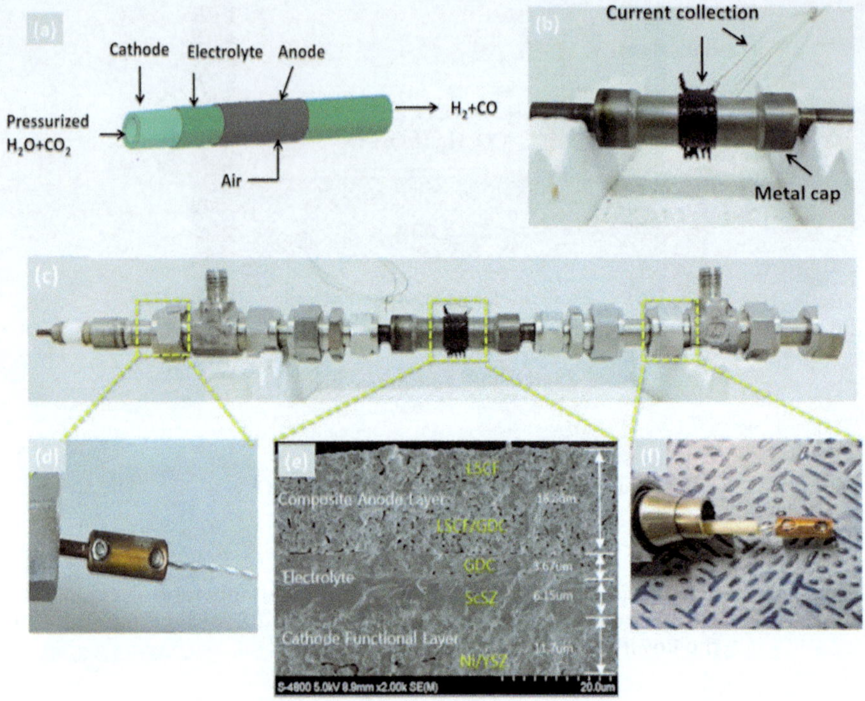

Figure 2.17 (a) Schematic illustration and (b) photograph of a tubular solid oxide co-electrolysis cell to be used for pressurised operation, (c) overall configuration of the tubular cell connected to sensors and stainless steel connectors for gas supply, (d) fuel electrode current collection, (e) SEM micrograph of the cross-section of the tubular cell showing the layers of the cell, (f) fuel electrode current collection. Reproduced from ref. 257 with permission from Elsevier, Copyright 2018.

of syngas. Their model confirmed that the improvement in the cell limiting current density is due to the gas diffusion resistance, which decreases when increasing the pressure of the system.

Recent modelling studies of pressurised systems[259] reported that an electrolyser stack with nominal power of 3.6 MW may be sufficient to assure the production of enough precursor syngas to further production of 150 L h^{-1} of Fischer–Tropsch diesel. In contradiction to other works, these authors found that the total efficiency of the system is lower at elevated operating pressure due to the internal production of methane from syngas, therefore it would be more beneficial to operate the system in post-compression mode rather than pressurised.

2.10 Durability Studies and Degradation Issues

Operational lifetime is one of the main actual concerns for the launch of co-electrolysis systems into the market. Thus, great efforts are being made in order to understand the degradation processes occurring upon long term operation and optimise the systems to achieve a durability of at least five years. In general, degradation processes in high-temperature cells are mainly related to the chemical and/or microstructural stability of the multicomponent device. The lack of chemical stability at high temperature can lead to the formation of secondary phases through diffusion of elements of the different components of the cell, or formation of different chemical species by reaction with the fuel constituents. The evolution of the cell microstructure can be produced by electrode delamination, formation of voids at the grain boundaries of the electrolyte, by migration or coarsening of Ni particles in the fuel electrode, or for example, by formation of carbon deposits on the surface of the electrodes.

In this sense, one of the first durability studies of solid oxide cells operating in the co-electrolysis mode was performed by Graves *et al.*[31] In this experiment, a Ni–YSZ/YSZ/LSM–YSZ cell was operated at different current densities for periods up to 400 h. When low current density was applied (-0.25 A cm^{-2}), there was a stabilisation period during the initial 100 h and then a very slow degradation rate (of about 0.008 mV h^{-1}) was observed during the last 100 h. This degradation rate clearly increases when increasing current density, up to 0.1 mV h^{-1} when -1 A cm^{-2} was applied. By distribution of relaxation times (DRT) from EIS studies, the authors identified up to five different electrode processes, which suggested different degradation mechanisms for high and low current density operation. At low current densities, the degradation at the Ni/YSZ electrode seems to dominate (accumulated impurities at the TPB), whereas at higher current densities the additional degradation of the LSM/YSZ electrode also contributes to the total resistance.[260] These findings are related to those for steam electrolysis operated at high current densities,[261,262] resulting in a build-up of oxygen pressure at the electrolyte/air electrode interface, and thus causing damage with the presence of pores, cracks and even delamination of the air electrode.

Several studies have shown that another usual source of degradation, the coarsening of nickel particles, can be strongly influenced by the steam partial pressures during cell operation. As this effect is already considered in SOFC mode, it is expected to be more significant in co-electrolysis mode, where high steam concentrations will be required. In the case of SOFCs, it has been shown that higher pH_2O leads to a more rapid degradation of the Ni–YSZ electrode. Holzer *et al.*[263] studied the degradation of Ni/CGO electrodes in SOFC under dry and humidified atmospheres and found that the kinetics of nickel grain growth cannot be explained solely by the typical Ostwald ripening mechanism. The grain growth of nickel is influenced, apart from the humidity, by the gas flow rate, electrode composition and many other factors. Besides, it is not independent of other degradation processes such as the evaporation of nickel or the formation of continuous ceramic layers on the surface of the nickel grains at longer exposure times. In the same trend, Hauch *et al.*[264] studied the degradation of anode-supported SOFC by impedance spectroscopy as a function of pH_2O and found an exponential relationship for anode degradation according to the eqn (2.12).

$$R_{Ni,TPB}(t) = R_{Ni,0} + \Delta R(1 - \exp(-t/\tau)) \quad (2.12)$$

The initial resistance and total increase for the Ni–YSZ charge-transfer resistance, $R_{Ni,0}$ and ΔR, were not directly correlated with pH_2O, but the characteristic time for the anode degradation, τ, was significantly higher for the lower pH_2O. As an example, τ varied from 134 h at $pH_2O = 0.2$ atm to 17 h at $pH_2O = 0.4$ atm.

Another degradation mechanism is typically related to the presence of impurities, especially coming from sealing materials and gases and even Zr from the electrolyte. For example, early durability studies performed at DTU in SOEC concluded that the initial performance passivation was associated with the Si impurities from the glass sealings.[265] This was later confirmed in another publication where FIB (focused ion beam)/TEM experiments revealed the build-up of impurities containing Si, Al and Na at the TPBs in the Ni–YSZ electrode close to electrolyte.[266] However, this degradation was prevented almost completely by cleaning the feed gas:[267] by comparing identical cells fed with either standard or purified gas, they concluded that the degradation occurring at the Ni/YSZ electrode can be completely eliminated by cleaning the inlet gas to the electrode.

Accelerated degradation tests were performed by Tao *et al.*[268] operating at high current densities using standard cells (in the range of −1.5 and −2.0 A cm^{-2} at ~875 °C for periods up to ~700 h). They found a significant increase in the ohmic resistance, the O^{2-} transport resistance at the Ni–YSZ electrode and the TPB reaction resistance of the Ni–YSZ electrode. This increase was explained by post-test observations that showed microstructural changes in the fuel electrode in the region close to the YSZ electrolyte. Percolation loss of Ni resulted in the formation of an inactive Ni/YSZ region and consequent increase of the polarisation resistance (Figure 2.18). Besides, zirconia nanoparticles were formed on the surface of Ni at the interface, contributing to the increase in the resistance.

Regarding the degradation issues involving the electrolyte, the electroreduction of the electrolyte at high voltage operation is particularly concerning when using YSZ. A decrease of oxygen activity at the Ni–YSZ/YSZ interface with increasing the cathodic overpotential could increase the risk of (partial) reduction of YSZ at the interface. As in the case of steam electrolysis, a fast degradation has been observed in co-electrolysis mode when operating at high current densities. Graves *et al.* found that high current densities in the range of −1 A cm^{-2}, result in a build-up of oxygen pressure at the LSM-YSZ/YSZ interface, leading to intergrain cracks in the electrolyte.[31]

The effect of different air electrodes on this electrolyte degradation has also been evaluated by several authors. Hjalmarsson *et al.*[269] studied co-electrolysis cells with two different air electrodes: LSM/YSZ and LSCF/CGO with a CGO barrier layer. The experiments were conducted at −1.5 A cm^{-2}, 800 °C and 62% conversion using a fuel composition of 45% CO_2 – 45% H_2O

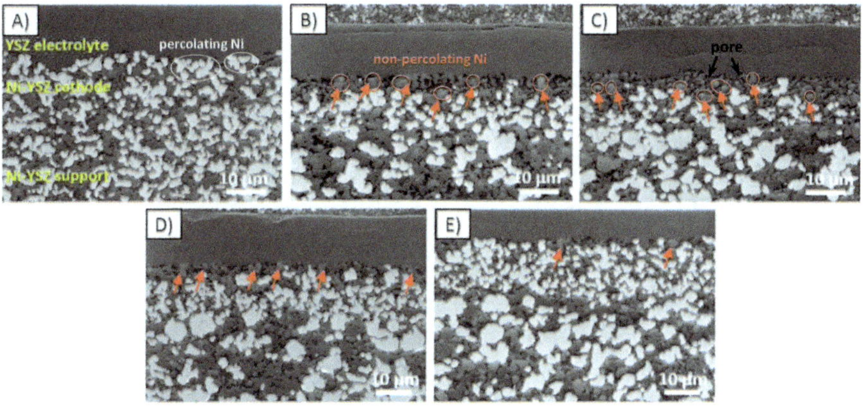

Figure 2.18 SEM micrographs of the Ni–YSZ electrodes after degradation experiments of SOC cells using a fuel composition of 45% H_2O + 45% CO_2 + 10% H_2 and current densities of −2.0 A cm^{-2}. (A) Reference cell without durability test, (B) cell 2 inlet and (C) outlet after test, (D) cell 3 inlet and (E) outlet after test. The bright and light grey colours indicate, respectively, the percolating nickel (white circles) and the non-percolating nickel (red circles). The non-percolating nickel particles are also indicated by the arrows. Reproduced from ref. 267 with permission from Elsevier, Copyright 2016.

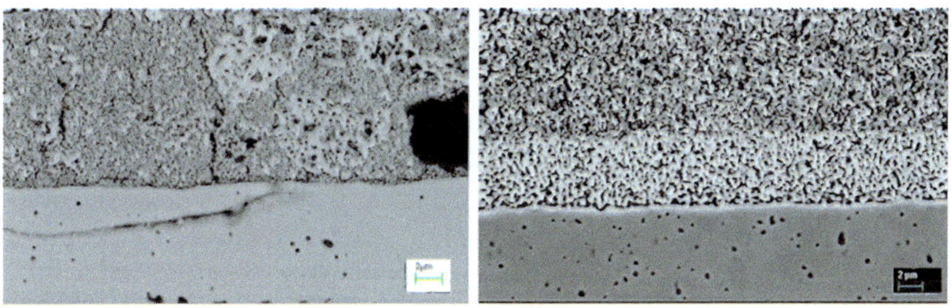

Figure 2.19 SEM image for operated cells with (left) LSM/YSZ air electrode without an interdiffusion barrier layer, and (right) with CGO barrier layer and LSCF-CGO electrode. Reproduced from ref. 268 with permission from Elsevier, Copyright 2013.

- 10% H_2. Their results showed a lower overall degradation for the cell with an LSCF:CGO air electrode, confirmed by impedance spectroscopy. However, the main reason for the lower degradation of the electrolyte (and fuel electrode in an indirect way) was the electrochemically more stable bi-layer electrolyte. It was proposed that the use of an inter-diffusion barrier and an electrochemically more active air electrode reduces the oxygen potential within the YSZ electrolyte, reducing the formation of the aforementioned inter-granular cracks within the YSZ that otherwise tend to show up under high polarisation (Figure 2.19).

Further studies by the same group[270] focused on cells with the aforementioned structure, including an inter-diffusion barrier layer of CGO between the YSZ electrolyte and the LSC–CGO air electrode. In this case long term durability studies were carried out up to 2700 h, claiming ASR values as low as 0.2 Ω cm^2, and a record of low degradation rates. Their experiments were conducted at −1 A cm^{-2}, 800 °C and 31% conversion and using a fuel composition of 45% CO_2 – 45% H_2O – 10% H_2. Although the cell degraded rapidly during the first 350 h, the degradation slowed down to an approximate rate of 5–10 mΩ cm^2 kh s^{-1}. The most interesting observation was that the cell partially reactivates its electrochemical performance when the electrolysis current was switched off and apparently the Ni–YSZ electrode was reactivated after an initial period of rapid degradation. From their EIS analysis it seems that the fuel conversion component is the most dominant contribution of the Ni–YSZ electrode.

2.11 Process Modelling

Due to the complexity of the different coupled processes involved in an electrochemical cell working in co-electrolysis mode, computational modelling of these systems needs to combine chemical, electrochemical and fluid dynamics models. These partial models should include (among others) the different chemical reactions taking place in the cell, electrical operating conditions (current densities, voltages) and processes of mass and heat transfer.

As previously mentioned, one of the first models to study the potential performance of large-scale high-temperature co-electrolysis plants for the production of syngas was developed by O'Brien *et al.*[29] An integral process model was developed, including all the components present in a real plant comprising a co-electrolyser coupled to a nuclear reactor. The predicted overall thermal-to-syngas efficiency values were generally within 5 percentage points of the reference-case power-cycle efficiency of 48.3%, decreasing with current density and operating voltage.

Ni[60,67] developed one-dimensional and two-dimensional thermal-fluid models of CO_2/H_2O co-electrolysis to study the distribution of gas composition and the behaviour of RWGS and methanation reactions in the fuel electrode. These studies concluded that the RWGS reaction could contribute to CO production at a low operating potential but consume CO at a high operating potential at an operating temperature of 800 °C and inlet gas composition of H_2O: 49.7%, CO_2: 25%, H_2: 25%, CO: 0.3%. Based on the elementary reaction models applied in SOFCs to describe the adsorption/desorption and surface reactions on the catalyst surface,[271,272] a detailed one-dimensional elementary reaction model for a CO_2/H_2O co-electrolysis cell was developed by Li *et al.*[273] considering the different reactions, electrode microstructure and the transport of mass and charge processes. The model, validated with experimental data for H_2O electrolysis, CO_2 electrolysis and CO_2/H_2O co-electrolysis at 700 °C, concluded the following order of adsorption/desorption rates: H_2 desorption > H_2O adsorption > CO_2 adsorption ≈ CO desorption. In addition, the electrochemical reduction rate of H_2O was found to be faster than that of CO_2.

Regarding the different regions of the cell, it was found that the heterogeneous reactions reach the equilibrium within 200 μm depths from the cathode's external surface, while the electrochemical reactions mainly occur near the interface with the electrolyte. Then, when the cathode is thick enough (*e.g.* more than 400 μm), the main zone of electrochemical reactions is far from that of the heterogeneous reactions. For example, the polarisation curve of CO_2/H_2O co-electrolysis lies between that of H_2O and CO_2 electrolysis when the cathode thickness is 700 μm, but gradually approaches that of H_2O electrolysis as the cathode thickness reduces from 700 μm to 30 μm, as also obtained from single cell experiments.

Cartographies were conducted by Aicart *et al.*[274] in order to find the optimal operation conditions. In concordance with experimental observations, co-electrolysis performances are found to be lower than those of H_2O electrolysis in similar conditions. Accordingly, the increase of temperature above the thermo-neutral voltage in electrolysis mode was much more pronounced compared to the one observed in co-electrolysis mode. As a consequence, higher mechanical stresses and material degradation rates are taking place in steam electrolysis mode. In addition, in the exothermic part of the co-electrolysis polarisation curves, the benefits from higher temperature are hindered by strong mass transport limitations. As also suggested by other authors, microstructure optimisation of the cell support to enhance CO/CO_2 gas transport is required. From their studies, CO_2 electroreduction was responsible for most of the CO production, as CO generated by the reversed WGS reaction was only in the 10–15% range at 1.3 V. Furthermore, the outlet ratio H_2/CO was shown to depend on current density, inlet flow rate and temperature.

As previously explained, the production and storage of methane is a very interesting way to convert excess electricity into a fuel that can be integrated with the existing natural-gas network. In this sense, Luo *et al.*[275] have recently completed a comprehensive exergy analysis using three different methane synthesis routes: (1) water electrolysis + Sabatier reaction, (2) H_2O/CO_2 co-electrolysis + methanation reactor (MR), and (3) a single SOEC-MR reactor. In their first route, a SOEC electrolyser was compared with a low-temperature electrolysis cell, concluding that using an SOEC is more efficient and has over 11% higher exergy efficiency at or below -0.8 A cm^{-2} due to the lower electrical consumption. Results from route (2) show a performance higher than that of route (1) at high current densities, being more suitable for high-yield methane production. The optimal operation conditions were found at a current density of -0.6 A cm^{-2}, a H/C ratio of 10.54, and at 650 °C. This intermediate temperature enables the system to achieve better thermal integration and improves the exergy efficiency to about 77% at 6 bar. In route (3), the use of a single SOEC-MR reactor adds some complexity due to the mismatch of the SOEC and MR optimal operating temperatures. By the use of a spatial temperature gradient, the exergy efficiency increases up to 81% in a compact system, while the energy-to-methane efficiency using a uniform-temperature SOEC-MR reactor only reaches about 50%.

In addition, parallel thermodynamic studies were performed by Samavati *et al.*[276] on the production of Fischer–Tropsch diesel from renewable sources using different pathways. In their work, SOEC and biomass entrained gasification (EG) were used to obtain the intermediate precursor syngas but comparisons were performed between operating the SOEC in co-electrolysis (scenario 1) or simply steam electrolysis (scenario 2). From models of the two different scenarios they analysed the effect of operating pressure (adverse in scenario 1 but positive in 2) and temperature (showing that it is possible to recover around 390 MW heat, increasing the system energy efficiency). Finally, although total efficiency of the system is higher in case of steam electrolysis, this integration would not allow the reuse of carbon dioxide. Besides, including co-electrolysis will increase the production potential of synthetic FT diesel compared to the electrolysis case.

Following this trend, reversible SOC systems have also been modelled. SOCs working alternately in both SOFC and SOEC modes are considered as a key technology for near future energy systems. For example, Wendel *et al.* studied how global efficiency is influenced by the stack temperature and pressure, and the fuel utilisation. Their strategy promotes exothermic methanation within the cell-stack to provide thermal energy for the endothermic steam/CO_2 electrolysis reactions during the fuel producing mode.[277] With their proposed thermal management, overall efficiencies greater than 70% were achieved at intermediate temperature (680 °C) and elevated stack pressure (20 bar). However, due to the reversible nature of the system, a trade-off between system efficiency and energy storage is required. For example, the conditions for optimal system efficiency correspond with sub-optimal stored methane content (<60 mol%), which reduces the storage energy density and adds challenges associated with storing hydrogen-rich gases.

Also relating to direct synthesis of methane, Cinti *et al.*[278] showed a conceptual design of an electricity-to-liquid fuel system consisted of a SOEC stack working in co-electrolysis and a Fischer–Tropsch reactor. A system-level model was built for three different system configurations: a basic open-loop layout (A), a closed-loop layout with reformate gases recirculation (B), and a closed-loop layout with direct recirculation of Fischer–Tropsch light gases (C). For the case (A), at thermoneutral conditions SOEC efficiency was estimated to be around 79%. Then, taking into account all of the syncrude components, the Fischer–Tropsch energy efficiency was around 53%, while considering only the crude fraction the efficiency drops to 41%. The highest efficiency was achieved in layout (C), where light gas fractions are recirculated directly to the SOEC stack. In addition, layout (C) achieves the following reductions: −73% H_2O, −51%

CO_2 and −100% H_2. In terms of the exergy analysis, layout (C) has a larger electrical consumption than layout (B), since its thermoneutral condition is achieved at higher values of power density.

Additional thermodynamic and/or electrochemical modelling studies can be found in the following ref. 279–289.

2.12 Concluding Remarks

Rising emissions of greenhouse effect gases have become a matter of concern in the past decade. Since one of the main sources of these gases, especially CO_2, is the consumption of fossil fuels to meet the major part of the global energy demand, a huge development of alternative fuels and energy systems has taken place. In the best case, this new scenario would be based on low carbon energy technologies coupled with carbon capture and storage systems.

Solid oxide fuel cells have been proposed to play an important role in future energy systems, not only because of their ability to use different fuels (from hydrogen to hydrocarbons) in an efficient and reasonably clean way, but also because they can be employed in reverse mode to electrolyse H_2O, CO_2 or to react them both to form syngas through a co-electrolysis process. Although syngas itself is widely used industrially, subsequent catalytic processes to form liquid hydrocarbon fuels could close a renewable energy cycle. Solid oxide fuel cells, which have been shown to work effectively with these fuels, could be a preferred method to convert the fuel back to electricity. In this proposal, a significant advantage is that the liquid fuels are very similar to current fossil fuels and thus present less technological and infrastructure barriers than, for example, an energy system based on hydrogen. This would allow an easier transition to a carbon-neutral and cost-competitive liquid fuel economy using renewable electricity. Furthermore, thermal synergies in the liquid fuel cell should allow substantially higher round-trip energy efficiency than a hydrogen cycle.

In this context, this chapter gathers some recent studies and developments covering various aspects of the co-electrolysis process, from the alternatives regarding the materials and operational parameters to durability tests, including economic considerations and scalability studies.

Abbreviations

AP-XPS –	Ambient-pressure X-ray photoelectron spectroscopy
ASR –	Area specific resistance
BEP –	Brønsted–Evans–Polanyi relationship
BZCYZn –	$BaCe_{0.5}Zr_{0.3}Y_{0.16}Zn_{0.04}O_{3-\delta}$
CEA –	Commissariat à l'Energie Atomique
CCS –	Carbon capture and storage
CGO –	Cerium gadolinium oxide

DFT – Density functional theory
DME – dimethyl ether
DRT – Distribution of relaxation times
DTU – Technical University of Denmark
EG – Entrained gasification
EIRC – European Research Institute of Catalysis
EIS – Electrochemical impedance spectroscopy
FIB – Focused ion beam
FT – Fischer–Tropsch
FVO – FeV_2O_4
GHG – Greenhouse effect gases
HRDC – Toshiba Group Hydrogen Energy Research & Development Center
HTSE – High temperature steam electrolysis
INL – Idaho National Laboratory
KIT – Karlsruhe Institute of Technology
LCFNb – $La_{0.9}Ca_{0.1}Fe_{0.9}Nb_{0.1}O_{3-\delta}$
LCNT – $La_{0.43}Ca_{0.37}Ni_{0.06}Ti_{0.94}O_3$
LSC – $La_{0.8}Sr_{0.2}CoO_3$
LSCF – Lanthanum strontium cobalt ferrite
LSCM – $La_{0.8}Sr_{0.2}Cr_{0.5}Mn_{0.5}O_{3-\delta}$ or $La_{0.75}Sr_{0.25}Cr_{0.5}Mn_{0.5}O_{3-\delta}$
LSCN – $La_{0.8}Sr_{0.2}Co_{0.8}Ni_{0.2}O_{3-\delta}$
LSF – $La_{0.8}Sr_{0.2}FeO_{3-\delta}$
LSFCr – $La_{0.3}Sr_{0.7}Fe_{0.7}Cr_{0.3}O_{3-\delta}$
LSFN – $La_{0.6}Sr_{0.4}Fe_{0.8}Ni_{0.2}O_{3-\delta}$
LSGM – $La_{0.9}Sr_{0.1}Ga_{0.8}Mg_{0.2}O_{3-\delta}$
LST – $La_xSr_{1-x}TiO_{3-\delta}$
$LSTMN_A$– $(La_{0.2}Sr_{0.8})_{0.95}Ti_{0.9}Mn_{0.1}O_{3+\delta}$
LSV – $La_xSr_{1-x}VO_{3-\delta}$
MEA – Monoethanolamine
MECC – Mixed electronic and carbonate-ion conductor
MIEC – Mixed ionic and electronic conductor
MOFs – Metal–organic frameworks
MR – Methanation reactor
NASA – National Aeronautics and Space Administration
P2G – Power-to-gas
P2L – Power-to-liquid
ReSOCs – Reversible solid oxide cells
RWGS – Reverse water gas shift reaction
10Sc1CeSZ – 10 mol% Sc_2O_3 and 1% CeO_2 stabilised zirconia
ScSZ – Scandia-stabilised zirconia
ScYZ – Scandia and yttria-stabilised zirconia
SEM – Scanning electron microscopy
SFM – $Sr_2Fe_{1.5}Mo_{0.5}O_{6-\delta}$
SFMn – $Sr_{0.5}Fe_{1.5}Mn_{0.5}O_{6-\delta}$
SFMo – $Sr_2Fe_{1.5}Mo_{0.5}O_{6-\delta}$

SOC –	Solid oxide cell
SOEC –	Solid oxide electrolyser cell
SOFC –	Solid oxide fuel cells
SMR –	Steam methane reforming
STN –	$SrTi_{0.9}Nb_{0.1}O_{3-\delta}$
TEC –	Thermal expansion coefficient
TEM –	Transmission electron microscopy
TOFC –	Topsoe fuel cell
TPBs –	Triple phase boundaries
WGS –	Water gas shift reaction
YbScSZ –	Ytterbia and scandia stabilised zirconia
3YSZ –	3 mol% yttria-stabilised zirconia
YSZ –	Yttria-stabilised zirconia

References

1. D. Y. C. Leung, G. Caramanna and M. M. Maroto-Valer, *Renewable Sustainable Energy Rev.*, 2014, **39**, 426.
2. S. Mekhilef, R. Saidur and A. Safari, *Renewable Sustainable Energy Rev.*, 2011, **15**, 1777.
3. J. A. Herron, J. Kim, A. A. Upadhye, G. W. Huber and C. T. Maravelias, *Energy Environ. Sci.*, 2015, **8**, 126.
4. F. Díaz-González, A. Sumper, O. Gomis-Bellmunt and R. Villafáfila-Robles, *Renewable Sustainable Energy Rev.*, 2012, **16**, 2154.
5. A. D. Azahin, *Prog. Energy Combust. Sci.*, 2004, **30**, 501.
6. A. Demirbas, *Prog. Energy Combust. Sci.*, 2005, **31**, 171.
7. E. Barbier, *Renewable Sustainable Energy Rev.*, 2002, **6**, 3.
8. C. R. Chamorro, M. E. Mondejar, R. Ramos, J. J. Segovia, M. C. Martin and M. A. Villamanan, *Energy*, 2012, **42**, 10.
9. C. W. Forsberg, *Prog. Nucl. Energy*, 2009, **51**, 192.
10. M. M. Abu-Khader, *Prog. Nucl. Energy*, 2009, **51**, 225.
11. L. Lin, Z. Cunshan, S. Vittayapadung, S. Xiangqian and D. Mingdong, *Appl. Energy*, 2011, **88**, 1020.
12. A. E. Atabani, A. S. Silitonga, I. A. Badruddin, T. M. I. Mahlia, H. H. Masjuki and S. Mekhilef, *Renewable Sustainable Energy Rev.*, 2012, **16**, 2070.
13. UK Renewable Energy Roadmap Update, 2013, https://www.gov.uk/government/collections/uk-renewable-energy-roadmap, accessed June 2018.
14. B. Sakintuna, F. Lamari-Darkrim and M. Hirscher, *Int. J. Hydrogen Energy*, 2007, **32**, 1121.
15. A. C. Dillon, K. M. Jones, T. A. Bekkedahl, C. H. Kiang, D. S. Bethune and M. J. Heben, *Nature*, 1997, **386**, 377.
16. J. L. C. Rowsell and O. M. Yaghi, *Angew. Chem., Int. Ed.*, 2005, **44**, 4670.
17. M. Mogensen and K. Kammer, *Annu. Rev. Mater. Res.*, 2003, **33**, 321.

18. X.-M. Ge, S.-H. Chan, Q.-L. Liu and Q. Sun, *Adv. Energy Mater.*, 2012, **2**, 1156.
19. M. A. Laguna-Bercero, in *Encyclopedia of Sustainable Technologies*, ed. M. Abraham, Elsevier, The Netherlands, 2017, p. 331.
20. W. Donitz and E. Erdle, *Int. J. Hydrogen Energy*, 1985, **10**, 291.
21. J. Laurencin and J. Mougin, in *Hydrogen Production: Electrolysis*, ed. A. Godula-Jopek, Wiley-VCH Verlag GmbH & Co. KGaA, 2015, p. 191.
22. A. Hauch, S. D. Ebbesen, S. H. Jensen and M. Mogensen, *J. Mater. Chem.*, 2008, **18**, 2331.
23. M. A. Laguna-Bercero, *J. Power Sources*, 2012, **203**, 4.
24. S. D. Ebbesen, S. H. Jensen, A. Hauch and M. B. Mogensen, *Chem. Rev.*, 2014, **114**, 10697.
25. M. Ni, M. K. H. Leung and D. Y. C. Leung, *Int. J. Hydrogen Energy*, 2008, **33**, 2337.
26. C. Zhao, W. Zhang, B. Yu, J. Wang and J. Chen, *J. Prog. Chem.*, 2016, **28**, 1265.
27. S. Y. Gómez and D. Hotza, *Renewable Sustainable Energy Rev.*, 2016, **61**, 155.
28. J. Carneiro and E. Nikolla, *Nano Res.*, 2019, **12**, 2081.
29. J. E. O'Brien, M. G. McKellar, C. M. Stoots, J. S. Herring and G. L. Hawkes, *Int. J. Hydrogen Energy*, 2009, **34**, 4216.
30. P. Kim-Lohsoontorn, N. Laosiripojana and J. Bae, *Curr. Appl. Phys.*, 2011, **11**, S223.
31. C. Graves, S. D. Ebbesen and M. Mogensen, *Solid State Ionics*, 2011, **192**, 398.
32. R. J. Gorte and J. M. Vohs, *J. Catal.*, 2003, **216**, 477.
33. H. Kim, C. Lu, W. L. Worrell, J. M. Vohs and R. J. Gorte, *J. Electrochem. Soc.*, 2002, **149**, A247.
34. K. Sasaki and Y. Taraoka, *J. Electrochem. Soc.*, 2003, **150**, A885.
35. J. P. Vanhook, *Catal. Rev.: Sci. Eng.*, 1980, **21**, 1.
36. J. A. Moulijn, M. Makkee and A. E. van Diepen, *Chemical Process Technology*, Wiley-VCH Verlag GmbH & Co. KGaA, 2nd edn, 2013, p. 127.
37. C. Van Der Giesen, R. Kleijn and G. J. Kramer, *Environ. Sci. Technol.*, 2014, **48**, 7111.
38. W. L. Becker, R. J. Braun, M. Penev and M. Melaina, *Energy*, 2012, **47**, 99.
39. E. V. Kondratenko, G. Mul, J. Baltrusaitis, G. O. Larrazábal and J. Pérez-Ramírez, *Energy Environ. Sci.*, 2013, **6**, 3112.
40. Y. Zheng, J. Wang, B. Yu, W. Zhang, J. Chen, J. Qiao and J. Zhang, *Chem. Soc. Rev.*, 2017, **46**, 1427.
41. A. Goeppert, M. Czaun, J.-P. Jones, G. K. Surya Prakash and G. A. Olah, *Chem. Soc. Rev.*, 2014, **43**, 7995.
42. Z. Wang, B. Yu, W. Zhang, J. Chen and J. Xu, *Prog. Chem.*, 2013, **25**, 1229.
43. G. Leonzio, *J. CO2 Util.*, 2018, **27**, 326.

44. G. Gahleitner, *Int. J. Hydrogen Energy*, 2013, **38**, 2039.
45. S. H. Jensen, C. Graves, M. Mogensen, C. Wendel, R. Braun, G. Hughes, Z. Gao and S. A. Barnett, *Energy Environ. Sci.*, 2015, **8**, 2471.
46. M. Bailera, P. Lisbona, L. M. Romeo and S. Espatolero, *Renewable Sustainable Energy Rev.*, 2017, **69**, 292.
47. A. Maroufmashat and M. Fowle, *Energies*, 2017, **10**, 1089.
48. K. Ghaib and F.-Z. Ben-Fares, *Renewable Sustainable Energy Rev.*, 2018, **81**, 433.
49. S. R. Foit, I. C. Vinke, L. G. J. de Haart and R. Eichel, *Angew. Chem.*, 2017, **56**, 5402.
50. S. Schemme, R. C. Samsun, R. Peters and D. Stolten, *Fuel*, 2017, **205**, 198.
51. A. Varone and M. Ferrari, *Renewable Sustainable Energy Rev.*, 2015, **45**, 207.
52. Y. Luo, Y. Shi and N. Cai, *J. Energy Eng.*, 2018, **144**, 04017079.
53. The Strategic Energy at the heart of Energy Research and Innovation in Europe Technology (SET) Plan, https://ec.europa.eu/energy/en/topics/technology-and-innovation/strategic-energy-technology-plan, accessed May 2019.
54. S. Paltsev, H. D. Jacoby, J. M. Reilly, Q. J. Ejaz, J. Morris, F. O'Sullivan, S. Rausch, N. Winchester and O. Kragha, *Energy Policy*, 2011, **39**, 5309.
55. http://www.dtu.dk/english/news/2017/03/dynamo-theme4-hydrogen-from-reverse-fuel-cells?id=e804ab15-4822-4f1c-92be-09a3e5bece1e, accessed May 2019.
56. D. M. Bierschenk, J. R. Wilson and S. A. Barnett, *Energy Environ. Sci.*, 2011, **4**, 944.
57. J. B. Hansen, *Faraday Discuss.*, 2015, **182**, 9.
58. A. O. Isenberg, *Solid State Ionics*, 1981, **3–4**, 431.
59. C. Stoots, J. O'Brien and J. Hartvigsen, *Int. J. Hydrogen Energy*, 2009, **34**, 4208.
60. M. Ni, *J. Power Sources*, 2012, **202**, 209.
61. F. Haber, *Z. Phys. Chem.*, 2017, **68**, 726.
62. Y. Wang, T. Liu, S. Fang, G. Xiao, H. Wang and F. Chen, *J. Power Sources*, 2015, **277**, 261.
63. Z. Zhan and L. Zhao, *J. Power Sources*, 2010, **195**, 7250.
64. X. Chen, C. Guan, G. Xiao, X. Du and J.-Q. Wang, *Faraday Discuss.*, 2015, **182**, 341.
65. S. D. Ebbesen, C. Graves and M. Mogensen, *Int. J. Green Energy*, 2009, **6**, 646.
66. Z. Zhan, W. Kobsiriphat, J. R. Wilson, M. Pillai, I. Kim and S. A. Barnett, *Energy Fuels*, 2009, **23**, 3089.
67. M. Ni, *Int. J. Hydrogen Energy*, 2012, **37**, 6389.
68. W. Li, Y. Shi, Y. Luo and N. Cai, *J. Power Sources*, 2015, **273**, 1.
69. W. Zhang, Y. Zheng, B. Yu, J. Wang and J. Chen, *Int. J. Hydrogen Energy*, 2017, **42**, 29911.

70. Z. A. Feng, M. L. Machala and W. C. Chueh, *Phys. Chem. Chem. Phys.*, 2015, **17**, 12273.
71. X. Sun, M. Chen, S. H. Jensen, S. D. Ebbesen, C. Graves and M. Mogensen, *Int. J. Hydrogen Energy*, 2012, **37**, 17101.
72. J. B. Goodenough and Y.-H. Huang, *J. Power Sources*, 2007, **173**, 1.
73. Y. Tao, S. D. Ebbesen and M. B. Mogensen, *J. Electrochem. Soc.*, 2014, **161**, F337.
74. Y. Tao, S. D. Ebbesen, W. Zhang and M. B. Mogensen, *ChemCatChem*, 2014, **6**, 1220.
75. S.-W. Kim, H. Kim, K. J. Yoon, J.-H. Lee, B.-K. Kim, W. Choi, J.-H. Lee and J. Hong, *J. Power Sources*, 2015, **280**, 630.
76. The Greater London Industrial Archaeology Society (GLIAS), http://www.glias.org.uk/Chemicals_from_Coal/PART_1.HTM, accessed May 2019.
77. H. W. Chandler, Isomet report no. 5907-PR4-61, 1961.
78. H. W. Chandler, Study of electrolytic reduction of carbon dioxide. Technical Documentary Report No. MRL-TDR-62–16, 1962.
79. J. Weissbart and W. H. Smart, NASA CR-680, 1967.
80. M. Steinberg, Electrolytic synthesis of methanol from CO_2, *US Pat.*, 3959094, 1976.
81. J. G. Lewis and A. J. Martin, Method for obtaining carbon dioxide from the atmosphere and for production of fuels, *US Pat.*, 4140602A, 1979.
82. M. J. Corbett and S. C. Salina, Production of synthetic hydrocarbons from air, water and low cost electrical power, *US Pat.*, 4282187, 1981.
83. Y. Yamauchi, Y. Tokita, N. Murakami, K. Takita, Y. Mori and K. Muraishi, Method for producing methanol by use of nuclear heat and power generating plant, *US Pat.*, 5479462, 1994.
84. R. J. Jensen, J. L. Lyman, J. D. King and R. D. Guettler, Solar reduction of CO_2, *US Pat.*, 6066187, 2000.
85. Danish Technological Institute GreenSynFuels, Final Project Report, EUDP project journal number: 64010-0011, 2011.
86. Biogas-SOEC, Electrochemical upgrading of biogas to pipeline quality by means of SOEC electrolysis, ForskNG, Project no. 10677, 2011.
87. http://www.nordicenergy.org/project/synthetic-fuels-for-heavy-transportation/, accessed May 2019.
88. Danish Energy Technology Development and Demonstration Programme (EUDP), number 64011-0036, Green Natural Gas.
89. H. Iskov and N. B. Rasmussen, Global screening of projects and technologies for Power-to-Gas and Bio-SNG, Project Report, November 2013.
90. M. Chen, J. V. T. Høgh, J. U. Nielsen, J. J. Bentzen, S. D. Ebbesen and P. V. Hendriksen, *Fuel Cells*, 2013, **13**, 638.
91. https://energiforskning.dk/en/node/7155, accessed May 2019.

92. http://www.helmeth.eu/, accessed May 2019.
93. Programme Review Report 2014. Fuel Cells and Hydrogen Joint Undertaking, Publications Office of the European Union, 2015.
94. http://www.sunfire.de/en/products-technology/hydrogen-generator, accessed May 2019.
95. G. Cinti, G. Discepoli, G. Bidini, A. Lanzini and M. Santarelli, *Int. J. Energy Res.*, 2016, **40**, 207.
96. M. Reytier, S. Di Iorio, A. Chatroux, M. Petitjean, J. Cren, M. De Saint Jean, J. Aicart and J. Mougin, *Int. J. Hydrogen Energy*, 2015, **40**, 11370.
97. S. Diethelm, J. van Herle, D. Montinaro and O. Bucheli, *Fuel Cells*, 2013, **13**, 631.
98. C. Stoots, J. O'Brien and J. Hartvigsen, Results of Recent High Temperature Co-Electrolysis Studies at the Idaho National Laboratory, 2007 AIChE Annual Meeting, INL/CON-07–12821.
99. M. G. McKellar, J. E. O'Brien, C. M. Stoots and G. L. Hawkes, ASME International Mechanical Engineering Congress and Exposition, *Proceedings of the ASME*, 2007, vol. 6, p. 691.
100. C. Stoots, J. O'Brien and J. Hartvigsen, *Int. J. Hydrogen Energy*, 2009, **34**, 4208.
101. M. McKellar, C. Stoots, M. Sohal, L. Mulloth, B. Luna and M. Abney, *American Institute of Aeronautics and Astronautics (AIAA)*, 2010, 6273.
102. L. Frost, E. Elangovan, J. Hartvigsen and J. Can, *J. Chem. Eng.*, 2016, **94**, 636.
103. https://www.fuelcellenergy.com/, accessed May 2019.
104. https://www.toshiba-energy.com/en/hydrogen/product/h2one.htm, accessed May 2019.
105. S. J. Noh, W. J. Byeon and S. K. Lee, *J. Korean Phys. Soc.*, 2015, **67**, 860.
106. J. E. O'Brien, M. G. McKellar, E. A. Harvego and C. M. Stoots, *Int. J. Hydrogen Energy*, 2010, **35**, 4808.
107. W. Zhang, B. Yu and J. Xu, *Int. J. Hydrogen Energy*, 2012, **37**, 12060.
108. Y. Bo, Z. Wenqiang, X. Jingming and C. Jing, *Int. J. Hydrogen Energy*, 2010, **35**, 2829.
109. H. Zhang, S. Su, X. Chen, G. Lin and J. Chen, *Int. J. Hydrogen Energy*, 2013, **38**, 4298.
110. J. Sanz-Bermejo, J. Muñoz-Antón, J. Gonzalez-Aguilar and M. Romero, *Appl. Energy*, 2014, **131**, 238.
111. J. Sigurvinsson, C. Mansilla, P. Lovera and F. Werkoff, *Int. J. Hydrogen Energy*, 2007, **32**, 1174.
112. Y. Shi, Y. Luo, W. Li, M. Ni and N. Cai, High Temperature Electrolysis for Hydrogen or Syngas Production from Nuclear or Renewable Energy, in *Handbook of Clean Energy Systems*, John Wiley & Sons, Ltd, 2015.
113. C. Graves, S. D. Ebbesen, M. Mongensen and K. S. Lackner, *Renewable Sustainable Energy Rev.*, 2011, **15**, 1.
114. M. Kauw, R. M. J. Benders and C. Visser, *Energy*, 2015, **90**, 208.

115. J. Fang, X. Jin and K. Huang, *J. Membr. Sci.*, 2018, **549**, 142.
116. K. L. Buchheit, J. D. Smith, U. Guntupalli and C. Chen, *Energy & Fuels*, 2016, **30**, 10721–10729.
117. P. Collet, E. Flottes, A. Favre, L. Raynal, H. Pierre, S. Capela and C. Peregrina, *Appl. Energy*, 2017, **192**, 282.
118. E. Giglio, A. Lanzini, M. Santarelli and P. Leone, *J. Energy Storage*, 2015, **1**, 22.
119. S. Chiuta, N. Engelbrecht, G. Human and D. G. Bessarabov, *J. CO2 Util.*, 2016, **16**, 399.
120. C. H. Wendel and R. Braun, *J. Appl. Energy*, 2016, **172**, 118.
121. Y. Redissi and C. Bouallou, *Energy Procedia*, 2013, **37**, 6667.
122. A. Buttler, R. Koltun, R. Wolf and H. Spliethoff, *Int. J. Hydrogen Energy*, 2015, **40**, 38.
123. I. Hannula, *Biomass Bioenergy*, 2015, **74**, 26.
124. M. Fasihi, D. Bogdanov and C. Breyer, *Energy Procedia*, 2016, **99**, 243.
125. E. Giglio, A. Lanzini, M. Santarelli and P. Leone, *J. Energy Storage*, 2015, **2**, 64.
126. R. Elder, D. Cumming and M. B. Mogensen, in *Carbon Dioxide Utilisation: Closing the Carbon Cycle*, ed. P. S. Elsje, A. Quadrelli and K. Armstrong, Elsevier, Nederland, 2014, p. 183.
127. V. A. Sadykov, S. N. Pavlova, G. M. Alikina, N. N. Sazonova, N. V. Mezentseva, M. V. Arapova, V. A. Rogov, T. A. Krieger, A. V. Ishchenko, R. V. Gulyaev, A. V. Zadesenets, A. C. Roger, C. E. Chan-Thaw and O. L. Smorygo, in *Perovskite: Crystallography, Chemistry and Catalytic Performance*, Nova Publishers, USA, 2013, p. 1.
128. Y. Wang, T. Liu, L. Lei and F. Chen, *Fuel-Process. Technol.*, 2017, **161**, 248.
129. S. Uhm and Y. D. Kim, *Curr. Appl. Phys.*, 2014, **14**, 672.
130. F. M. Sapountzi, J. M. Gracia, C. J. K.-J. Weststrate, H. O. A. Fredriksson and J. W. H. Niemantsverdriet, *Prog. Energy Combust. Sci.*, 2017, **58**, 1.
131. V. N. Nguyen and L. Blum, *Chem.-Ing.-Tech.*, 2015, **87**, 354.
132. P. A. Lessing, *J. Mater. Sci.*, 2007, **42**, 3465.
133. S. H. Jensen, P. H. Larsen and M. Mogensen, *Int. J. Hydrogen Energy*, 2007, **32**, 3253.
134. A. Brisse, J. Schefold and M. Zahid, *Int. J. Hydrogen Energy*, 2008, **33**, 5375.
135. A. Momma, T. Kato, Y. Kaga, S. Nagata and J. Ceram, *Soc. Jpn.*, 1997, **105**, 369.
136. P. Kim-Lohsoontorn, D. J. L. Brett, N. Laosiripojana, Y.-M. Kim and J.-M. Bae, *Int. J. Hydrogen Energy*, 2010, **35**, 3958.
137. M. A. Laguna-Bercero, R. Campana, A. Larrea, J. A. Kilner and V. M. Orera, *Fuel Cells*, 2011, **11**, 116.
138. A. V. Virkar, *Int. J. Hydrogen Energy*, 2010, **35**, 9527.
139. J. Schefold, A. Brisse and M. Zahid, *J. Electrochem. Soc.*, 2009, **156**, B897.

140. M. A. Laguna-Bercero, R. Campana, A. Larrea, J. A. Kilner and V. M. Orera, *J. Power Sources*, 2011, **196**, 8942.
141. J. R. Mawdsley, J. D. Carter, A. J. Kropf, B. Yildiz and V. A. Maroni, *Int. J. Hydrogen Energy*, 2009, **34**, 4198.
142. C. Graves, S. D. Ebbesen, S. H. Jensen, S. B. Simonsen and M. B. Mogensen, *Nat. Mater.*, 2015, **14**, 239.
143. M. A. Laguna-Bercero, S. J. Skinner and J. A. Kilner, *J. Power Sources*, 2009, **192**, 126.
144. M. A. Laguna-Bercero, J. A. Kilner and S. J. Skinner, *Chem. Mater.*, 2010, **22**, 1134.
145. J. S. Herring, J. E. O'Brien, C. M. Stoots, G. L. Hawkes, J. J. Hartvigsen and M. Shahnam, *Int. J. Hydrogen Energy*, 2007, **32**, 440.
146. T. Ishihara, N. Jirathiwathanakul and H. Zhong, *Energy Environ. Sci.*, 2010, **3**, 665.
147. S. Elangovan, J. J. Hartvigsen and L. J. Frost, *Int. J. Appl. Ceram. Technol.*, 2007, **4**, 109.
148. M. A. Laguna-Bercero, R. D. Bayliss and S. J. Skinner, *Solid State Ionics*, 2014, **262**, 398.
149. F. Chauveau, J. Mougin, J. M. Bassat, F. Mauvy and J. C. Grenier, *J. Power Sources*, 2010, **195**, 744.
150. M. A. Laguna-Bercero, N. Kinadjian, R. Sayers, H. El Shinawi, C. Greaves and S. J. Skinner, *Fuel Cells*, 2011, **11**, 102.
151. F. Chauveau, J. Mougin, F. Mauvy, J.-M. Bassat and J.-C. Grenier, *Int. J. Hydrogen Energy*, 2011, **36**, 7785.
152. M. A. Laguna-Bercero, H. Monzón, A. Larrea and V. M. Orera, *J. Mater. Chem. A*, 2016, **4**, 1446.
153. X. Tong, F. Zhou, S. Yang, S. Zhong, M. Wei and Y. Liu, *Ceram. Int.*, 2017, **43**, 10927–10933.
154. E. Kravchenko, D. Khalyavin, K. Zakharchuk, J. Grins, G. Svensson, V. Pankov and A. Yaremchenko, *J. Mater. Chem. A*, 2015, **3**, 23852.
155. M. A. Laguna-Bercero, A. R. Hanifi, L. Menand, N. K. Sandhu, N. E. Anderson, T. H. Etsell and P. Sarkar, *Electrochim. Acta*, 2018, **268**, 195.
156. S. D. Ebbesen and M. Mogensen, *J. Power Sources*, 2009, **193**, 349.
157. T.-J. Huang, X.-D. Shen and C.-L. Chou, *J. Power Sources*, 2009, **187**, 348.
158. A. Gadalla and J. White, *Br. Ceram. Soc.*, 1966, **65**, 383.
159. N. Kiratzis, P. Holtappels, C. E. Hatchwell, M. Mogensen and J. T. S. Irvine, *Fuel Cells*, 2001, **1**, 211.
160. C.-Y. Cheng, G. H. Kelsall and L. Kleiminger, *J. Appl. Electrochem.*, 2013, **43**, 1131.
161. D. L. Trimm, *Catal. Today*, 1999, **49**, 3.
162. R. J. Gorte, H. Kim and J. M. Vohs, *J. Power Sources*, 2002, **106**, 10.
163. T. Takeguchi, R. Kikuchi, T. Yano, K. Eguchi and K. Murata, *Catal. Today*, 2003, **84**, 217.
164. I. Gavrielatos, V. Drakopoulos and S. G. Neophytides, *J. Catal.*, 2008, **259**, 75.

165. H. Kan and H. Lee, *Appl. Catal., B*, 2010, **97**, 108.
166. E. Nikolla, J. Schwank and S. Linic, *J. Catal.*, 2007, **250**, 85.
167. O. A. Marina, C. A. Coyle, M. H. Engelhard and L. R. Pederson, *J. Electrochem. Soc.*, 2011, **158**, B424.
168. W. M. Haynes, in *CRC Handbook of Chemistry and Physics*, 2016.
169. H. Zhu, W. Wang, R. Ran, C. Su, H. Shi and Z. Shao, *Int. J. Hydrogen Energy*, 2012, **37**, 9801.
170. T. Hibino, A. Hashimoto, M. Yano, M. Suzuki and M. Sano, *Electrochim. Acta*, 2003, **48**, 2531.
171. Z. Zhan and S. A. Barnett, *Solid State Ionics*, 2005, **176**, 871.
172. F. Bidrawn, G. Kim, G. Corre, J. T. S. Irvine, J. M. Vohs and R. J. Gorte, *Electrochem. Solid-State Lett.*, 2008, **11**, B167.
173. X. Zhang, Y. Song, F. Guan, Y. Zhou, H. Lv, Q. Liu, G. Wang and X. Bao, *J. Power Sources*, 2018, **400**, 104.
174. H. Li, G. Sun, K. Xie, W. Qi, Q. Qin, H. Wei, S. Chen, Y. Wang, Y. Zhang and Y. Wu, *Int. J. Hydrogen Energy*, 2014, **39**, 20888.
175. C. Ruan, K. Xie, L. Yang, B. Ding and Y. Wu, *Int. J. Hydrogen Energy*, 2014, **39**, 10338.
176. W. Qi, Y. Gan, D. Yin, Z. Li, G. Wu, K. Xie and Y. Wu, *J. Mater. Chem. A*, 2014, **2**, 6904.
177. L. Yang, K. Xie, L. Wu, Q. Qin, J. Zhang, Y. Zhang, T. Xie and Y. Wu, *Phys. Chem. Chem. Phys.*, 2014, **16**, 21417.
178. L. Ye, X. Hu, X. Wang, F. Chen, D. Tang, D. Dong and K. Xie, *J. Mater. Chem. A*, 2019, **7**, 2764.
179. S. Wang and T. Ishihara, *ECS Trans.*, 2017, **57**, 3171.
180. P. K. Addo, B. Molero-Sanchez, M. Chen, S. Paulson and V. Birss, *Fuel Cells*, 2015, **15**, 689.
181. L. Ye, C. Pan, M. Zhang, C. Li, F. Chen, L. Gan and K. Xie, *ACS Appl. Mater. Interfaces*, 2017, **9**, 25350.
182. Y. Li, X. Chen, Y. Yang, Y. Jiang and C. Xia, *ACS Sustainable Chem. Eng.*, 2017, **5**, 11403.
183. A. K. Opitz, A. Nenning, C. Rameshan, M. Kubicek, T. Götsch, R. Blume, M. Hävecker, A. Knop-Gericke, G. Rupprechter, B. Klötzer and J. Fleig, *ACS Appl. Mater. Interfaces*, 2017, **9**, 35847.
184. V. Singh, S. Hashigami, H. Muroyama, T. Matsui, T. Inagaki and K. Eguchi, *J. Electrochem. Soc.*, 2016, **163**, F3084.
185. T. H. Shin, J.-H. Myung, M. Verbraeken, G. Kim and J. T. S. Irvine, *Faraday Discuss.*, 2015, **182**, 227.
186. C. Stoots, J. O'Brien and J. Hartvigsen, *Int. J. Hydrogen Energy*, 2009, **34**, 4208.
187. Z. Zhan, W. Kobsiriphat, J. R. Wilson, M. Pillai, I. Kim and S. A. Barnett, *Energy Fuels*, 2009, **23**, 3089.
188. W. Wang, J. M. Vohs and R. J. Gorte, *Top. Catal.*, 2007, **46**, 380.
189. M. Torrell, S. García-Rodríguez, A. Morata, G. Penelas and A. Tarancón, *Faraday Discuss.*, 2015, **182**, 241.

190. X. Zhang, Y. Song, G. Wang and X. Bao, *J. Energy Chem.*, 2017, **26**, 839.
191. F. Bidrawn, G. Kim, G. Corre, J. T. S. Irvine, J. M. Vohs and R. J. Gorte, *Electrochem. Solid-State Lett.*, 2008, **11**, 167.
192. S. D. Ebbesen, R. Knibbe and M. Mogensen, *J. Electrochem. Soc.*, 2012, **159**, F482.
193. W. Li, H. Wang, Y. Shi and N. Cai, *Int. J. Hydrogen Energy*, 2013, **38**, 11104.
194. P. Kim-Lohsoontorn and J. Bae, *J. Power Sources*, 2011, **196**, 7161.
195. C.-A. Thieu, J. Hong, H. Kim, K. J. Yoon, J.-H. Lee, B.-K. Kim and J.-W. Son, *J. Mater. Chem. A*, 2017, **5**, 7433.
196. A. Cho, J. Ko, B.-K. Kim and J. W. Han, *ACS Catal.*, 2019, **9**, 967.
197. E. T. Ioannidou, C. S. Neofytides, S. G. Neophytides and D. K. Niakolas, *ECS Trans.*, 2017, **78**, 3267.
198. Ch. Neofytidis, E. Ioannidou, L. Sygellou, M. Kollia and D. K. Niakolas, *J. Catal.*, 2019, **373**, 260.
199. E. Ioannidou, S. Neophytides and D. K. Niakolas, *Catalysts*, 2019, **9**, 151.
200. S.-W. Kim, M. Park, H. Kim, K. J. Yoon, J.-W. Son, J.-H. Lee, B.-K. Kim, J.-H. Lee and J. Hong, *Appl. Catal., B*, 2017, **200**, 265.
201. S.-W. Kim, M. Park, H. Kim, K. J. Yoon, J.-W. Son, J.-H. Lee, B.-K. Kim, J.-H. Lee and J. Hong, *Appl. J. Electrochem. Soc.*, 2016, **163**, F3171.
202. C. Jin, C. Yang, F. Zhao, D. Cui and F. Chen, *Int. J. Hydrogen Energy*, 2011, **36**, 3340.
203. R. Xing, Y. Wang, Y. Zhu, S. Liu and C. Jin, *J. Power Sources*, 2015, **274**, 260.
204. Y. Ye, T. He, Y. Li, E. H. Tang, T. L. Reitz and S. P. Jiang, *J. Electrochem. Soc.*, 2008, **155**, B811.
205. X. Yue and J. T. S. Irvine, *J. Mater. Chem. A*, 2017, **5**, 7081.
206. Y. Li, Y. J. Zhou, D. Dong, Y. Wang, J. Z. Jiang, H. Xiang and K. Xie, *Phys. Chem. Chem. Phys.*, 2015, **14**, 15547.
207. K. Xie, Y. Zhang, G. Meng and J. T. S. Irvine, *Energy Environ. Sci.*, 2011, **4**, 2218.
208. Y. Li, P. Li, B. Hu and C. Xia, *J. Mater. Chem. A*, 2016, **4**, 9236.
209. L. Gan, L. Ye, C. Ruan, S. Chen and K. Xie, *Adv. Sci.*, 2016, **3**, 1500186.
210. A. K. Opitz, A. Nenning, C. Rameshan, R. Rameshan, R. Blume, M. Hävecker, A. Knop-Gericke, G. Rupprechter, J. Fleig and B. Klötzer, *Angew. Chem.*, 2015, **54**, 2628.
211. K. Hosoi, H. Hagiwara, S. Ida and T. Ishihara, *J. Phys. Chem. C*, 2016, **120**, 16110.
212. S.-E. Yoon, S.-H. Song, J. Choi, J.-Y. Ahn, B.-K. Kim and J.-S. Park, *Int. J. Hydrogen Energy*, 2014, **39**, 5497.
213. S.-E. Yoon, J.-Y. Ahn, B.-K. Kim and J.-S. Park, *Int. J. Hydrogen Energy*, 2015, **40**, 13558–13565.
214. X. Ge, L. Zhang, Y. Fang, J. Zeng and S. H. Chan, *RSC Adv.*, 2011, **1**, 715.

215. Z. Huang, Z. Zhao, H. Qi, X. Wang, B. Tu and M. Cheng, *J. Energy Chem.*, 2020, **40**, 46.
216. X. Hou, K. Zhao, J.-I. Yang and S. Ha, *Int. J. Hydrogen Energy*, 2016, **41**, 11895.
217. L. Liu, Y. Wang, X. Zhou, Y. Yang, C. Ma, Y. Li, C. Wang, X. Zhao and B. Li, *Electrochim. Acta*, 2017, **235**, 365.
218. D. Dogu, S. Gunduz, K. E. Meyer, D. J. Deka, A. C. Co and U. S. Ozkan, *Catal. Lett.*, 2019, **149**, 1743.
219. Y. Yu, B. Mao, A. Geller, R. Chang, K. Gaskell, Z. Liu and B. W. Eichhorn, *Phys. Chem. Chem. Phys.*, 2014, **16**, 11633.
220. Z. A. Feng, F. El Gabaly, X. Ye, Z. X. Shen and W. C. Chueh, *Nat. Commun.*, 2014, **5**, 4374.
221. C. Graves, C. Chatzichristodoulou and M. B. Mogensen, *Faraday Discuss.*, 2015, **182**, 75.
222. N. Tsvetkov, Q. Lu and B. Yildiz, *Faraday Discuss.*, 2015, **182**, 257.
223. M. Lo Faro, S. C. Zignani, S. Trocino, V. Antonucci and A. S. Aricò, *Electrochim. Acta*, 2019, **296**, 458.
224. X. Chen, C. Guan, G. Xiao, X. Du and J.-Q. Wang, *Faraday Discuss.*, 2015, **182**, 341.
225. H.-N. Im, S.-Y. Jeon, D.-K. Lim, B. Singh, M. Choi, Y.-S. Yoo and S.-J. Song, *J. Electrochem. Soc.*, 2015, **162**, F54.
226. C. Gaudillere, L. Navarrete and J. M. Serra, *Int. J. Hydrogen Energy*, 2014, **39**, 3047.
227. H. Zheng, Y. Tian, L. Zhang, B. Chi, J. Pu and L. Jian, *J. Power Sources*, 2018, **383**, 93.
228. D. J. Deka, S. Gunduz, T. Fitzgerald, J. T. Miller, A. C. Co and U. S. Ozkan, *Appl. Catal., B*, 2019, **248**, 487.
229. Z. Yang, N. Wang, C. Ma, X. Jin, Z. Lei, X. Xiong and S. Peng, *J. Electroanal. Chem.*, 2019, **836**, 107.
230. L. Liu, X. Zhou, Y. Wang, S. Li, R. Yin, P. Guo, J. Zhao, X. Zhao and B. Li, *Int. J. Hydrogen Energy*, 2017, **42**, 14905.
231. J. T. S. Irvine, D. Neagu, M. C. Verbraeken, C. Chatzichristodoulou, C. Graves and M. B. Mogensen, *Nat. Energy*, 2016, **1**, 15014.
232. K. Xie, Y. Zhang, G. Meng and J. T. Irvine, *J. Mater. Chem.*, 2011, **21**, 195.
233. E. Ruiz-Trejo and J. T. Irvine, *Solid State Ionics*, 2013, **252**, 157.
234. G. Wu, K. Xie, Y. Wu, W. Yao and J. Zhou, *J. Power Sources*, 2013, **232**, 187.
235. N. Shi, Y. Xie, D. Huan, Y. Yang, S. Xue, Z. Qi, Y. Pan, R. Peng, C. Xia and Y. Lu, *J. Mater. Chem. A*, 2019, **7**, 4855.
236. J. M. Vohs and R. J. Gorte, *Adv. Mater.*, 2009, **21**, 943.
237. S. P. Jiang, *Int. J. Hydrogen Energy*, 2012, **37**, 449.
238. D. Ding, X. Li, S. Y. Lai, K. Gerdes and M. Liu, *Energy Environ. Sci.*, 2014, **7**, 552.

239. M. A. Laguna-Bercero, A. R. Hanifi, T. H. Etsell, P. Sarkar and V. M. Orera, *Int. J. Hydrogen Energy*, 2015, **40**, 5469.
240. H.-Y. Jeong, K. J. Yoon, J.-H. Lee, Y.-C. Chun and J. Hong, *J. Korean Ceram. Soc.*, 2018, **55**, 50.
241. J.-Y. Ahn, B.-K. Kim and J.-S. Park, *J. Electrochem. Soc.*, 2016, **163**, F1288.
242. J. Hong, C. Balamurugan, H.-N. Im, S.-Y. Jeon, Y.-S. Yoo and S.-J. Song, *J. Electrochem. Soc.*, 2018, **165**, F132.
243. H. Fan and M. Han, *J. Power Sources*, 2016, **336**, 179.
244. D. J. Cumming, A. R. Thompson and R. H. Rothman, *J. Electrochem. Soc.*, 2016, **163**, F3057.
245. B. Hua, M. Li, Y.-F. Sun, J.-H. Li and J.-L. Luo, *ChemSusChem*, 2017, **10**, 3333.
246. J.-H. Myung, D. Neagu, D. N. Miller and J. T. S. Irvine, *Nature*, 2016, **537**, 528.
247. S. Liu, Q. Liu and J.-L. Luo, *ACS Catal.*, 2016, **6**, 6219.
248. L. Ye, M. Zhang, P. Huang, G. Guo, M. Hong, C. Li, J. T. S. Irvine and K. Xie, *Nat. Commun.*, 2017, **8**, 14785.
249. J. Lu, C. Zhu, C. Pan, W. Lin, J. P. Lemmon, F. Chen, C. Li and K. Xie, *Sci. Adv.*, 2018, **4**, 5100.
250. V. M. Orera, M. A. Laguna-Bercero and A. Larrea, *Front. Energy Res.*, 2014, **2**, 22.
251. S.-H. Lee, J.-W. Lee, S.-B. Lee, S.-J. Park, R.-H. Song, U.-J. Yun and T.-H. Lim, *Int. J. Hydrogen Energy*, 2016, **41**, 7530.
252. S.-B. Yu, S.-H. Lee, M. T. Mehran, J.-E. Hong, J.-W. Lee, S.-B. Lee, S.-J. Park, R.-H. Song, J. H. Shim, Y.-G. Shul and T.-H. Lim, *Chem. Eng. J.*, 2018, **335**, 41.
253. L. Chen, F. Chen and C. Xia, *Energy Environ. Sci.*, 2014, **7**, 4018.
254. L. Lei, T. Liu, S. Fang, J. P. Lemmon and F. Chen, *J. Mater. Chem. A*, 2017, **5**, 2904.
255. B. Chen, H. Xu, L. Chen, Y. Li, C. Xia and M. Ni, *J. Electrochem. Soc.*, 2016, **163**, F3001.
256. L. Song, J. Du, Y. Wang, W. Zhao and T. Liu, *Int. J. Electrochem. Sci.*, 2017, **12**, 2949.
257. M. T. Mehran, S.-B. Yu, D.-Y. Lee, J.-E. Hong, S.-B. Lee, S.-J. Park, R.-H. Song and T.-H. Lim, *Appl. Energy*, 2018, **212**, 759.
258. L. Bernadet, J. Laurencin, G. Roux, D. Montinaro, F. Mauvy and M. Reytier, *Electrochim. Acta*, 2017, **253**, 114.
259. M. Samavati, M. Santarelli, A. Martin and V. Nemanova, *Energy*, 2017, **122**, 37.
260. X. Sun, M. Chen, Y.-L. Liu, P. Hjalmarsson, S. D. Ebbesen, S. H. Jensen, M. B. Mogensen and P. V. Hendriksen, *J. Electrochem. Soc.*, 2013, **60**, F1074.
261. M. A. Laguna-Bercero, R. Campana, A. Larrea, J. A. Kilner and V. M. Orera, *J. Power Sources*, 2011, **196**, 8942.

262. A. V. Virkar, *Int. J. Hydrogen Energy*, 2010, **35**, 9527.
263. L. Holzer, B. Iwanschitz, Th. Hocker, B. Münch, M. Prestat, D. Wiedenmann, U. Vogt, P. Holtappels, J. Sfeir, A. Mai and Th. Graule, *J. Power Sources*, 2011, **196**, 1279.
264. A. Hauch, M. Mogensen and A. Hagen, *Solid State Ionics*, 2011, **192**, 547.
265. A. Hauch, S. D. Ebbesen, S. H. Jensen and M. Mogensen, *J. Electrochem. Soc.*, 2008, **155**, B1184.
266. A. Hauch, J. R. Bowen, L. T. Kuhn and M. Mogensen, *Electrochem. Solid-State Lett.*, 2008, **11**, B38.
267. S. D. Ebbesen, C. Graves, A. Hauch, S. H. Jensen and M. Mogensen, *J. Electrochem. Soc.*, 2010, **157**, B1419.
268. Y. Tao, S. D. Ebbesen and M. B. Mogensen, *J. Power Sources*, 2016, **328**, 452.
269. P. Hjalmarsson, X. Sun, Y.-L. Liu and M. Chen, *J. Power Sources*, 2013, **223**, 349.
270. P. Hjalmarsson, X. Sun, Y.-L. Liu and M. Chen, *J. Power Sources*, 2014, **262**, 316.
271. W. G. Bessler, S. Gewies and M. Vogler, *Electrochim. Acta*, 2007, **53**, 1782.
272. G. M. Goldin, H. Zhu, R. J. Kee, D. Bierschenk and S. A. Barnett, *J. Power Sources*, 2009, **187**, 123.
273. W. Li, Y. Shi, Y. Luo and N. Cai, *J. Power Sources*, 2013, **243**, 118.
274. J. Aicart, F. Usseglio-Viretta, J. Laurencin, M. Petitjean, G. Delette and L. Dessemond, *Int. J. Hydrogen Energy*, 2016, **41**, 17233.
275. Y. Luo, X.-Y. Wu, Y. Shi, A. F. Ghoniem and N. Cai, *Appl. Energy*, 2018, **215**, 371.
276. M. Samavati, A. Martin, V. Nemanova and M. Santarelli, *Int. J. Hydrogen Energy*, 2018, **43**, 4785.
277. C. H. Wendel, P. Kazempoor and R. J. Braun, *J. Power Sources*, 2015, **276**, 133.
278. G. Cinti, A. Baldinelli, A. Di Michele and U. Desideri, *Appl. Energy*, 2016, **162**, 308.
279. W. Zhang, B. Yu, X. Wang and J. Chen, *Int. J. Hydrogen Energy*, 2016, **41**, 15960.
280. M. J. Bos and D. W. F. Brilman, *Chem. Eng. J.*, 2015, **278**, 527.
281. C. H. Wendel, P. Kazempoor and R. J. Braun, *J. Power Sources*, 2016, **301**, 93.
282. Y. Shi, Y. Luo, N. Cai, J. Qian, S. Wang, W. Li and S. Wang, *Electrochim. Acta*, 2013, **88**, 644.
283. D. Klotz, A. Leonide, A. Weber and E. Ivers-Tiffée, *Int. J. Hydrogen Energy*, 2014, **39**, 20844.
284. J. Aicart, J. Laurencin, M. Petitjean and L. Dessemond, *Fuel Cells*, 2014, **14**, 430.

285. J. P. Stempien, M. Ni, Q. Sun and S. H. Chan, *Energy*, 2015, **82**, 714.
286. M. R. Singh, E. L. Clark and A. T. Bell, *Proc. Natl. Acad. Sci. U. S. A.*, 2015, E6111.
287. D. H. König, N. Baucks, R.-U. Dietrich and A. Wörner, *Energy*, 2015, **91**, 833.
288. V. Menon, Q. Fu, V. M. Janardhanan and O. Deutschmann, *J. Power Sources*, 2015, **274**, 768.
289. S. Diethelm, F. Maréchal and J. Van herle, *Appl. Energy*, 2019, **247**, 572.

CHAPTER 3

Electrolyte Development for Solid-state Lithium Batteries

SOURAV BAG AND VENKATARAMAN THANGADURAI*

Department of Chemistry, University of Calgary, 2500 University Drive NW, Calgary, Alberta, Canada T2N 1N4
*E-mail: vthangad@ucalgary.ca

3.1 Introduction

The current total power consumption of the world is around 13 276.3 Mtoe.[1] In 2015, oil accounted for more than 30% of the total energy consumption worldwide, whereas coal and natural gas provided 28% and 21% of energy consumption, respectively.[2] It is clear that fossil fuel consumption presents a significant global challenge due to the environmental pollution resulting from their continued use. A large quantity of greenhouse gases such as CO_2 are produced because of the current utilisation of fossil fuels without carbon capture and storage facilities.[3] So, there is an urgent need for alternative energy sources to address the depletion of fossil fuels and the environmental impact of their continued use. However, the large-scale development of renewable energy technologies such as wind or solar, that are intermittent and unpredictable, requires highly reliable electrical energy storage (EES) devices to store the excess produced electricity and to release it on demand.[4] However, an appropriate EES technology is required to store the electricity

Inorganic Materials Series No. 5
Energy Storage and Conversion Materials
Edited by Stephen Skinner

Published by the Royal Society of Chemistry, www.rsc.org

generated from these renewable sources and deliver it efficiently when it is needed. Moreover, electrical energy storage is also vital for the smart grid and distributed power generation development. Among the present energy storage technologies, electrochemical energy storage (ECES) devices, including batteries, electrolysis cells and capacitors have recently attracted intensive interest.[5]

3.2 Electrochemical Energy Storage (ECES)

The major advantage of ECES systems is the direct conversion of chemical energy into electrical energy with high efficiency. Batteries are the most promising ECES technologies due to their wide range of energy density and ease of transportation, unlike other power conversion devices; also, they can be easily integrated with renewable energy sources such as solar and wind. Generally, like fuel cells, batteries are an assembly of two electrodes separated by a suitable electrolyte and/or separator, which allow ions to migrate from one electrode to another, forming a single cell. The ions move through the electrolyte while electrons travel through an external circuit between the two electrodes. The differences in the electrochemical potentials of the two electrodes determine the voltage of a battery, *i.e.*,

$$E_{\text{cell}} = E_{\text{cathode}} - E_{\text{anode}} \tag{3.1}$$

Hence, the operating voltage of the cell is directly related to the electrochemical potential of the electro-active species in the electrodes. Determining suitable electrode materials for both electrodes is crucial if higher voltage and energy densities are to be achieved. The Gibbs free energy of a chemical reaction is related to the voltage of a cell by the expression:

$$G_{\text{cell}} = -nFE_{\text{cell}} \tag{3.2}$$

where n is the number of electrons transferred for the associated electrochemical reactions and F is the Faraday constant. Both electrodes and electrolytes are equally important components in developing a highly energy-efficient electrochemical cell or battery. To date, it has been challenging to fully access the theoretical energy of any ECES device due to some associated side reactions, electrode–electrolyte interfacial resistance, electrode reaction rates and the nature of the electrolyte. Loss of electrical energy is generally manifested in the form of heat in a battery.

Capacity is one of the most important and commonly used parameters among battery consumers and developers, which is defined as the total amount of electrical charge stored in the electrochemical reactions of cell. It is a function of the number of electrons transferred for the associated electrochemical reaction and of the available active material in the cell. Theoretical capacity per mol of electro-active material is defined as the specific capacity (C_S):[6]

$$C_s = nF/M \tag{3.3}$$

where M is the molecular weight (g mol^{-1}) of the active material. The capacity of the electrode is defined in units of *ampere-hour* (Ah). As the capacity is directly related to the mass of the active material in the electrode, scientists often use specific capacity (gravimetric and volumetric) to understand the charge storage and electrochemical properties of the battery electrode materials.

Specific capacity is defined by the capacity of unit mass of the material having a unit of *ampere-hour per gram* (A h g^{-1}). The theoretical energy of a battery is equal to the work done by the cell due to the change in the free energy associated with the electrochemical reaction of the cell. In practice, the theoretical energy is not achieved by any of the present ECES systems due to energy loss. The practical/actual energy is calculated by multiplying the obtained capacity during discharge ($C_{\text{discharge}}$) and the actual operational voltage window (V) of the cell. The energy is defined in units of *Watt-hour* (Wh). The practical or actual energy density (E) can be simply calculated by the following equation;[6,7]

$$\textit{Actual energy density}, E = C_{\text{discharge}} \times V \tag{3.4}$$

Specific energy is often used and is the most effective parameter used to evaluate the performance of various battery technologies. Specific energy is expressed with respect to gravimetric (Wh kg^{-1}) or volumetric (Wh L^{-1}) energy densities.[6–8] ECES systems that are lightweight and of smaller volume offer higher specific energy density systems. For example, Li-based chemistry provides the highest energy density compared to corresponding Na and K batteries. *Power* (W) is defined as the amount of delivered energy per unit time. Specific power and power density are usually expressed in W kg^{-1} and W L^{-1}, respectively.

Batteries are classified into two categories: (i) primary or non-rechargeable and (ii) secondary or rechargeable. Figure 3.1 shows a schematic representation and typical example of primary and secondary batteries. Primary batteries can only be used once (*i.e.* are not rechargeable), while secondary batteries can be re-used repeatedly and are extensively popular among consumers. Though various concepts of battery technologies have been proven, only a handful of them have been commercialised because of the high manufacturing cost, low energy density, poor recyclability and safety concerns. Currently, lead-acid, Li-ion, Ni–Cd, Alkaline, and Ni-metal hydride are the most promising and widely used secondary batteries. Figure 3.2 compares the specific energy density

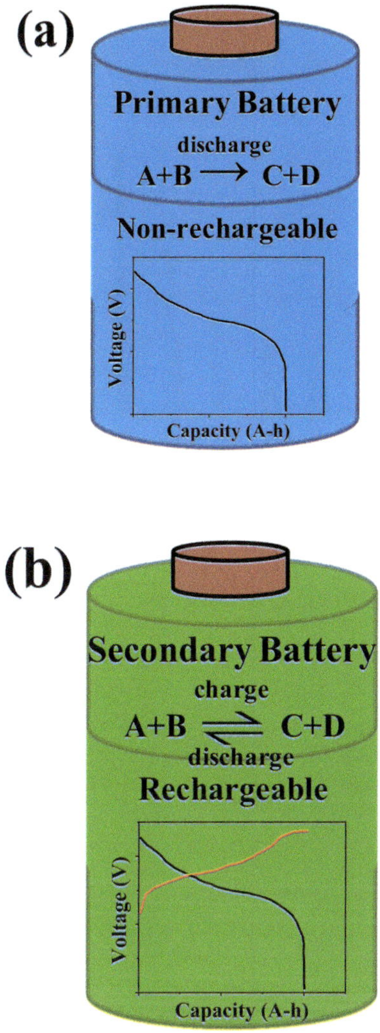

Figure 3.1 Scheme illustrating the characteristic features of (a) primary and (b) secondary batteries.

of various secondary batteries. Li-ion battery technology has attracted a great deal of attention due to the high energy density, safe operation and cost-effective production of these devices.[6–8]

3.3 Li-ion Batteries

Rechargeable Li-ion technologies were first commercialised in 1991 by the Sony corporation and afterwards have been widely employed for portable electronics and hybrid electric vehicles. Figure 3.3 shows a

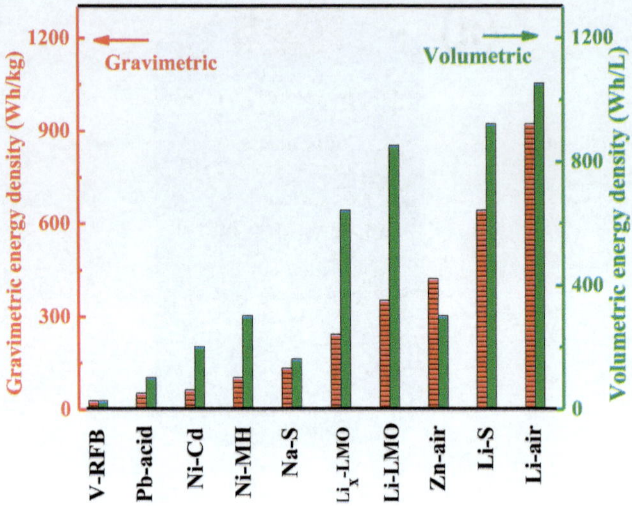

Figure 3.2 Specific energy densities of different electrochemical energy storage devices.

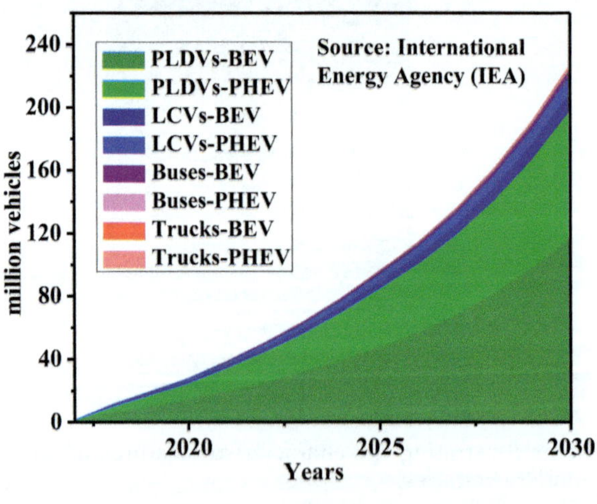

Figure 3.3 Projection of electric vehicle sale in 2017–3030. Based on IEA Analysis from the Global EV Outlook and Mobility Model© OECD/IEA 2018, IEA Publishing. Licence: http://www.iea.org/t&c, as modified by the authors. Adapted from ref. 9.

projection for sales of electric vehicles based on Li-ion battery storage for the period 2017 to 2030.[9] The present technologies of Li-ion batteries are based on Li intercalation anode and cathode materials, that are separated by an organic polymer-based Li-ion conducting membrane. Figure 3.4 shows the half-cell reaction of various battery electrodes, including

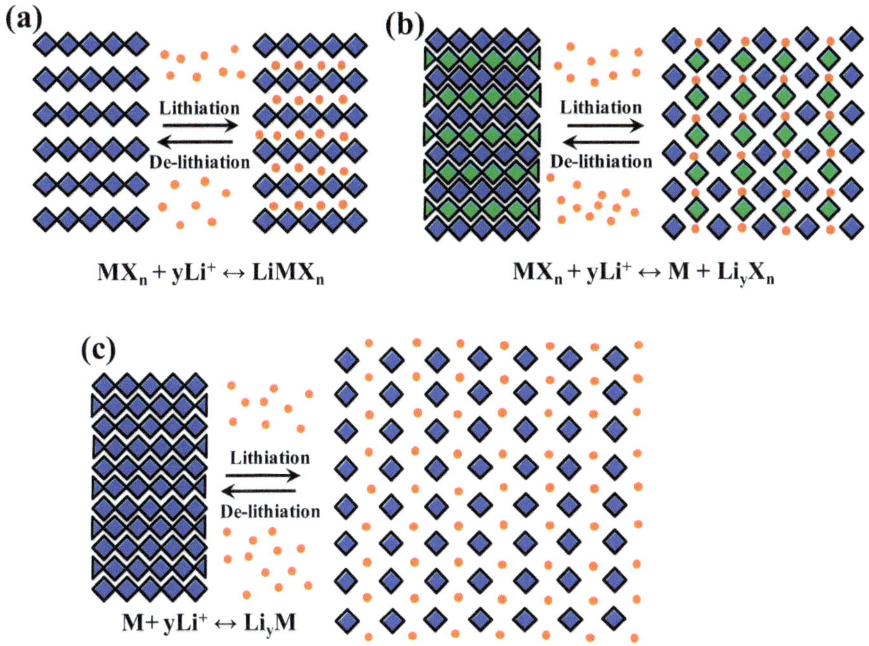

Figure 3.4 Schematic illustrations of three types of reaction mechanisms: (a) intercalation, (b) conversion and (c) alloy associated with a Li battery electrode.

intercalation, conversion and alloying.[10,11] The electrode materials and the electrolyte play a crucial role in determining the overall performance of a Li-ion battery. Typical examples of various Li-based electrodes are listed in Table 3.1.[12–40]

During charge–discharge, intercalation electrodes accommodate the Li-ion inside the host lattice without any abrupt change of the primary structure. On the other hand, an electrochemical reaction for conversion and alloying electrode materials involves complete bond breaking and forming in the crystal structures. Metal oxides, sulphides, nitrides and phosphides are mostly investigated as conversion electrodes in batteries. Though they possess high capacity, the major limitations of conversion electrodes are low conductivity, volume expansion and voltage hysteresis. Silicon is the most promising alloying anode and it exhibits a specific capacity of 4000 mA h g^{-1}.[10,41] Anodes with alloying reaction such as Sn, Sb and Bi have also shown high capacity. The major problems associated with commercialisation of alloying-type electrode materials are the parasitic and irreversible reactions during cycling. Graphite is a state-of-the-art intercalation anode material, which exhibits a theoretical specific capacity of ~372 mA h g^{-1}.[41]

Cathodes are mainly transition metal-based materials having layered $LiMO_2$ (M = Co, Ni), olivine $LiMPO_4$ (M = Fe, Co, Ni) and spinel $LiMn_2O_4$

Table 3.1 Theoretical and practical capacity for Li-ion battery electrode materials[12–40].

Electrode	Theoretical specific capacity (mA h g^{-1})	Practical specific capacity (mA h g^{-1})	Reference
$LiTiS_2$	225	210	12
$LiCoO_2$	274	148	13
$LiNiO_2$	275	150	14
$LiMnO_2$	285	140	15
$LiNi_{0.33}Mn_{0.33}Co_{0.33}O_2$	280	160	16
$LiNi_{0.8}Co_{0.15}Al_{0.05}O_2$	279	199	17
Li_2MnO_3	458	180	18
$LiMn_2O_4$	148	120	19
$LiCo_2O_4$	142	84	20
$LiFePO_4$	170	165	21
$LiMnPO_4$	171	168	22
$LiCoPO_4$	142	84	23
$LiFeSO_4F$	167	125	24
$LiVOPO_4F$	156	129	25
Hard carbon	1116	530	26 and 27
CNTs	780	1050	28 and 29
Graphite	372	360	30
Graphene	—	780–1100	31
$LiTi_4O_5$	175	170	32
TiO_2	330	300	32
Si	4212	2000–3500	33
Ge	1624	980–1130	34
Tin	993	510	35
Antimony	660	460	36
Tin oxide	790	761	37
SiO	1600	700–900	38
Fe_2O_3	1007	700–911	39
Fe_3O_4	926	830	40
CoO	715	474	37

structures, which accommodate Li^+ inside the lattice.[42] Most of the cathode materials undergo intercalation/de-intercalation type reactions during the charge–discharge process. Layered structured $LiCoO_2$ is the most commonly used cathode material with a specific practical capacity of ~140 mA h g^{-1}.[43] During charge–discharge, Li ions, the working ionic component for the electrochemical reactions, travel through the electrolyte towards both the electrodes. Electrical charge is stored by means of storing Li-ions in the electrodes. The energy density of a Li-ion battery is determined as the Li^+ storage capacity of the electrode material and the potential of the cell. Figure 3.5 shows the specific capacity and operational voltage of different Li electrode materials. However, the Li-ion storage capacity of the intercalation electrode depends on the ability of the material to change the valence state of the central element of the layered electrode, interlayer spacing and the reversibility of the intercalation reaction.[42]

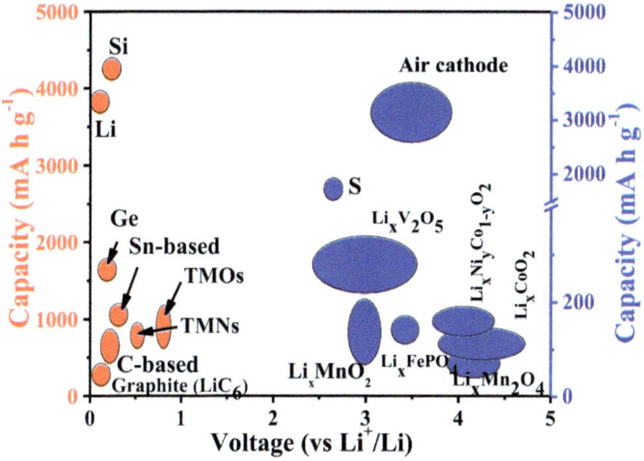

Figure 3.5 Specific capacity and operational voltage of electrode material for Li batteries.

The electrolyte determines the operating voltage, power density, safety and cycle-life stability of the batteries. Presently, non-aqueous electrolytes are mostly used in Li-ion batteries. Alkyl ether and carbonate are mainly employed as the solvent for electrolyte preparation. Ethylene carbonate (EC), propylene carbonate (PC), dimethyl carbonate (DMC), tetraethyl glycol dimethyl ether (TEGDME) and diethyl carbonate (DC) are used as solvents.[44] Lithium salts such as lithium tetrafluroborate ($LiBF_4$), lithium hexafluoro phosphate ($LiPF_6$), lithium perchlorate ($LiClO_4$) and lithium trifluoromethansulfonate ($LiCF_3SO_2$) are utilised as the electrolyte salt for the Li-ion battery.[45] In the most common cases, most of the Li-ion batteries are made of a graphite anode with $LiPF_6$ in an EC-DMC electrolyte with a $LiCoO_2$ cathode. Fluro-organic compounds such as fluoroethylene carbonate (FEC) are also employed as an additive to the electrolyte to enhance the stability of the electrolyte. To improve the safety and prevent internal short circuits, a polymeric membrane separator soaked in electrolyte is employed between the cathode and anode. The separator membranes should have less thickness, optimal Li^+ ion permeability and porosity, with good wettability in the electrolyte solvent and chemical, thermal and mechanical stability. The most widely employed separators are polypropylene, polyethylene, polyethylene oxide (PEO), polyacrylonitrile (PAN), poly(methyl-methacrylate) (PMMA) and polyvinylidene fluoride (PVDF).[46]

3.4 Solid-state Li Batteries

Li-ion battery technology has been proven as an efficient ECES system exhibiting high power and energy density with good recyclability. The state-of-the-art Li-ion battery delivers a volumetric and gravimetric energy density of up to 750 Wh L^{-1} and 250 Wh kg^{-1}, respectively.[47] The increasing demand for higher energy density and high power density for electric vehicle applications requires innovative technology. Solid-state batteries composed of high voltage and thermally stable solid electrolytes instead of liquid electrolytes, are one of the most promising technologies to improve energy densities. The advantages of employing solid electrolytes in all-solid-state Li-ion batteries are listed below.

- High energy density can be obtained by employing a Li-metal anode.
- Mechanical integrity of the solid electrolyte improves the stability/recyclability and safety by prevention of the dendrite formation.
- No chemical crossover of the electrode materials (leached metal ions/polysulphide in Li–S battery) mitigates self-discharge.
- Unlike liquid electrolytes, no concentration polarisation is observed in solid inorganic electrolytes, as they are only Li-ion conductors.
- Solid inorganic/ceramic electrolytes show higher Li-ion conductivity at elevated temperatures.

Solid Li-ion electrolytes determine the performance of solid-state batteries and can be classified into (i) solid organic polymeric and (ii) inorganic/ceramic electrolytes.

3.5 Electrolytes for All-solid-state Li Batteries

The term 'solid-state battery' generally refers only to ceramic (lithium metal oxides and or non-oxides) Li-ion membrane-based batteries. However, a large number of high-temperature stable solid polymer-based membranes are being discovered to replace conventional liquid electrolytes in Li-ion batteries. These solid polymer membrane-based batteries are also known as 'solid-state batteries'. In the following sections, we briefly report the advances in solid polymer and ceramic membranes (oxides) for application in next-generation all-solid-state-Li batteries.

3.5.1 Polymer Electrolytes

Fenton and co-workers first discovered polymer-based electrolytes in 1973.[49] They have shown that the complex salt formed by alkali salts and polyethylene oxide (PEO) exhibits excellent ionic conductivity. Since then, intensive research on exploring new and efficient polymer-based electrolytes, their ion transport mechanism, corresponding electrode/electrolyte interface and theoretical modelling has been carried out.[49–51] Polymer electrolytes have several advantages over the liquid organic and solid inorganic electrolytes as they are resistant to volume expansion of the electrode material during charge/discharge and provide flexibility of the cell design. Moreover, dendrite growth associated with the Li anode could also be mitigated by the use of a solid polymer electrolyte. In 1979, the all-solid-state battery was demonstrated

Table 3.2 Ionic conductivity of selected polymer electrolytes (those reported at room temperature are marked with *).

Electrolyte type	Electrolyte composition	Ionic conductivity (S cm^{-1})	Temperature (°C)	Reference
Dry SPE	P(EO)20/$LiClO_4$	2.78×10^{-7}	27	54
	P(EO)20/$LiBF_4$	6.32×10^{-7}	27	54
	PEO/5 wt% $LiPF_6$	1.20×10^{-6}	25	49
	PEO/11.1 wt% $LiAsF_6$	1.43×10^{-4}	25	55
	PEO/15 wt% $LiCF_3SO_3$	1.0×10^{-6}	27	54
	$P(EO)_{24}/LiN(CF_3SO_2)_2$*	3.84×10^{-4}	—	56
Plasticised polymer–salt complex	$(PEO)_8$-$LiClO_4$ (EC : PC)	1×10^{-3}	20	49
	$(PEO)_8$-$LiClO_4$ (PC, 50 mol%)	8×10^{-4}	20	49
	PEO-$LiCF_3SO_3$ plasticised with PEG	1×10^{-3}	25	57
	PEO-$LiBF_4$ with 12-crown-4*	7×10^{-4}	—	58
Gel PE	PAN-EC/PC/DMF-$LiClO_4$	4×10^{-4}	22	59
	PMMA-EC/PC-$LiClO_4$	1×10^{-3}	25	60
	PAN-EC/PC-$LiClO_4$	4×10^{-3}	25	61
	PVC-EC/PC-$LiClO_4$	1×10^{-3}	20	61
	PVDF-EC/DEC-$LiBF_4$*	6×10^{-3}	—	62
CPE	PEO-LiI-Al_2O_3*	1×10^{-4}	—	63
	PEO-$LiBF_4$-superacid-ZrO_2*	1×10^{-6}	—	64
	PEO-$LiClO_4$-α-Al_2O_3	1×10^{-5}	25	65
	PEO-$LiClO_4$-SiO_2*	1×10^{-5}	—	66
	PEO-$LiClO_4$-TiO_2	2×10^{-5}	30	67

using a solid polymer electrolyte based on a PEO–Li salt complex. Presently, a large number of Li^+ conducting solid polymer organic electrolytes are being investigated.[49,50]

The glass transition temperature (T_g), is a crucial parameter for polymer electrolytes below which the polymer becomes hard and brittle, like glassy materials. Polymer matrices having a low glass transition temperature maintain good flexibility and faster ion transport. The glass transition is related to amorphous polymers whose chains are not orderly arranged in crystals, but are just spread around in any fashion, even in the solid state. The ion transport of the polymer–salt complex largely depends upon reducing the mobility of the counter anion of the salt by anchoring on the polymer backbone, which enhances the single ion conduction of the electrolytes.[52] Gadjourova *et al.* have newly proposed that the ionic conduction of polymers is related to the ordered environment of the crystalline phase above the glass transition temperature.[53] The size, shape and charge of the anions influences the conductivity of the electrolyte. However, the ion transport properties of the polymer are still not understood properly due to the lack of information on the structure–property relationship.[49] Table 3.2 lists the ionic conductivity of typical polymer organic electrolytes that were developed for Li-ion batteries.[49,54–67] Table 3.3 details the molecular formula and physical properties of some commonly used polymer membranes.[49,50]

Table 3.3 Chemical and physical properties of commonly used polymer matrices in Li-ion batteries.[49,50]

Polymer	Repeat unit	Glass transition temperature (°C)	Melting point (°C)
Poly(ethylene oxide)	$-(CH_2CH_2O)_n-$	−64	65
Poly(propylene oxide)	$-(CH(-CH_3)CH_2O)_n-$	−60	—
Poly(dimethylsiloxane)	$-[SiO(-CH_3)_2]_n-$	−127	−40
Poly(acrylonitrile)	$-(CH_2CH(-CN))_n-$	125	317
Poly(methylemethylacrylate)	$(CH_2C(-CH_3)(-COOCH_3))_n-$	105	160
Poly(vinyl chloride)	$-(CH_2CHCl)_n-$	82	220
Poly(vinylidene fluoride)	$-(CH_2CF_2)_n-$	−35	177
Poly(vinylidene fluoride-hexafluoropropylene)	$-(CH_2CF_2)_n-[CF_2CF(CF_3)]_m-$	−90	135
Polyvinyl alcohol	$-(CH_2CHOH)_n-$	85	240

Dry solid polymer electrolyte: Dry solid polymer electrolytes (SPEs) are conventional polymer–salt complexes formed by the addition of Li-salts in the higher molecular weight polymer matrix. PEO dissolves Li-based salts by complexation of the Li ions with polar oxygen. Polypropylene oxide, poly[bis(methoxy-ethoxy-ethoxy)phosphazene] and polysiloxane have also been investigated for application in Li-ion batteries.[7] Li salts such as $LiClO_4$, $LiBF_4$, $LiPF_6$, $LiAsF_6$, $LiCF_3SO_3$ and $LiN(CF_3SO_2)_2$ are generally used in polymer electrolytes.[49] The ability to dissolve salts in PEO and SPE is determined by the solvation and lattice energy. Low lattice energy of both the polymer and inorganic salt enhance the stability of the resultant SPE. Higher ionic conductivity is obtained at a lower salt to PEO ratio. Both conductivity and Li^+-ion transference number have been found to decrease with increase of the Li-salt concentration. The conductivity is mainly determined by the amorphous nature and lower T_g of the matrices in polymer electrolytes.[51] Modifications of polymer matrices by blending, cross-linking and co-polymerisation have been carried out to enhance the amorphous phase and lower the T_g value. PEO crystallizes below 70 °C, above which it becomes amorphous. Hence, a practical useful conductivity in the PEO-based polymer–salt complex could be achieved in the temperature range of 70–90 °C.

Plasticised polymer–salt complex: Several efforts have been made to enhance the conductivity of PEO-based dry SPE at ambient temperature. One of the most common approaches to increase conductivity is mixing of a substantial amount of plasticisers or additives, such as poly(ethylene glycol) and dimethyl sulfoxide with dry SPE.[49] Plasticisers facilitate the salt dissociation and increase the segmental motion of polymer chain, which enhance the charge transport and conductivity. Crown ether (*e.g.* 12-crown-4) and ethylene carbonate are also added as plasticisers, which reduce crystallinity and increase free volume for ion transport. Though conductivity is enhanced with the addition of plasticisers, the electrolyte loses its mechanical rigidity and compatibility towards Li electrodes.

Gel polymer electrolyte (GPE): The gel polymer electrolyte (GPE) was first reported in 1975.[50] In a GPE, a polymer–salt complex is incorporated with a certain amount of liquid plasticiser and/or solvent. The polymer network is swollen with solvents, which enhance cohesive and diffusive transport properties. The plasticiser solutions of the Li-based salts are confined within the matrix of the polymer and transform the whole system into gels with a very high ionic conduction comparable to the liquid electrolytes. 'GPE' is also known as a 'polymer hybrid' and 'gel-ionic'.[49,50] Generally, the plasticisers used in GPEs are lower molar mass and low-evaporation solvents such as ethylene carbonate, propylene carbonate, dimethyl formamide, diethyl carbonate, dimethyl carbonate and ionic liquids. Plasticisers also facilitate the dissociation of the Li-salt. A large number of polymer matrices, including polyethylene oxide, polyvinyl chloride (PVC), polyacrylonitrile (PAN), poly(methyl methacrylate) (PMMA), polyvinylidene fluoride (PVDF), and poly-(vinylidene fluoride-hexafluoropropylene) co-polymer P(VDF-HFP) have been intensively studied as matrices in GPEs.[52] GPEs show high mechanical strength, ionic conductivity and electrochemical stability toward the electrodes. For example, (i) PMMA-based electrolytes show poor mechanical strength; (ii) PAN-based electrolytes undergo severe passivation upon contact with lithium metal; (iii) PVC-based electrolytes exhibit poor ionic conductivity and are reactive towards Li; and (iv) PVDF-based electrolytes show poor interfacial properties with lithium.[49,50]

Composite PE: Though the addition of a plasticiser increases the conductivity, it compromises the mechanical robustness and increases resistance at the Li interface through passivation. Addition of inorganic fillers such as Al_2O_3, TiO_2 and SiO_2 enhance the mechanical and chemical stability and Li^+ transport. The size, shape and amount of the dispersed filler particles play a crucial role in determining the overall electrochemical properties of the composite polymer electrolyte (CPE).[50,51] Nanostructured ceramic material dispersed in the polymer electrolyte enhances the electrochemical properties, mechanical integrity and the electrolyte/electrode interface under operation at room temperature by up to two orders of magnitude when compared with the polymer electrolyte without filler particles. Weston and Stile have investigated the effect of α-Al_2O_3 fillers on a PEO polymer matrix and on the mechanical integrity and conductivity of a CPE for the first time. It is believed that nanostructured fillers reduce the crystallinity of polymer matrices and so increase the ionic transport.[68] Tarascon *et al.* demonstrated that dispersed SiO_2 nanoparticles on P(VDF-HFP) polymer absorbs solvents on the surface and increases the conductivity.[69] Stephan *et al.* studied the effect on the electrochemical properties of a P(VDF-HFP) polymer electrolyte incorporated with $[AlO(OH)_n]$ fillers that reduced the crystallinity of the polymer matrix and increased the ionic conduction to an extent of up to 12.5%.[70]

Inorganic/organic hybrid: This is a novel type of polymer-based electrolyte studied recently and polyhedral oligomericsilsesquioxanes (POSS) are the most important examples of the organic/inorganic hybrid building block polymers. POSS have silicon/oxygen nanostructured cores encircled by organic functional groups. POSS have a general formula of $(R–SiO_{1.5})_n$, where R is an organic functional group, *n* is usually 6, 8 or 10 and the ratio of Si/O is 1.5.[71] The surface functionalities of the POSS nanostructures facilitate copolymerisation, grafting or blending with other polymers. The advantage of POSS is the high chain mobility of the polymer due to the available additional free volume of the polymer matrix for steric effects. This suggests that the ionic conductivity of electrolytes could be increased by one to two orders of magnitude with inorganic/organic hybrid polymer matrices.[72]

Albeit several attempts have been made to improve the ionic conductivity and electrochemical stability of organic polymer-based Li-ion electrolytes, their thermal stability and electrochemical decomposition voltage have not been adequate for commercial application due to their flammability and low energy densities. Hence, replacement of the organic polymer electrolyte with non-flammable inorganic solid-state (ceramic) membranes can overcome long-term issues of safety and lifetime.[7] The following section briefly highlights recent developments in ceramic Li-ion electrolytes for all-solid-state Li-ion batteries.

3.5.2 Solid-state (Ceramic) Li-ion Electrolytes

Figure 3.6 shows a schematic representation of the all-solid-state cell utilising a solid-state ceramic membrane. Various crystalline and amorphous oxides and non-oxides have been investigated as solid Li-ion electrolytes for all-solid-state Li-ion batteries.[73] An ideal ceramic electrolyte should have the following physical and chemical properties to be used in advanced Li-ion batteries:[74,75]

- An ionic conductivity of $\sim 10^{-2}$ S cm^{-1} with negligibly small electronic and anionic conductivities at the battery operating temperature.
- Negligible grain-boundary impedance to total (bulk + grain-boundary) ionic conductivity.
- Chemically stable against reaction with high voltage Li cathodes.
- Chemically stable against reaction with high capacity Li anodes such as elemental Li and Li-metal alloys.
- Wide electrochemical potential window (up to 6 V *vs.* Li^+/Li) at the operating condition.
- Should exhibit negligibly small charge transfer impedance between the Li anode and or Li cathode and solid Li-ion electrolyte.
- Solid electrolyte should be able to prepare a thin film using current ceramic processing technologies.

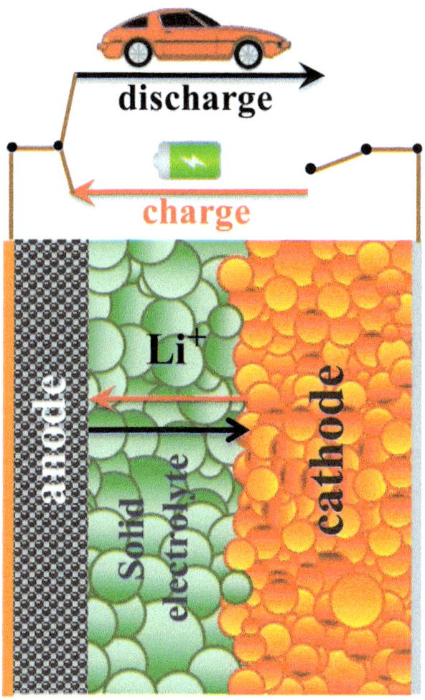

Figure 3.6 Schematic illustration of an all-solid-state Li battery.

- Easy to scale the synthesis and cost effective.
- Environmentally safe.

None of the known solid electrolytes simultaneously meet all of the above requirements. In the following sections, we discuss some of the most popularly known ceramic (solid state) Li-ion electrolytes that have been developed for application in all-solid-state Li-ion batteries.

3.5.2.1 Li_3N

Li_3N is one of the most well-known solid Li-ion electrolytes that were developed in the early 1970s that exhibit anisotropic Li-ionic conductivity. Lithium nitride (Li_3N) crystallises in a layered structure, like beta-alumina, $LiCoO_2$ and graphite.[76] Figure 3.7 shows the crystal structure of Li_3N along with the ionic conductivity of polycrystalline and single-crystal Li_3N.[76] In this structure N is surrounded by eight Li ions in a hexagonal bi-pyramidal geometry. Li-ion conduction along the *ab* plane is much higher than along the *c*-axis. High ionic conduction in the *ab* plane is due to the presence of Li vacancies in the Li_2N layers. Various elements such as H^+, Ca^{2+}, Mg^{2+} and Al^{3+} have been substituted in the Li_2N layer to improve the ionic conductivity.[76,77] Lithium nitride halides are also attractive ternary compounds due to their thermodynamic stability. $Li_{1.8}N_{0.4}Cl_{0.6}$ is the most interesting among all the ternary

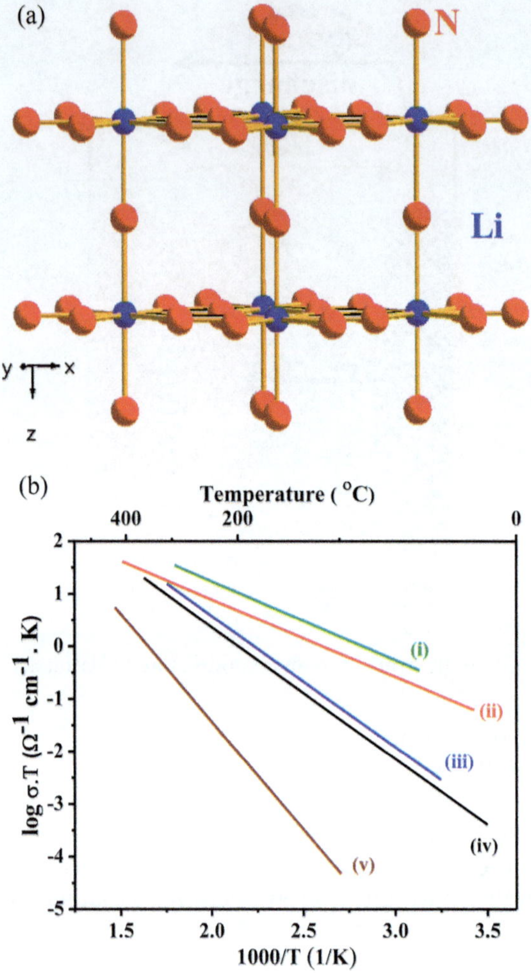

Figure 3.7 (a) Idealised crystal structure of Li_3N and (b) Li-ion conductivity of polycrystalline and single-crystal Li_3N and doped Li_3N: (i) single crystal Li_3N perpendicular to the *c*-axis, (ii) polycrystalline Li_3N, (iii) single crystal Li_3N parallel to the *c*-axis, (iv) $Li_{1.8}N_{0.4}Cl_{0.6}$, and (v) $Li_{9.11}N_{2.7}I$. Adapted from ref. 76 with permission from Elsevier, Copyright 1982, and ref. 78 with permission from Elsevier, Copyright 1981.

halides because it exhibits low activation energy and high ionic conductivity.[78] Although Li_3N is chemically in contact with elemental Li, it decomposes at potentials of 0.44 V *versus* Li^+/Li. Therefore, it cannot be used as an electrolyte in high power density all-solid-state Li batteries.[77]

3.5.2.2 Li_4SiO_4 and $LiAlCl_4$

Stoichiometric Li_4SiO_4 shows poor Li-ion conductivity and substitution of P for Si, which leads to Li-ion vacancies in the structure that enhances Li-ion conductivity.[79–81] P doped for Si or Al doped for Si in Li_4SiO_4 show improved lithium ion conductivity in the group of silicate structures. The Li_4SiO_4 structure consists of isolated tetrahedral SiO_4 units and lithium ions distributed over the interstitial sites. Figure 3.8 shows the structure and conductivity of Li_3PO_4–Li_4SiO_4 solid solution.[82] Furthermore, substitution of trivalent ions for tetravalent Si in Li_4SiO_4 increases the Li-ion concentration at the interstitial sites and also shows moderate improvement in Li-ionic conductivity.[80] A further experiment on solid solutions of γ-Li_3PO_4 and Li_4SiO_4 shows a maximum conductivity (Figure 3.8) at 1 : 1 composition, which was found to be independent of the type of structure.[83]

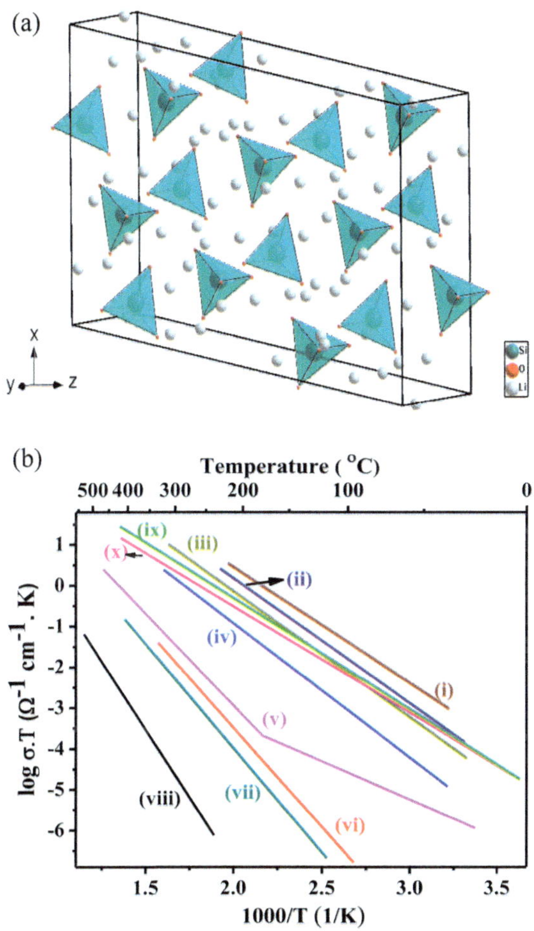

Figure 3.8 (a) Idealised crystal structure of Li_4SiO_4 (tetrahedral: SiO_4; closed spheres: Li ions) and (b) electrical conductivity of (i) Li_3PO_4, (ii) 50 mol% Li_3PO_4, (iii) Li_4SiO_4, (iv) Li_4SiO_4 (West), (v) 9.1% Li_3PO_4, (vi) 20% Li_3PO_4, (vii) 30% Li_3PO_4, (viii) 40% Li_3PO_4, (ix) 50% Li_3PO_4 in metastable Li_4SiO_4 solid solution, and (x) conversion of (ix) into γ-Li_3PO_4 at 900 °C. Adapted from ref. 82 with permission from Elsevier, Copyright 1977, and ref. 83 with permission from Elsevier, Copyright 1981.

Another family of compounds that consists of isolated $AlCl_4$, like SiO_4 in Li_4SiO_4, shows fast ionic conduction at the melting temperature. For example, the ionic conductivity value at room temperature was reported as 1.2×10^{-6} S cm^{-1} for $LiAlCl_4$ and it reaches a value ~0.1 S cm^{-1} at 130 °C.[84] Both Li_4SiO_4 and $LiAlCl_4$ are chemically stable against reaction with elemental Li but later exhibit fast conduction in the liquid state while the former shows moderate total (bulk + grain-boundary) conductivity at low temperatures. Hence, the practical applications of these materials are limited in all-solid-state Li-ion batteries.

3.5.2.3 *LISICON*

The term LISICON stands for a lithium super ionic conductor and is named after the nominal chemical composition $Li_{14}Zn(GeO_4)$. The prototypical LISICON phase crystallises with the Li_3PO_4 structure and shows a Li-ion conductivity of 0.125 S cm^{-1} at 300 °C.[85] This is due to an immobile sub-lattice trapping the mobile Li^+ ions at lower temperatures due to defect complexes. The ionic conductivity of LISICON-type solid electrolytes is further improved by the replacement of oxide by S. These sulphide solid electrolytes are called as thio-LISICONs. As the radius of S^{2-} is larger than O^{2-}, replacement of O^{2-} with S^{2-} increases the size of Li^+ transport bottlenecks. Furthermore, higher polarisation capability of S^{2-} as compared to O^{2-} weakens the Li interaction. LISICON-type $Li_{3.25}Ge_{0.25}P_{0.75}S_4$ showed a high conductivity of 2.2×10^{-3} S cm^{-1} at room temperature with negligible electronic conductivity and high electrochemical stability.[86] Recently, Kamaya *et al.* reported a thio-LISICON with a nominal composition $Li_{10}GeP_2S_{12}$, which showed the highest Li-ionic conductivity of 12 mS cm^{-1} at 27 °C.[87] Ionic conductivity of some LISCON structure compounds is shown in Figure 3.9.[85]

3.5.2.4 *NASICON*

Li-based super ionic conductors with composition $LiM_2(PO_4)_3$ (M = Zr, Ti, Ge), based on the sodium analogue NASICON structure, show moderate Li-ion conductivity. The size of the transition metal plays an important role in the ionic conductivity of NASICON phases. Figure 3.10 shows the idealised crystal structure of NASICON that consists of MO_6 octahedra and PO_4 tetrahedra that are connected three-dimensionally (3D) through edge sharing.[74] Co-substitution of trivalent ions and Li^+ for M^{4+}-sites led to improved conductivity. Among the $LiM_2(PO_4)_3$ phases, the Ti-based phases show higher ionic conductivity with Al-doped $Li_{1.3}Al_{0.3}Ti_{1.7}(PO_4)_3$ showing the highest ionic conductivity at room temperature.[74] Also, attempts have been made to substitute the M site in $LiM_2(PO_4)_3$ with Nb and In, $LiInNb(PO_4)_3$, which resulted in even higher conductivity.[88] Also, Ca-doped $LiZr_2(PO_4)_3$ shows appreciable Li-ion conductivity of 1.2×10^{-4} S cm^{-1} at room temperature with an activation energy of 0.48 eV.[89]

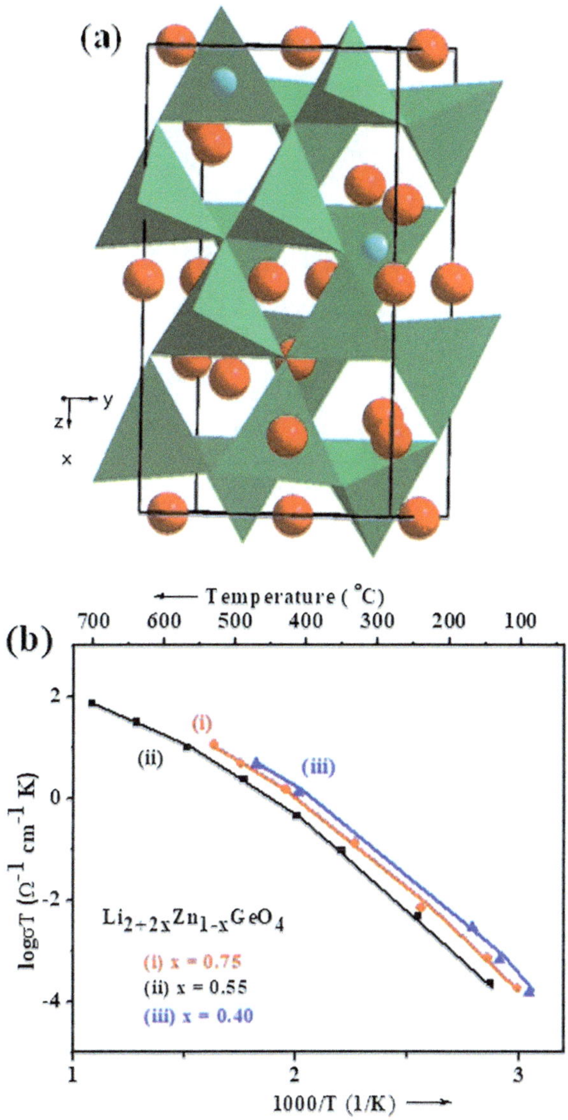

Figure 3.9 (a) Idealised crystal structure of LISICON (tetrahedral: $Zn(Ge)O_4$; closed spheres: Li ions) and (b) Li-ion conductivities (bulk + grain boundary) of $Li_{2+2x}Zn_{1-x}GeO_4$ solid solutions with x values: (i) 0.75, (ii) 0.55, and (iii) 0.40. Adapted from ref. 85 with permission from Elsevier, Copyright 1982.

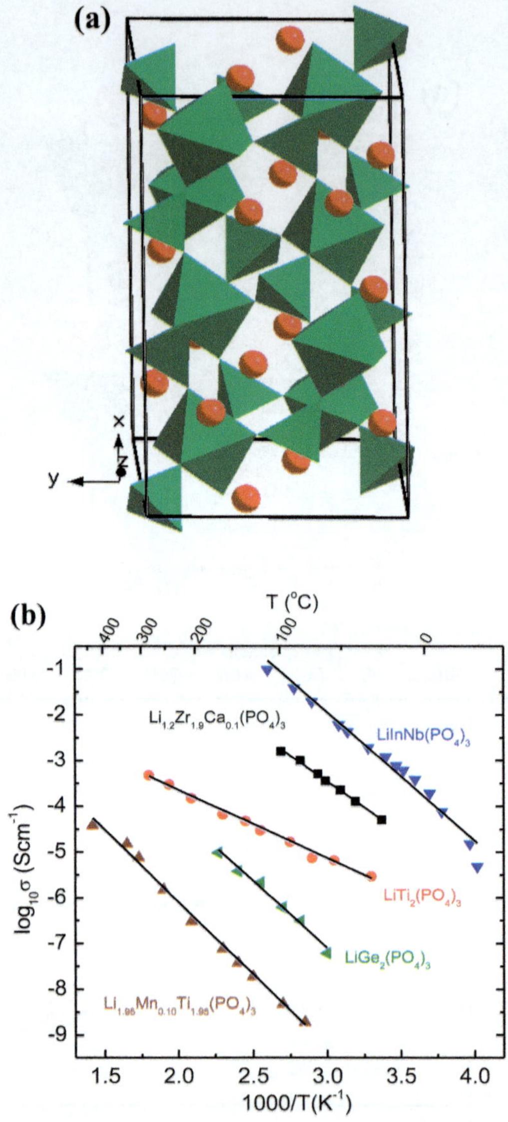

Figure 3.10 (a) Ideal crystal structure of NASICON $LiM_2(PO_4)_3$ (tetrahedral: PO_4; octahedra: MO_6; closed spheres: Li ions) and (b) Li-ion conductivity of different NASICON structures.

3.5.2.5 A-site Deficient Perovskites

A stoichiometric perovskite-type structure adopts a general formula of ABO_3 where A and B represent twelve and six-fold coordinated cations, respectively. Several A-site deficient perovskite-type Li-ion electrolytes have been studied following the discovery of fast Li-ion conduction in $La_{(2/3)-x}Li_{3x}TiO_3$ (LLTO) in 1994.[90] The high ionic conductivity in the structure was explained using A-site

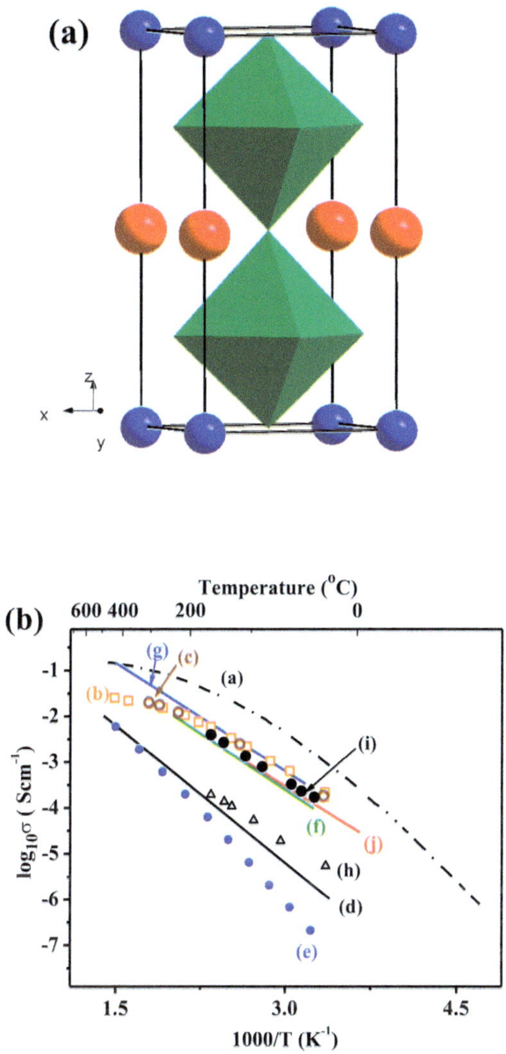

Figure 3.11 (a) Idealised crystal structure of A-site deficient perovskite $La_{(2/3)-x}Li_{3x}TiO_3$ (LLTO) (octahedra: TiO_6; closed spheres: Li or La ions) and (b) Arrhenius plot for Li-ion conductivity of different perovskites; (a) $Li_{0.34}La_{0.51}TiO_{2.94}$, (b) $Li_{0.31}La_{0.63}Ti_{0.9}Co_{0.1}O_3$, (c) $Li_{0.31}La_{0.63}Ti_{0.9}Ni_{0.1}O_3$, (d) $Li_{0.1}La_{0.63}Mg_{0.5}W_{0.5}O_3$, (e) $Li_{0.34}Nd_{0.55}TiO_3$, (f) $Li_{0.1}La_{0.3}NbO_3$, (g) $Li_{0.25}La_{0.25}TaO_3$, (h) $Li_{0.3}Sr_{0.6}Nb_{0.5}Ti_{0.5}O_3$, (i) $Li_{0.3}Sr_{0.6}Ta_{0.5}Ti_{0.5}O_3$ and (j) $Li_{0.5}Sr_{0.5}Fe_{0.25}Ta_{0.75}O_3$. Adapted from ref. 91 with permission from American Chemical Society, Copyright 2003.

deficiencies in the structure. Figure 3.11 shows the idealised crystal structure of A-site deficient perovskite $La_{(2/3)-x}Li_{3x}TiO_3$ (LLTO).[91] The $x = 0.1$ member of the LLTO family shows the highest bulk ion conductivity of 10^{-3} S cm^{-1} at room temperature, which is about two orders of magnitude higher than that of total (bulk + grain-boundary (σ_{gb})) conductivity. A detailed review on the structure–composition–ionic conductivity relationship in the perovskite-type structure was reported by Stramare *et al.*[91] Although LLTO shows the highest bulk ionic conductivity among the oxide-based Li-ion electrolytes, its stability with Li metal is a problem and limits its use as a membrane in all-solid-state Li batteries.

3.5.2.6 *Anti-perovskites*

Very recently, anti-perovskites with a general formula of $X^{+}{}_{3}B^{2-}A^{-}$ where X is the highly electropositive element, and A and B are the monovalent and divalent anions, respectively, were reported as fast Li-ion conductors.[92] The crystal structure of Li_3OX (X = Cl, Br) is shown in Figure 3.12.[93] The conductivity of the mixed member $Li_3O(Cl, Br)$ is higher compared to both end members Li_3OBr and Li_3OCl because of its topological nature as it accommodates both Br^- and Cl^- in the structure. The Br^- anion is large and fills the very small dodecahedral site completely and it causes insufficient A-site filling and this results in lower conductivity in Li_3OBr. In contrast, the smaller Cl^- anion is not large enough to substantially fill the dodecahedral site, which causes structural distortion and shows low conductivity.[93] A super ionic conductivity of 10 S cm^{-1} was observed for $Li_3O(Cl, Br)$ at 266 °C, which is comparable with that of high conducting electrolytes such as $Li_7La_3Zr_2O_{12}$, thio-LISICON ($Li_{10}GeP_2S_{12}$) and some liquid electrolytes.[87,93,94] Anti-perovskite-type structure Li-ion conductors show high ionic conductivity at high temperatures, poor electronic conductivity, and have a wide potential stability window; thus, these materials can contribute to a new era in solid-state Li electrolyte research.

3.5.2.7 *Garnet-type Li-ion Conductors*

The ideal garnet structure has a general formula of $A_3B_2C_3O_{12}$, where A (Ca^{2+}, Mg^{2+}, or Fe^{2+}), B (Al^{3+}, Cr^{3+}, or Fe^{3+}) and C (Si^{4+}, Ge^{4+}, or Al^{4+}) cations occupy the eight, six, and four coordination sites in the lattice, respectively, and stabilises the face-centered cubic structure with space group *Ia* –3*d*. The crystal structure of an ideal garnet is shown in Figure 3.13a. Due to these various coordination sites, garnet materials have a variety of substitution options by alkali metals, alkali earth, transition and rare earth elements. $Li_5La_3M_2O_{12}$ (M = Nb, Ta) was first reported by Mazza in 1988 to crystallise in the garnet-like oxygen framework.[95] The Li-ion sites are not fully occupied, which suggests that these compounds may be capable of conducting Li ions. Thangadurai *et al.* reported a Li-ion conductivity of 10^{-6} S cm^{-1}, at room temperature, for the garnet-type $Li_5La_3M_2O_{12}$.[96] Since their discovery as Li-ion conductors, there has been increased interest in the development of garnet-type Li-ion solid electrolytes for all-solid-state LIBs. In an attempt to improve the electrical properties of the garnet materials, the M site has been fully replaced with other transition metal ions and La by both divalent ions and Li ions.

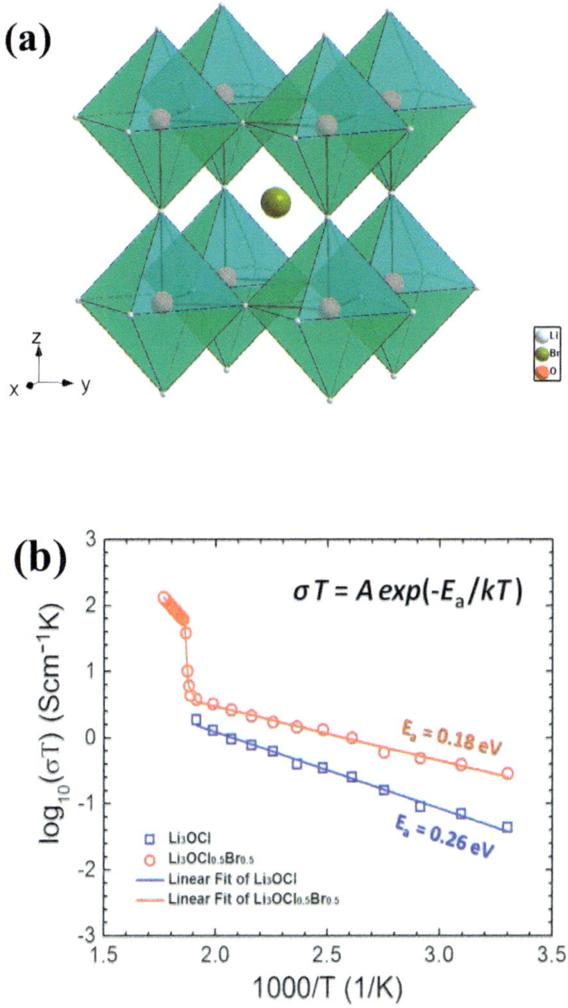

Figure 3.12 (a) Crystal structure of anti-perovskite Li_3OX (octahedra: OLi_6; closed spheres: X ions) and (b) Li-ion conductivity of anti-perovskites. Reproduced from ref. 93 with permission from American Chemical Society, Copyright 2012.

Partial substitution at the M site of the pentavalent ion in $Li_5La_3M_2O_{12}$ (M = Nb, Ta) with a trivalent ion yields new members of the garnet-like family of compounds with increased Li content. The larger ionic radius of Y^{3+} (0.90 Å) and In^{3+} (0.79 Å), compared to the radii of Nb^{5+} (0.64 Å) and Ta^{5+} (0.64 Å), resulted in an increased lattice parameter of the garnet samples. The higher Li content, together with the larger lattice parameters of the doped members, is expected to increase the total Li-ion conductivity of the materials. For example, $Li_{5.5}La_3Nb_{1.5}In_{0.5}O_{12}$ showed a total Li-ion conductivity of 10^{-4} S cm^{-1} at 50 °C, while the reported conductivity for $Li_{6.25}La_3Nb_{1.375}Sc_{0.625}O_{12}$ is 3.7×10^{-4} S cm^{-1} at the same temperature.[97,98]

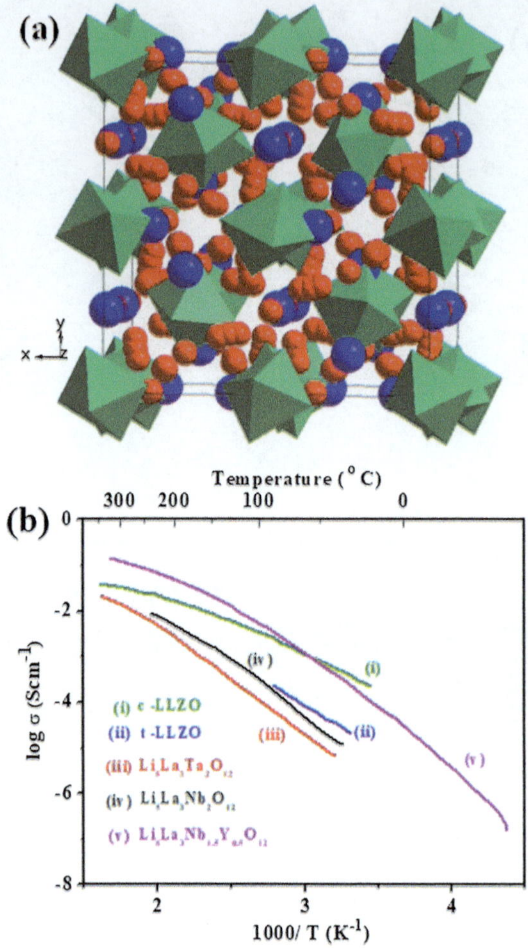

Figure 3.13 (a) Crystal structure of garnet-type Li-ion conductor $Li_5La_3M_2O_{12}$ (octahedra: MO_6; small closed spheres: Li ions; large closed spheres: La ions) and (b) conductivity of selected members of garnet-type Li stuffed metal oxides. Adapted from ref. 105 with permission from the Royal Society of Chemistry.

Murugan *et al.* reported highly conducting $Li_7La_3Zr_2O_{12}$ (LLZO) with a cubic garnet structure in 2007.[94] In 2009 Awaka *et al.* reported the same material crystallised in a tetragonal framework, with a lower Li-ion conductivity compared to the cubic structure.[99] Subsequently, Geiger *et al.* showed that Al contamination from the alumina crucibles used in the high temperature synthesis aids in the stabilisation of the cubic phase.[97] Since the cubic structure is highly conductive, various attempts have been made to gain a better understanding on the effect of Al and final sintering temperature on the formation of cubic $Li_7La_3Zr_2O_{12}$.[100]

A Hf analogue, $Li_7La_3Hf_2O_{12}$, was synthesised in a tetragonal garnet-type structure, which showed Li-ion conductivity of 10^{-7} S cm^{-1} at room temperature. Zaiß *et al.* prepared a cubic Hf-phase at a temperature of 1250 °C that showed bulk conductivity of 2.4×10^{-4} S cm^{-1} at room temperature.[101] A further study by Ohta *et al.* showed the formation of the cubic phase in Nb-doped LLZO $Li_{7-x}La_3Zr_{2-x}Nb_xO_{12}$.[102] An ionic conductivity of 8×10^{-4} S cm^{-1} at room temperature was observed for the $x = 0.25$ member of Li_{7-x} $La_3Zr_{2-x}Nb_xO_{12}$. Li *et al.* prepared the same family of materials under slightly different laboratory conditions, and obtained a conductivity of 1×10^{-3} S cm^{-1} for the $x = 0.6$ member. A comparison of the best conducting members of each of the mentioned families of doped garnets is shown in Figure 3.13b.[103–105]

3.6 Application of Solid Li-ion Electrolytes in All-solid-state Batteries

The performance of the all-solid-state Li batteries (ASSLBs) is mainly determined by the properties of the solid electrolyte.[104,105] The ionic conductivities of different electrolytes are shown in Figure 3.14.[87] Lack of a suitable solid electrolyte is the major limitation hindering practical realisation of ASSLBs. Li_4SiO_4–Li_3PO_4, LiPON and garnet are the most promising solid electrolyte families for ASSLBs due to their high ionic conductivity and electrochemical stability. Though Li^+ conduction of sulphide-based thio-LISICON type electrolytes at ambient temperature are higher than the garnet-type oxide, their mechanical stability is far inferior to the garnet family members. Several studies on ASSLBs have been carried out using different solid electrolytes. Most of the studied ASSLBs were fabricated using a $LiCoO_2$ or $LiFePO_4$ cathode, metallic Li or $Li_4Ti_5O_{12}$ anode.[106] Moreover, Li–S and Li-air batteries have also been fabricated to evaluate the opportunities for solid electrolytes. Tables 3.4–3.6 illustrate recently studied ASSLBs, Li–S and Li-air batteries, respectively.[107–147] The interfacial resistance of the electrode and electrolyte is a major concern for the practical utilisation of ASSLBs. The poor contact between the Li metal anode and the inorganic electrolyte results in a focused local ionic current and hence restricts the applied current density.

Despite having good ionic conductivities, A-site deficient perovskite-type $La_{(2/3)-x}Li_{3x}TiO_3$ (LLTO) and NASICON-type $Li_{1.5}Al_{0.5}Ge_{1.5}(PO_4)_3$ (LAGP) are unstable in the presence of metallic Li at the interface. Ti^{4+} (LLTO, LATP) and Ge^{4+} (LAGP) ions in the crystal structures of the electrolyte are reduced to Ti^{3+} and Ge^{3+} by the metallic Li. Li is converted to Li_2O or Li_2O_2 and increases interfacial impedance.[148,149] Thio-LISICON, $Li_{10}GeP_2S_{12}$ phases have potential risks as they may produce toxic H_2S or SO_2 gas on oxidation.[150] Garnet-based oxides are highly stable towards metallic Li but only point contacts could be made, which limits the Li^+ conduction through the interfaces.

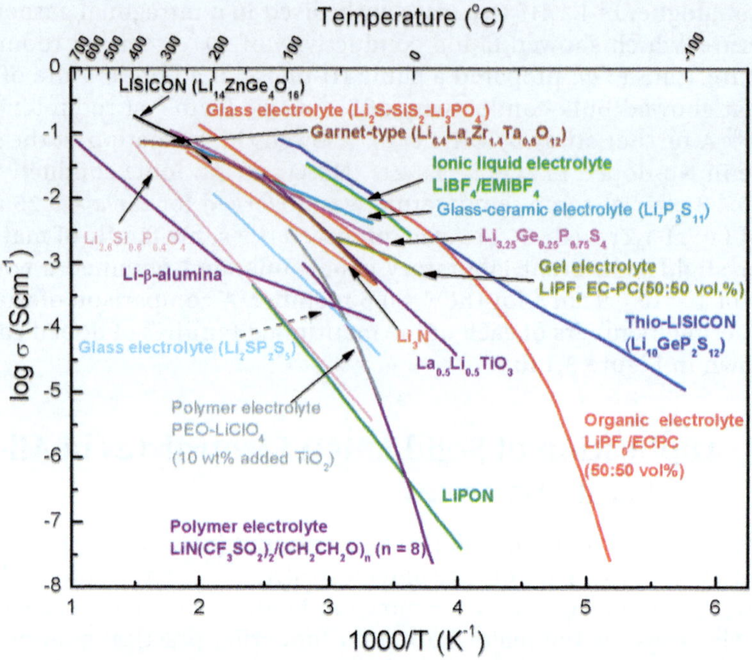

Figure 3.14 Ionic conductivity of some known typical Li-ion electrolytes. Reproduced from ref. 87 with permission from Springer Nature, Copyright 2011.

Several efforts have been made to overcome the interfacial issues for these electrolytes to enable the fabrication of ASSLBs. Coating of the electrolyte surface with metal, metal oxide and polymer was reported to reduce the interfacial impedance and also stabilise the electrolyte. It has been observed that coating with Si, Ge, Au, ZnO and Al_2O_3 on a garnet-based oxide electrolyte reduced the interfacial resistance.[151–153] For example, atomic layer deposited Al_2O_3 on the $Li_7La_{2.75}Ca_{0.25}Zr_{1.75}Nb_{0.25}O_{12}$ electrolyte reduced the interfacial resistance abruptly (1710 to 1 ohm cm^2 at room temperature under the open circuit condition) and exhibited a stable capacity against a $Li_2FeMn_3O_8$ cathode.[154] Goodenough *et al.* introduced a polymer coating over the electrolyte surface that improved the contacts and enhanced the performance of the ASSLB.[155] Cathode material ($LiCoO_2$) coated with $Li_4Ti_5O_{12}$ and $LiNbO_3$ also enhanced the stability and contact with electrolytes.[106] A coating of LAGP on a $LiCoO_2$ particle surface led to good reversibility with excellent capacity retention up to 200 cycles.[156] ASSLBs with exceptional power density (>10 kW kg^{-1}) and good rate capability were recently reported by Kato *et al.* using a $Li_{9.6}P_3S_{12}$ electrolyte and $LiNbO_3$-coated $LiCoO_2$ cathode and In–Li alloy anode.[106]

The electrodes of the ASSLBs are fabricated by mixing the active material and electrolyte particles to improve electronic and ionic percolation. Though the composite electrode enhances the cell performance, rigid

Table 3.4 Some examples of recently investigated all-solid-state Li batteries.

Anode	Cathode	Electrolyte	Cell voltage (*V*)	Capacity (mA h g^{-1})	Reference
Li	$LiFePO_4$	poly(triethylene glycol carbonate) (PTEC)	4	129.3	107
Li	$LiFe_{0.2}Mn_{0.8}PO_4$	poly(triethylene glycol carbonate) (PTEC)	4.35	124.8	107
Li	$Li_4Ti_5O_{12}$	Silica-PEO-LiTFSI	2	3.6	108
Li	$LiFePO_4$	PEO18-LiTFSI-1%LGPS-10 Succinonitrile	4	160.6	109
Li–In	Li_3PO_4 thin film coated $LiNi_{0.5}Mn_{1.5}O_4$	$80Li_2S$-$20P_2S_5$	4.8	62	110
Li	$LiFePO_4$	PEO-LiTFSI-20% $Li_{1.5}Al_{0.5}Ge_{1.5}(PO_4)_3$	3.8	166	111
Li	$LiMn_{0.5}Ni_{0.5}O_2$	PEO18-$Li(CF_3SO_2)_2N$-$Li_{1+x+y}Al_xTi_{2-x}Si_yP_{3-y}O_{12}$	4.5	70	112
Li	$LiMn_2O_4$	Porous honeycomb $Li_{0.55}La_{0.35}TiO_3$/Li-PMMA	4.3	71	113
Li	$LiCoO_2$	$78Li_2S$-$22P_2S_5$	3.6	139	114
Li	$LiFePO_4$	PEO18-LiTFSI-1%LGPS	3.8	158	115
Li	$LiFePO_4$ and $LiFe_{0.15}Mn_{0.85}PO_4$	PEO : LLZTO	4.2	141.5	116
Li	$LiFePO_4$	(LiTNFSI/PEO, EO/Li^+ = 20)	3.8	146.0	117
TiO_2	$LiFePO_4$	PEO-20% $LiCF_3SO_3$-20% GFs	2.5	3.7	118
$Li_4Ti_5O_{12}$	$LiCoO_2$ was coated with $LiNb_{0.5}Ta_{0.5}O_3$	$Li_{10}GeP_2S_{12}$	2.7	90	119
3D Li-rGO	$LiFePO_4$	PEG-LiTFSI-LLZTO	4	110	120
Li	$LiCoO_2$	PVDF/LLZTO	4.2	140	121
Li	$LiFePO_4$	Mg-BTC MOF-PEO-LiTFSI	3.8	110	122
Li	$Li(Ni_{0.6}Co_{0.2}Mn_{0.2})O_2$	LATP–Bi_2O_3 with interlayer based on polyethylene oxide	4.3	122	123
Li	$LiFePO_4$	Sandwiched garnet with Ge (20 nm) and PVDF gel layer	4	140	124
Si	$LiCoO_2$	Li_3PO_4	1	3500	125
Li	Al_2O_3 coated $LiCoO_2$	$Li_{3.15}Ge_{0.15}P_{0.85}S_4$ and $77.5Li_2S$-$22.5P_2S_5$	4.3	120	126

coupling between the electrolyte and active material will lead to local strain during charge–discharge. The local strain is due to the volume changes of the electrodes in the confined cell structure. This local strain may lead to crack and pore formation on the particle. Soft ionic conductors like sulphide

Table 3.5 Selected examples of Li–S batteries.

Anode	Cathode	Electrolyte	Cell voltage (*V*)	Capacity (mA h g^{-1})	Reference
Li	S	Li_2S-P_2S_5	3	605	127
Li	S	PEO–LiTFSI in metal–organic framework	2.8	1225	128
Li	S	$Li_7La_{2.75}Ca_{0.25}Zr_{1.75}Nb_{0.25}O_{12}$ (LLCZN)	3.5	1000	129
Li/Li–In/Li–Al	S	Li_6PS_5Cl	3.6	1300	130
$Li_{4.4}Ge$	CuS and Cu_9S_8	Amorphous-60Li_2S–40SiS_2	2.5	980	131
Li	S	(LiTNFSI/PEO, EO/Li^+ = 20)	3	851.4	117
Li	S	$Li_{10}GeP_2S_{12}$ and 75%Li_2S–24%P_2S_5–1%P_2O_5	2.8	1384.5	132
Li	S	PEO-$LiCF_3SO_3$ + 10 mass% ZrO_2	3	900	133
Li	S	PEO-Al^{3+}/Nb^{5+} doped $Li_7La_3Zr_2O_{12}$	2.8	>900	134
Li	Li_2S	Li_6PS_5Cl	3	830	135
Li	S	Li_2S–P_2S_5 glass–ceramic	2	850	136
Li	Li_2S	78Li_2S-22P_2S_5	4	600	137
Li–Al	S	$Li_{3.25}Ge_{0.25}P_{0.75}S_4$	3	786	138
Li	S	$Li_7P_{2.9}Mn_{0.1}S_{10.7}I_{0.3}$	3	791.6	139
Li	S	$Li_{1.5}Al_{0.5}Ge_{1.5}(PO_4)_3$	3	1510	140
Li	S	$Li_7La_{2.75}Ca_{0.25}Zr_{1.75}Nb_{0.25}O_{12}$	3.5	645	141

Table 3.6 Examples of solid-state Li-air or Li–O_2 batteries.

Anode	Cathode	Electrolyte	Cell Voltage (*V*)	Capacity (mA h g^{-1})	Reference
Li	Carbon	$Li_{6.4}La_3Zr_{1.4}Ta_{0.6}O_{12}$	4	2184	142
Li	$TiC_{0.7}N_{0.3}$	Polyacrylonitrile-$LiNO_3$	4.5	400	143
Li	*N*-rGO/RuO*x*	Gel polymer electrolyte	4.5	6000	144
Li	Single-wall carbon nanotube	$Li_{1+x}Al_yGe_{2-y}(PO_4)_3$ (LAGP)	5	1200	145
Li	Carbon nanofibre	$Li_{1.3}Al_{0.3}Ti_{1.7}(PO_4)_3$ (LATP)	3.5	6300	146
Li	3D-Graphene hydrogel	Lithium iodine (LiI), 3-hydroxypropionitrile	3.0	2100	147

and thio-phosphate-based materials are suitable to sustain these mechanical effects. Though this is an important issue it has still not properly been addressed.[119]

A solid-state Li–S battery was first demonstrated in the late 1980s using PEO-based polymer electrolytes.[157] Later, several fillers such as ZrO_2 and SiO_2 were introduced to enhance performance.[49] As ceramic inorganic electrolytes can mitigate polysulphide dissolution along with the Li dendrite growth, Li_2S–P_2S_5, thio-LISICON and garnet-type electrolytes were explored for application in solid-state Li–S batteries.[127,130,134,135] As glass–ceramic electrolytes such as $Li_2S–P_2S_5$ contain sulphides, compatible CuS cathodes were investigated for Li–S systems.[131] Thio-LISICON-based electrolytes $Li_2S\text{-}GeS_2\text{-}P_2S_5$ and $Li_{3.25}Ge_{0.25}O_{0.75}S_4$ were explored as potential electrolytes for solid-state Li–S batteries.[158] A sulfur-encapsulated bilayer $Li_7La_{2.75}Ca_{0.25}Zr_{1.7}Nb_{0.25}O_{12}$ showed a high specific capacity of 645 mA h g^{-1}.[141]

The first solid-state Li-air battery comprised a Li metal anode, Li^+ ion conducting glass ceramic (GC) sandwiched between two polymer electrolyte membranes and carbon-GC powder composite air cathode and was demonstrated by Kumar *et al.*[159] Several polymer-based (PEO, PAN, PVDF) electrolytes were investigated for Li-air battery applications and good performance was observed.[49] Zhou *et al.* obtained a first capacity of 1700 mA h g^{-1} with an operational voltage of 2 to 4.2 V using a $Li_{1+x}Al_yGe_{2-y}(PO_4)_3$ electrolyte with a carbon nanotube cathode.[145] A further all-solid-state Li-air cell based on a garnet $Li_{6.4}La_3Zr_{1.4}Ta_{0.6}O_{12}$ was demonstrated by Sun *et al.* with an impressive specific capacity of 2184 mA h g^{-1} at 20 μA cm^{-2} at 200 °C.[142]

3.7 Conclusions

The fast-growing interest in all-solid-state-Li batteries is due to several advantages over conventional mature state-of-the-art Li-ion battery technology in terms of energy density, power density and safety. Several Li-based solid-state batteries have been designed and their performances have been evaluated even at high current densities. The cell architecture, fundamental understanding of interfacial reactions and ion transport inside ASSLBs are still to be improved for commercialisation. The major factors, which should be addressed for practical applications of ASSLBs are: (i) development of state-of-the-art solid electrolytes with excellent ionic conductivities, mechanical and chemical stability in air; (ii) a decrease of the area-specific polarisation resistance between solid Li-ion electrolyte and electrodes, which is a major obstacle for the development of ASSLBs; (iii) a deep understanding of the electrode–electrolyte interfacial reaction and kinetics for further improvement of interfacial impedance. The present literature is inadequate in providing a clear perspective on the interface issue. Detailed studies on the interface using *in situ* and *ex situ* techniques are required; (v) development of cost-effective material processing and cell manufacture for competing conventional LIB technology; and (vi) development of *in situ* and *ex situ* methods to determine volume expansion of electrode materials due to intercalation/de-intercalation during charging and discharging cycles. This volume expansion can produce mechanical pressure, which could reduce the stability of the ASSLBs. The development of ASSLBs is still in its infancy. Several positive and negative perspectives were discussed and analysed regarding the ASSLB concept. Recent experiments on mitigating interfacial resistance and achieving high power and energy density ASSLBs are promising for their practical utilisation in the near future.

Abbreviations

A h	Ampere hour (unit of charge)
ASSLBs	All-solid-state-Li batteries
BEV	Battery electric vehicles
$C_{discharge}$	Practical discharge capacity
CPE	Composite polymer electrolyte
C_s	Specific capacity
DC	Diethyl carbonate
DMC	Dimethyl carbonate
E	Energy density
EC	Ethylene carbonate
ECES	Electrochemical energy storage
EES	Electrical energy storage
E_i	Electrochemical potential of the *i*th system
FEC	Fluoroethylene carbonate
GC	Glass ceramic
G_j	Gibbs free energy for the *j*th chemical reaction
GPE	Gel polymer electrolyte
LAGP	$Li_{1.5}Al_{0.5}Ge_{1.5}(PO_4)_3$ NASICON structured solid electrolyte
LATP	$LiAlTi(PO_4)_3$ NASICON-type ceramic electrolyte
LCVs	Light commercial vehicles
LISICON	Li super ionic conductor
LLTO	$La_{(2/3)-x}Li_{3x}TiO_3$ perovskite-type solid electrolyte
LLZO	$Li_7La_3Zr_2O_{12}$ garnet-type solid ceramic electrolyte
M	Molecular weight
Mtoe	Million tonnes of oil equivalent
NASICON	Na super ionic conductor
PAN	Polyacrylonitrile
PC	Propylene carbonate
PEO	Polyethylene oxide
PHEVs	Plug-in hybrid electric vehicles
PLDVs	Passenger light duty vehicles
PMMA	Poly(methyl-methacrylate)
POSS	Polyhedral oligomericsilsesquioxanes
PVC	Polyvinyl chloride
PVDF	Polyvinylidene fluoride
P(VDF-HFP)	Poly(vinylidene fluoride-*co*-hexafluoropropylene)
SPE	Solid polymer electrolyte
TEGDME	Tetraethyl glycol dimethyl ether
T_g	Glass transition temperature
V	Voltage
W	Watt (unit of *power*)
Wh	Watt-hour (unit of *energy*)
Wh kg^{-1}	Watt-hour per kilogram (unit of gravimetric *energy density*)

Wh L^{-1} Watt-hour per litre (unit of volumetric *energy density*)
W kg^{-1} Watt per kilogram (unit of gravimetric *power density*)
W L^{-1} Watt per litre (unit of volumetric *power density*)

Acknowledgements

This work is supported through discovery grants from the Natural Sciences and Engineering Research Council of Canada (NSERC) awarded to V.T (Award number: RGPIN-2016-03853).

References

1. BP, Statistical Review of World Energy 2017, https://www.bp.com/content/dam/bp/en/corporate/pdf/energy-economics/statistical-review-2017/bp-statistical-review-of-world-energy-2017-full-report.pdf, accessed November 2017.
2. Key World Energy Statistic, International Energy Agency, 2017, https://www.iea.org/publications/freepublications/publication/Key-World2017.pdf, accessed November 2017.
3. M. Tran, D. Banister, J. D. K. Bishop and M. D. McCulloch, *Nat. Clim. Change*, 2012, **2**, 328.
4. G. N. Tiwari and R. K. Mishra, *Advanced Renewable Energy Sources*, Royal Society of Chemistry, 2012.
5. J. Zhang, L. Zhang, H. Liu, A. Sun and R.-S. Liu, *Electrochemical Technologies for Energy Storage and Conversion*, John Wiley & Sons, 2012.
6. D. Linden and T. B. Reddy, *Handbook of Batteries*, McGraw Hill, New York, 3rd edn, 2001.
7. G.-A. Nazri and G. Pistoia, *Lithium Batteries: Science and Technology*, Springer Science & Business Media, 2008.
8. J. B. Goodenough, in *Lithium Ion Batteries*, Wiley-VCH Verlag GmbH, 2007, pp. 1–25.
9. Global EV Outlook 2018, International Energy Agency, https://www.iea.org/gevo2018/, accessed October 2018.
10. S. Goriparti, E. Miele, F. De Angelis, E. Di Fabrizio, R. Proietti Zaccaria and C. Capiglia, *J. Power Sources*, 2014, **257**, 421–443.
11. A. Grimaud, *Nat. Energy*, 2017, **2**, 17003.
12. G. Che, K. B. Jirage, E. R. Fisher, C. R. Martin and H. Yoneyama, *J. Electrochem. Soc.*, 1997, **144**, 4296–4302.
13. J. Cho, Y.-W. Kim, B. Kim, J.-G. Lee and B. Park, *Angew. Chem., Int. Ed.*, 2003, **42**, 1618–1621.
14. T. Ohzuku, A. Ueda and M. Nagayama, *J. Electrochem. Soc.*, 1993, **140**, 1862–1870.
15. P. G. Bruce, A. Robert Armstrong and R. L. Gitzendanner, *J. Mater. Chem.*, 1999, **9**, 193–198.
16. F. Lin, I. M. Markus, D. Nordlund, T.-C. Weng, M. D. Asta, H. L. Xin and M. M. Doeff, *Nat. Commun.*, 2014, **5**, 3529.

17. S. K. Martha, O. Haik, E. Zinigrad, I. Exnar, T. Drezen, J. H. Miners and D. Aurbach, *J. Electrochem. Soc.*, 2011, **158**, A1115–A1122.
18. R. Wang, X. He, L. He, F. Wang, R. Xiao, L. Gu, H. Li and L. Chen, *Adv. Energy Mater.*, 2013, **3**, 1358–1367.
19. M.-J. Lee, S. Lee, P. Oh, Y. Kim and J. Cho, *Nano Lett.*, 2014, **14**, 993–999.
20. S. Choi and A. Manthiram, *J. Electrochem. Soc.*, 2002, **149**, A162–A166.
21. A. Yamada, S. C. Chung and K. Hinokuma, *J. Electrochem. Soc.*, 2001, **148**, A224–A229.
22. D. Choi, D. Wang, I.-T. Bae, J. Xiao, Z. Nie, W. Wang, V. V. Viswanathan, Y. J. Lee, J.-G. Zhang, G. L. Graff, Z. Yang and J. Liu, *Nano Lett.*, 2010, **10**, 2799–2805.
23. K. Amine, H. Yasuda and M. Yamachi, *Electrochem. Solid-State Lett.*, 2000, **3**, 178–179.
24. A. Sobkowiak, M. R. Roberts, R. Younesi, T. Ericsson, L. Häggström, C.-W. Tai, A. M. Andersson, K. Edström, T. Gustafsson and F. Björefors, *Chem. Mater.*, 2013, **25**, 3020–3029.
25. J. Barker, R. K. B. Gover, P. Burns, A. Bryan, M. Y. Saidi and J. L. Swoyer, *J. Power Sources*, 2005, **146**, 516–520.
26. H. Fujimoto, K. Tokumitsu, A. Mabuchi, N. Chinnasamy and T. Kasuh, *J. Power Sources*, 2010, **195**, 7452–7456.
27. Y. Liu, J. Xue, T. Zheng and J. Dahn, *Carbon*, 1996, **34**, 193–200.
28. V. Meunier, J. Kephart, C. Roland and J. Bernholc, *Phys. Rev. Lett.*, 2002, **88**, 075506.
29. C. M. Schauerman, M. J. Ganter, G. Gaustad, C. W. Babbitt, R. P. Raffaelle and B. J. Landi, *J. Mater. Chem.*, 2012, **22**, 12008–12015.
30. H. Buqa, D. Goers, M. Holzapfel, M. E. Spahr and P. Novák, *J. Electrochem. Soc.*, 2005, **152**, A474–A481.
31. J. Hou, Y. Shao, M. W. Ellis, R. B. Moore and B. Yi, *Phys. Chem. Chem. Phys.*, 2011, **13**, 15384–15402.
32. Z. Chen, I. Belharouak, Y. K. Sun and K. Amine, *Adv. Funct. Mater.*, 2013, **23**, 959–969.
33. J. R. Szczech and S. Jin, *Energy Environ. Sci.*, 2011, **4**, 56–72.
34. N. G. Rudawski, B. R. Yates, M. R. Holzworth, K. S. Jones, R. G. Elliman and A. A. Volinsky, *J. Power Sources*, 2013, **223**, 336–340.
35. P. G. Bruce, B. Scrosati and J.-M. Tarascon, *Angew. Chem., Int. Ed.*, 2008, **47**, 2930–2946.
36. C.-M. Park, J.-H. Kim, H. Kim and H.-J. Sohn, *Chem. Soc. Rev.*, 2010, **39**, 3115–3141.
37. Z. Wang, L. Zhou and X. W. Lou, *Adv. Mater.*, 2012, **24**, 1903–1911.
38. J. Yang, Y. Takeda, N. Imanishi, C. Capiglia, J. Y. Xie and O. Yamamoto, *Solid State Ionics*, 2002, **152–153**, 125–129.
39. J. Jiang, Y. Li, J. Liu, X. Huang, C. Yuan and X. W. Lou, *Adv. Mater.*, 2012, **24**, 5166–5180.
40. L. Ji, Z. Lin, M. Alcoutlabi and X. Zhang, *Energy Environ. Sci.*, 2011, **4**, 2682–2699.

41. W. Qi, J. G. Shapter, Q. Wu, T. Yin, G. Gao and D. Cui, *J. Mater. Chem. A*, 2017, **5**, 19521–19540.
42. N. Nitta, F. Wu, J. T. Lee and G. Yushin, *Mater. Today*, 2015, **18**, 252–264.
43. V. Etacheri, R. Marom, R. Elazari, G. Salitra and D. Aurbach, *Energy Environ. Sci.*, 2011, **4**, 3243–3262.
44. K. Xu, *Chem. Rev.*, 2004, **104**, 4303–4418.
45. D. Aurbach, E. Zinigrad, Y. Cohen and H. Teller, *Solid State Ionics*, 2002, **148**, 405–416.
46. H. Lee, M. Yanilmaz, O. Toprakci, K. Fu and X. Zhang, *Energy Environ. Sci.*, 2014, **7**, 3857–3886.
47. K. Kerman, A. Luntz, V. Viswanathan, Y.-M. Chiang and Z. Chen, *J. Electrochem. Soc.*, 2017, **164**, A1731–A1744.
48. D. E. Fenton, J. M. Parker and P. V. Wright, *Polymer*, 1973, **14**, 589.
49. L. Long, S. Wang, M. Xiao and Y. Meng, *J. Mater. Chem. A*, 2016, **4**, 10038–10069.
50. R. C. Agrawal and G. P. Pandey, *J. Phys. D: Appl. Phys.*, 2008, **41**, 223001–223018.
51. W. H. Meyer, *Adv. Mater.*, 1998, **10**, 439–448.
52. P. G. Bruce, *Solid State Electrochemistry*, Cambridge University Press, 1997.
53. Z. Gadjourova, Y. G. Andreev, D. P. Tunstall and P. G. Bruce, *Nature*, 2001, **412**, 520.
54. S. Choudhary and R. J. Sengwa, *Ionics*, 2012, **18**, 379–384.
55. C. V. Subba Reddy, G. P. Wu, C. X. Zhao, W. Jin, Q. Y. Zhu, W. Chen and S.-i. Mho, *Curr. Appl. Phys.*, 2007, **7**, 655–661.
56. S. Klongkan and J. Pumchusak, *Electrochim. Acta*, 2015, **161**, 171–176.
57. S. Lascaud, M. Perrier, A. Vallee, S. Besner, J. Prud'homme and M. Armand, *Macromolecules*, 1994, **27**, 7469–7477.
58. S. Murakami, K. Ueda, T. Kitade, Y. Ikeda and S. Kohjiya, *Solid State Ionics*, 2002, **154–155**, 399–406.
59. Y. Kang, J. Lee, J.-i. Lee and C. Lee, *J. Power Sources*, 2007, **165**, 92–96.
60. J. Shim, K. Y. Bae, H. J. Kim, J. H. Lee, D.-G. Kim, W. Y. Yoon and J.-C. Lee, *ChemSusChem*, 2015, **8**, 4133–4138.
61. D.-G. Kim, J. Shim, J. H. Lee, S.-J. Kwon, J.-H. Baik and J.-C. Lee, *Polymer*, 2013, **54**, 5812–5820.
62. A. Ghosh and P. Kofinas, *J. Electrochem. Soc.*, 2008, **155**, A428–A431.
63. W. A. Henderson and S. Passerini, *Electrochem. Commun.*, 2003, **5**, 575–578.
64. A. R. Lee, Y. D. Kim, S. K. Lee and N. J. Jo, *J. Nanosci. Nanotechnol.*, 2013, **13**, 7208–7211.
65. D. J. Bannister, G. R. Davies, I. M. Ward and J. E. McIntyre, *Polymer*, 1984, **25**, 1291–1296.
66. W.-W. Cui and D.-Y. Tang, *J. Appl. Polym. Sci.*, 2012, **126**, 510–518.
67. D. R. Sadoway, *J. Power Sources*, 2004, **129**, 1–3.

68. F. Croce, L. Persi, B. Scrosati, F. Serraino-Fiory, E. Plichta and M. A. Hendrickson, *Electrochim. Acta*, 2001, **46**, 2457–2461.
69. J. M. Tarascon, A. S. Gozdz, C. Schmutz, F. Shokoohi and P. C. Warren, *Solid State Ionics*, 1996, **86–88**, 49–54.
70. A. Manuel Stephan and K. S. Nahm, *Polymer*, 2006, **47**, 5952–5964.
71. S. Kudaibergenov, W. Jaeger and A. Laschewsky, in *Supramolecular Polymers Polymeric Betains Oligomers*, Springer, 2006, pp. 157–224.
72. J. Shim, D.-G. Kim, H. J. Kim, J. H. Lee, J.-H. Baik and J.-C. Lee, *J. Mater. Chem. A*, 2014, **2**, 13873–13883.
73. S. P. Kammampata and V. Thangadurai, *Ionics*, 2018, **24**, 639–660.
74. V. Thangadurai and W. Weppner, *Ionics*, 2002, **8**, 281–292.
75. A. D. Robertson, A. R. West and A. G. Ritchie, *Solid State Ionics*, 1997, **104**, 1–11.
76. A. Rabenau, *Solid State Ionics*, 1982, **6**, 277–293.
77. T. Lapp, S. Skaarup and A. Hooper, *Solid State Ionics*, 1983, **11**, 97–103.
78. W. Weppner, P. Hartwig and A. Rabenau, *J. Power Sources*, 1981, **6**, 251–259.
79. A. West, *Ber. Bunsenges. Phys. Chem.*, 1989, **93**, 1235–1241.
80. A. Garcia, G. Torres-Trevino and A. R. West, *Solid State Ionics*, 1990, **40**, 13–17.
81. P. Quintana, F. Velasco and A. West, *Solid State Ionics*, 1989, **34**, 149–155.
82. R. A. Huggins, *Electrochim. Acta*, 1977, **22**, 773–781.
83. A. Khorassani, G. Izquierdo and A. R. West, *Mater. Res. Bull.*, 1981, **16**, 1561–1567.
84. W. Weppner and R. Huggins, *Solid State Ionics*, 1980, **1**, 3–14.
85. P. G. Bruce and A. R. West, *J. Solid State Chem.*, 1982, **44**, 354–365.
86. R. Kanno and M. Murayama, *J. Electrochem. Soc.*, 2001, **1148**, A742–A746.
87. N. Kamaya, K. Homma, Y. Yamakawa, M. Hirayama, R. Kanno, M. Yonemura, T. Kamiyama, Y. Kato, S. Hama and K. Kawamoto, *Nat. Mater.*, 2011, **10**, 682–686.
88. M. Manickam, K. Minato and M. Takata, *J. Electrochem. Soc.*, 2003, **150**, A1085–A1089.
89. H. Xie, J. B. Goodenough and Y. Li, *J. Power Sources*, 2011, **196**, 7760–7762.
90. H. Kawai and J. Kuwano, *J. Electrochem. Soc.*, 1994, **141**, L78–L79.
91. S. Stramare, V. Thangadurai and W. Weppner, *Chem. Mater.*, 2003, **15**, 3974–3990.
92. P. Hartwig, A. Rabenau and W. Weppner, *J. Less-Common Met.*, 1981, **78**, 227–233.
93. Y. Zhao and L. L. Daemen, *J. Am. Chem. Soc.*, 2012, **134**, 15042–15047.
94. R. Murugan, V. Thangadurai and W. Weppner, *Angew. Chem., Int. Ed.*, 2007, **46**, 7778–7781.
95. D. Mazza, *Mater. Lett.*, 1988, **7**, 205–207.
96. V. Thangadurai, H. Kaack and W. J. Weppner, *J. Am. Ceram. Soc.*, 2003, **86**, 437–440.

97. H. Nemori, Y. Matsuda, M. Matsui, O. Yamamoto, Y. Takeda and N. Imanishi, *Solid State Ionics*, 2014, **266**, 9–12.
98. V. Thangadurai and W. Weppner, *J. Solid State Chem.*, 2006, **179**, 974–984.
99. J. Awaka, N. Kijima, H. Hayakawa and J. Akimoto, *J. Solid State Chem.*, 2009, **182**, 2046–2052.
100. C. A. Geiger, E. Alekseev, B. Lazic, M. Fisch, T. Armbruster, R. Langner, M. Fechtelkord, N. Kim, T. Pettke and W. Weppner, *Inorg. Chem.*, 2011, **50**, 1089–1097.
101. T. Zaiß, M. Ortner, R. Murugan and W. Weppner, *Ionics*, 2010, **16**, 855–858.
102. S. Ohta, T. Kobayashi, J. Seki and T. Asaoka, *J. Power Sources*, 2012, **202**, 332–335.
103. D. I. Pinzaru, Optimization of Li-ion Conductivity of Garnet-type $Li_5La_3Nb_2O_{12}$ by Nb-site Substitution Approach, MSc thesis, University of Calgary, 2015.
104. J. C. Bachman, S. Muy, A. Grimaud, H.-H. Chang, N. Pour, S. F. Lux, O. Paschos, F. Maglia, S. Lupart, P. Lamp, L. Giordano and Y. Shao-Horn, *Chem. Rev.*, 2016, **116**, 140–162.
105. V. Thangadurai, S. Narayanan and D. Pinzaru, *Chem. Soc. Rev.*, 2014, **43**, 4714–4727.
106. Y. Kato, S. Hori, T. Saito, K. Suzuki, M. Hirayama, A. Mitsui, M. Yonemura, H. Iba and R. Kanno, *Nat. Energy*, 2016, **1**, 16030.
107. W. He, Z. Cui, X. Liu, Y. Cui, J. Chai, X. Zhou, Z. Liu and G. Cui, *Electrochim. Acta*, 2017, **225**, 151–159.
108. J. F. Vélez, M. Aparicio and J. Mosa, *Electrochim. Acta*, 2016, **213**, 831–841.
109. B. Chen, Z. Huang, X. Chen, Y. Zhao, Q. Xu, P. Long, S. Chen and X. Xu, *Electrochim. Acta*, 2016, **210**, 905–914.
110. S. Yubuchi, Y. Ito, T. Matsuyama, A. Hayashi and M. Tatsumisago, *Solid State Ionics*, 2016, **285**, 79–82.
111. Y. Zhao, Z. Huang, S. Chen, B. Chen, J. Yang, Q. Zhang, F. Ding, Y. Chen and X. Xu, *Solid State Ionics*, 2016, **295**, 65–71.
112. J. Xie, N. Imanishi, T. Zhang, A. Hirano, Y. Takeda, O. Yamamoto, X. B. Zhao and G. S. Cao, *J. Power Sources*, 2010, **195**, 8341–8346.
113. M. Kotobuki, Y. Suzuki, K. Kanamura, Y. Sato, K. Yamamoto and T. Yoshida, *J. Power Sources*, 2011, **196**, 9815–9819.
114. J. Kim, M. Eom, S. Noh and D. Shin, *J. Power Sources*, 2013, **244**, 476–481.
115. Y. Zhao, C. Wu, G. Peng, X. Chen, X. Yao, Y. Bai, F. Wu, S. Chen and X. Xu, *J. Power Sources*, 2016, **301**, 47–53.
116. J. Zhang, N. Zhao, M. Zhang, Y. Li, P. K. Chu, X. Guo, Z. Di, X. Wang and H. Li, *Nano Energy*, 2016, **28**, 447–454.
117. Q. Ma, X. Qi, B. Tong, Y. Zheng, W. Feng, J. Nie, Y.-S. Hu, H. Li, X. Huang, L. Chen and Z. Zhou, *ACS Appl. Mater. Interfaces*, 2016, **8**, 29705–29712.

118. B. De, A. Yadav, S. Khan and K. K. Kar, *ACS Appl. Mater. Interfaces*, 2017, **9**, 19870–19880.
119. W. Zhang, D. Schroder, T. Arlt, I. Manke, R. Koerver, R. Pinedo, D. A. Weber, J. Sann, W. G. Zeier and J. Janek, *J. Mater. Chem. A*, 2017, **5**, 9929–9936.
120. Y. Liu, D. Lin, Y. Jin, K. Liu, X. Tao, Q. Zhang, X. Zhang and Y. Cui, *Sci. Adv.*, 2017, **3**, eaao0713–eaao0722.
121. X. Zhang, T. Liu, S. Zhang, X. Huang, B. Xu, Y. Lin, B. Xu, L. Li, C.-W. Nan and Y. Shen, *J. Am. Chem. Soc.*, 2017, **139**, 13779–13785.
122. N. Angulakshmi, R. S. Kumar, M. A. Kulandainathan and A. M. Stephan, *J. Phys. Chem. C*, 2014, **118**, 24240–24247.
123. S.-D. Lee, K.-N. Jung, H. Kim, H.-S. Shin, S.-W. Song, M.-S. Park and J.-W. Lee, *ChemSusChem*, 2017, **10**, 2175–2181.
124. W. Luo, Y. Gong, Y. Zhu, Y. Li, Y. Yao, Y. Zhang, K. Fu, G. Pastel, C.-F. Lin, Y. Mo, E. D. Wachsman and L. Hu, *Adv. Mater.*, 2017, **29**, 1606042.
125. P. H. L. Notten, F. Roozeboom, R. A. H. Niessen and L. Baggetto, *Adv. Mater.*, 2007, **19**, 4564–4567.
126. J. H. Woo, J. E. Trevey, A. S. Cavanagh, Y. S. Choi, S. C. Kim, S. M. George, K. H. Oh and S.-H. Lee, *J. Electrochem. Soc.*, 2012, **159**, A1120–A1124.
127. J. E. Trevey, J. R. Gilsdorf, C. R. Stoldt, S.-H. Lee and P. Liu, *J. Electrochem. Soc.*, 2012, **159**, A1019–A1022.
128. C. Zhang, Y. Lin, Y. Zhu, Z. Zhang and J. Liu, *RSC Adv.*, 2017, **7**, 19231–19236.
129. K. K. Fu, Y. Gong, S. Xu, Y. Zhu, Y. Li, J. Dai, C. Wang, B. Liu, G. Pastel, H. Xie, Y. Yao, Y. Mo, E. Wachsman and L. Hu, *Chem. Mater.*, 2017, **29**, 8037–8041.
130. C. Yu, L. van Eijck, S. Ganapathy and M. Wagemaker, *Electrochim. Acta*, 2016, **215**, 93–99.
131. N. Machida, K. Kobayashi, Y. Nishikawa and T. Shigematsu, *Solid State Ionics*, 2004, **175**, 247–250.
132. X. Yao, N. Huang, F. Han, Q. Zhang, H. Wan, J. P. Mwizerwa, C. Wang and X. Xu, *Adv. Energy Mater.*, 2017, **7**, 1602923–1602931.
133. J. Hassoun and B. Scrosati, *Adv. Mater.*, 2010, **22**, 5198–5201.
134. X. Tao, Y. Liu, W. Liu, G. Zhou, J. Zhao, D. Lin, C. Zu, O. Sheng, W. Zhang, H.-W. Lee and Y. Cui, *Nano Lett.*, 2017, **17**, 2967–2972.
135. F. Han, J. Yue, X. Fan, T. Gao, C. Luo, Z. Ma, L. Suo and C. Wang, *Nano Lett.*, 2016, **16**, 4521–4527.
136. M. Nagao, A. Hayashi and M. Tatsumisago, *Electrochim. Acta*, 2011, **56**, 6055–6059.
137. M. Eom, S. Son, C. Park, S. Noh, W. T. Nichols, and D. Shin, *Electrochim. Acta*, 2017, **230**, 279–284.
138. M. Nagao, Y. Imade, H. Narisawa, T. Kobayashi, R. Watanabe, T. Yokoi, T. Tatsumi and R. Kanno, *J. Power Sources*, 2013, **222**, 237–242.
139. R.-c. Xu, X.-h. Xia, S.-h. Li, S.-z. Zhang, X.-l. Wang and J.-p. Tu, *J. Mater. Chem. A*, 2017, **5**, 6310–6317.

140. Y. Hao, S. Wang, F. Xu, Y. Liu, N. Feng, P. He and H. Zhou, *ACS Appl. Mater. Interfaces*, 2017, **9**, 33735–33739.
141. K. Fu, Y. Gong, G. T. Hitz, D. W. McOwen, Y. Li, S. Xu, Y. Wen, L. Zhang, C. Wang, G. Pastel, J. Dai, B. Liu, H. Xie, Y. Yao, E. D. Wachsman and L. Hu, *Energy Environ. Sci.*, 2017, **10**, 1568–1575.
142. J. Sun, N. Zhao, Y. Li, X. Guo, X. Feng, X. Liu, Z. Liu, G. Cui, H. Zheng, L. Gu and H. Li, *Sci. Rep.*, 2017, **7**, 41217.
143. V. G. Anju and S. Sampath, *Electrochim. Acta*, 2017, **252**, 119–126.
144. T. Liu, Q.-C. Liu, J.-J. Xu and X.-B. Zhang, *Small*, 2016, **12**, 3101–3105.
145. Y. Liu, B. Li, H. Kitaura, X. Zhang, M. Han, P. He and H. Zhou, *ACS Appl. Mater. Interfaces*, 2015, **7**, 17307–17310.
146. X. Zhu, T. Zhao, P. Tan, Z. Wei and M. Wu, *Nano Energy*, 2016, **26**, 565–576.
147. F.-C. Liu, Z. Shadike, X.-F. Wang, S.-Q. Shi, Y.-N. Zhou, G.-Y. Chen, X.-Q. Yang, L.-H. Weng, J.-T. Zhao and Z.-W. Fu, *Inorg. Chem.*, 2016, **55**, 6504–6510.
148. H. Chung and B. Kang, *Chem. Mater.*, 2017, **29**, 8611–8619.
149. Z. Zhang, Y. Zhao, S. Chen, D. Xie, X. Yao, P. Cui and X. Xu, *J. Mater. Chem. A*, 2017, **5**, 16984–16993.
150. J. Janek and W. G. Zeier, *Nat. Energy*, 2016, **1**, 16141.
151. C. Wang, Y. Gong, B. Liu, K. Fu, Y. Yao, E. Hitz, Y. Li, J. Dai, S. Xu and W. Luo, *Nano Lett.*, 2016, **17**, 565–571.
152. K. Fu, Y. Gong, B. Liu, Y. Zhu, S. Xu, Y. Yao, W. Luo, C. Wang, S. D. Lacey, J. Dai, Y. Chen, Y. Mo, E. Wachsman and L. Hu, *Sci. Adv.*, 2017, 3.
153. W. Luo, Y. Gong, Y. Zhu, K. K. Fu, J. Dai, S. D. Lacey, C. Wang, B. Liu, X. Han, Y. Mo, E. D. Wachsman and L. Hu, *J. Am. Chem. Soc.*, 2016, **138**, 12258–12262.
154. X. Han, Y. Gong, K. K. Fu, X. He, G. T. Hitz, J. Dai, A. Pearse, B. Liu, H. Wang and G. Rubloff, *Nat. Mater.*, 2017, **16**, 572–579.
155. W. Zhou, S. Wang, Y. Li, S. Xin, A. Manthiram and J. B. Goodenough, *J. Am. Chem. Soc.*, 2016, **138**, 9385–9388.
156. Y. Kobayashi, S. Seki, M. Tabuchi, H. Miyashiro, Y. Mita and T. Iwahori, *J. Electrochem. Soc.*, 2005, **152**, A1985–A1988.
157. X. Judez, H. Zhang, C. Li, G. G. Eshetu, J. A. González-Marcos, M. Armand and L. M. Rodriguez-Martinez, *J. Electrochem. Soc.*, 2018, **165**, A6008–A6016.
158. R. Kanno and M. Murayama, *J. Electrochem. Soc.*, 2001, **148**, A742–A746.
159. B. Kumar, J. Kumar, R. Leese, J. P. Fellner, S. J. Rodrigues and K. M. Abraham, *J. Electrochem. Soc.*, 2010, **157**, A50–A54.

CHAPTER 4

Redox Oxides for Thermochemical Energy Storage

ALFONSO J. CARRILLO[†a], JENNIFER L. M. RUPP[a,b] AND JUAN M. CORONADO*[c]

[a]Department of Materials Science and Engineering, Massachusetts Institute of Technology, 77 Massachusetts Ave Cambridge, MA 02139, USA; [b]Department of Electrical Engineering and Computer Science, Massachusetts Institute of Technology, 77 Massachusetts Ave Cambridge, MA 02139, USA; [c]Instituto de Catálisis y Petroleoquímica, CSIC, Marie Curie, 2, Cantoblanco, 28049 Madrid, Spain
*E-mail: jm.coronado@csic.es

4.1 Thermochemical Storage in a Dynamic Energy Landscape: Niches, Limitations and Challenges

Every day we take advantage of the chemical energy accumulated aeons ago in the molecules that constitute the fossil fuels we use for transportation and heating. The high calorific power of coal and hydrocarbons (*e.g.* the higher heating value (HHV) of gasoline is 48 MJ kg^{-1}) provides a ready available supply of energy, which can be used whenever is needed by means

[†]Current address: Instituto de Tecnología Química (Universitat Politècnica de València – CSIC), Avda. Los Naranjos s/n, 46022 Valencia, Spain.

Inorganic Materials Series No. 5
Energy Storage and Conversion Materials
Edited by Stephen Skinner

Published by the Royal Society of Chemistry, www.rsc.org

of mature technologies that transform heat into mechanical energy (combustion engines) or electricity (power block of thermal plants) using different thermodynamic cycles (such as Brayton, Rankine, *etc.*). Although this easy access to energy is a staple of our society and has contributed to the economic growth and welfare of developed countries, now it is widely recognised that a profound change of paradigm is needed in order to develop a fully sustainable model. In contrast with the existing technology, any reasonable alternative should not threaten the very existence of our way of life due to unforeseen cascade effects that the massive injection of CO_2 into the atmosphere is causing on the global climate. International summits such as that held in Paris on November 2015 on Climate Change are providing the political momentum for this necessary profound change of the whole of society.[1]

The past few decades have witnessed remarkable advances in the development of renewable energy technologies, with an unprecedented growth of commercial deployment of new plants in many countries during the past ten years. However, the current energy system still relies mostly in the utilisation of fossil fuels, which constitute more than 81% of the overall primary energy demand.[2] That elevated contribution is even higher if we consider the transportation sector because the main advances on the use of more sustainable sources correspond to the electricity generation, while the penetration of biofuels, despite being important in countries such as Brazil or USA, is globally modest.[3] Furthermore, the use of this class of alternative fuels has been strongly contested due to environmental and social reasons, prompting the current focus on the utilisation of residual feedstocks, the management of which is of socio-economic concern and its energy valorisation can contribute to the establishment of a circular economy.[4]

Within this panorama, there is a clear interest in the development of new routes for the direct conversion of energy from wind or sun into fuels. Furthermore, considering the inherent variability of power production of these renewable resources in short (day/night) and long cycles (seasons), storing energy is a key point for the success of these technologies as they cannot be switched on and off on-demand to satisfy the end-user necessities. Energy storage based on batteries can provide a buffer for these fluctuations; however, restrictions in material storage capacities and scarcity of the cathode materials may pose techno-economic challenges.[5–7] In this respect, an approach known as “power-to-gas“ has recently gained relevance, because it intends to take advantage of the overproduction of renewable electricity when consumers' demand is low for producing energy-rich chemicals such as H_2 and CH_4.[8] Usually this implies water splitting in an alkaline electrolyser, which is powered by the surplus electricity, and the hydrogen so-generated can be stored and eventually fed into a fuel cell to recover electricity when is needed, or used as a reagent for the catalytic conversion of CO_2. This last route has the added benefit of contributing to reduce the emissions of this greenhouse gas, and it usually takes place according to either the Sabatier process

$$4H_2 + CO_2 \rightleftharpoons CH_4 + 2H_2O \quad \Delta H = -165.0 \text{ kJ mol}^{-1} \tag{4.1}$$

or by the production of methanol through:

$$3H_2 + CO_2 \rightleftharpoons CH_3OH + H_2O \quad \Delta H = -41.2 \text{ kJ mol}^{-1} \tag{4.2}$$

Methane generated in the first process is sometimes referred to as synthetic natural gas (SNG) and it can be readily used as the fossil counterpart.[9,10] On the other hand, production of methanol can be more challenging as it requires higher pressures but, being a liquid, provides higher volumetric energy density.[11] Recently, progress has been made for achieving methanol production at atmospheric pressure using Ni–Ga catalysts.[12] Besides, methanol can be used as an intermediate for the synthesis of other relevant chemicals (*e.g.* dimethylether), including fuels. This is the case of the production of gasoline through Mobil's process using zeolite-based catalysts.[13]

Using these processes for storing energy in chemical bonds allows one to adjust renewable production and consumer demand, and in that way can contribute to the stability of the electrical grid.[8] This strategy has been frequently proposed to complement the deployment of huge wind farms in the North Sea basin, but the use of photovoltaics has been also considered.[14] In the past few years, pilot plants of different characteristics have been deployed to demonstrate the viability of implementing this scheme in the existing energy system.[15] Among them ETOGAS, which was built in Werlte (Low Saxony, Germany) with the participation of the automobile company Audi AG, was established to demonstrate commercial-scale power generation of 6.3 MW.[16]

The energy yield that can be achieved for methanation is estimated to be around 64% without considering gas compression. However, a more direct transformation may increment the efficiency of energy transferred to the carrier molecule.[9] With this aim, co-electrolysis at high temperature (800 °C) coupled to a catalytic reactor for methanation has been also proposed as an alternative but, despite the promising production of methane (up to 11.8% of yield) these systems are still on the lab scale.[17] On the other hand, low temperature aqueous-based CO_2 electroreduction, despite having achieved significant advances recently, still has to enhance the selectivity to the products of interest such as methane or other light hydrocarbons ($C_{\geq 2}$).[18]

Taking into account the enormous amount of energy delivered to the Earth by sunlight (about 4.3×10^{20} J in 1 h, equivalent to the world energy consumption in one year), the development of routes for its storage is particularly attractive. A scheme showing the main value chains for solar fuels' production is displayed in Figure 4.1. This graph provides a global picture, which also highlights the requirement of a previous stage for the capture of CO_2 and H_2O as ideal feedstocks, as well as the need of further catalytic upgrading of the some of the initially produced solar fuels, such as syngas (*i.e.* $CO + H_2$).

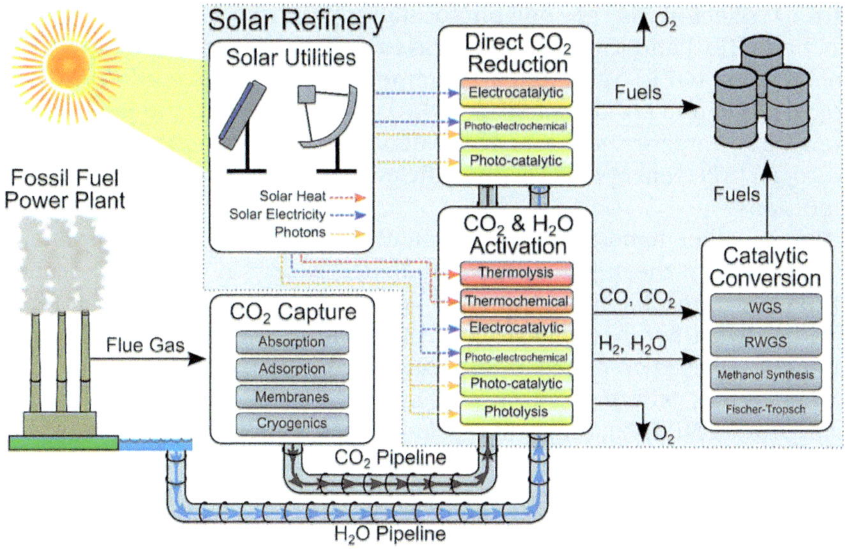

Figure 4.1 Schematic of the main routes for the production of solar fuels, showing all the value chain, from the CO_2 capture, solar-activated transformation and subsequent catalytic upgrading. WGS stands for water gas shift reaction ($CO+H_2O \rightleftharpoons CO_2 + H_2$), while RWGS refers to reverse water gas shift. Reproduced from ref. 19 with permission from the Royal Society of Chemistry.

Direct approaches for solar energy conversion to molecular energy carriers at mild conditions are based on either photoelectrocatalytic or photocatalysis processes.[19] Considering that these technologies operate at ambient conditions, if the required efficiency is achieved, they hold the promise of inexhaustible, carbon neutral and widely distributed fuel production. Accordingly, a huge scientific effort has been undertaken in the past decade, with initiatives such as the Joint Center for Artificial Photosynthesis, a DOE Energy Innovation Hub among others. Although remarkable progress related to the synthesis of photoactive nanomaterials and organic–inorganic hybrids have been reported, the final goal of efficient conversion of sunlight into chemical energy still does not appear on the near horizon.[20] The main limitations of photocatalysis are related to incomplete overlapping of the solar spectrum with the materials' optical absorption, and the high electron-hole recombination rates, which lead to unproductive photon interactions. In this respect, although the threshold for commercial application is estimated to be around 15% of conversion of photon energy into chemical energy, the best results obtained up to now only reached a solar to hydrogen conversion efficiency of 0.2%.[21] Those results were achieved under visible light illumination of the semiconductor $Ga_{1-x}Zn_xN_{1-x}O_x$ coupled with a mixed Rh–Cr oxide co-catalyst. Higher efficiencies have been obtained recently using a so-called Z-scheme, where two semiconductors operate in a coupling by a current collector, whereby one photocatalyst performs a hydrogen evolution reaction (HER) and the other converts to oxygen through an oxygen evolution reaction (OER). With that arrangement, and using La- and Rh-codoped $SrTiO_3$ (HER) and Mo-doped $BiVO_4$ (OER) powders with Ru and RuO_x, respectively, as co-catalysts, and everything deposited on a carbon film, a solar-to-hydrogen (STH) energy conversion efficiency of 1% was obtained at ambient conditions.[22]

On the other hand, photoelectrochemical processes of fuel production are limited by the high overpotential of the oxygen reduction reaction (ORR), which results in slow kinetics. However, promising results have been obtained with autonomous devices that combine smartly a photovoltaic component based on doped-amorphous silicon to directly provide an *in situ* generated bias, with the hydrogen- and oxygen-evolving catalysts, which are made from a NiMoZn alloy and a cobalt borate catalyst, respectively. These devices have been denominated "artificial leaf" and they can drive a water-splitting reaction under simulated solar illumination at efficiencies of up to 2.5% in the simplest wireless configuration.[23]

The heating power of sunlight can be also exploited for relevant energy applications. Low temperature solar heat is commercially utilised by means of dedicated panels worldwide for providing hot water for domestic and also for industrial uses, while large concentrating solar power (CSP) plants operating at high temperature (400–600 °C) for electricity generation have been actively deployed in the past decade in countries such as USA, Spain, Israel and China.[24] These facilities accounted for a total of 4.8 GW by 2016, although the share of this technology to global electricity production was less than 0.4%, which is less than one third of the contribution of photovoltaics (PV).[25] One of the reasons for these differences in the degree of development of these two solar technologies is that CSP is more complex and requires more extensive utilisation of land than a PV modular arrangement. The most common configuration of a thermosolar plant is based on a parabolic trough receiver (PTR) and it consists of these basic components: a field of solar collectors, a circuit with the heat transfer fluid, which may also include thermal energy storage, and a power block that converts high-temperature solar heat into electricity. Nevertheless, concentrating technologies provide clear advantages, because they facilitate an optimal exploitation of the full spectrum of solar light from the ultraviolet to the infrared. In addition, the simple and cost-effective heat accumulation in molten salts provides a competitive edge to CSP in comparison to PV, which relies on costly batteries for dealing with the mismatch between production and demand of electricity.[20] Currently, commercial plants like SOLANA in Gila Bend, Arizona (USA) can store up to 1000 MWh of energy, equivalent to 6 h of operation. In summary, the potential for growing CSP is considerable, especially in the belt of the world's arid lands at medium latitudes, as they present high annual irradiance (*ca.* 6 kW h m^{-2} per day). Furthermore, the expected progressive reduction of commissioning costs could aid motivation for the development of future CSP plants.[26]

Beyond electricity production, solar heat provides an ideal niche for the development of a myriad of emerging technologies for thermochemically-driven energy storage. Accordingly, future CSP plants are expected to achieve higher operation temperatures for improved efficiency and this fact can prompt the commercial development of thermochemical heat storage (TCS) because the solar salts currently used are not stable under those extreme conditions. Although these advanced TCS technologies are still under research, the intrinsic advantages in terms of high energy density and reduced energy losses have gathered interest in these systems.[27] In particular, as described in detail in Section 4.4, metal oxides can play an important role for this application.

Although thermochemical cycles for solar fuel production have been investigated since the late 1970s, interest in them has escalated in recent years, and as a consequence very encouraging results have recently been achieved with simple metal oxides.[28,29] In contrast with photonic processes, which are possible at ambient conditions, thermochemical cycles operate at very high radiation fluxes that result in extreme temperature windows (>1400 °C) and gradients. Another peculiarity is that, as suggested by the name, thermochemical cycles operate in at least two stages that have to be run alternately in order to achieve the expected products, hydrogen or syngas ($CO + H_2$) in one step and oxygen in the other. Although this way of functioning adds complexity to the system, it can be advantageous because it facilitates separation of O_2 and H_2 mixtures. More than 250 different cycles have been evaluated theoretically considering thermodynamic properties, but only a small percentage have been experimentally tested.[30] Some of them imply three or more stages (*e.g.* sulphur–iodine cycle), but for operational simplicity and energy integration two-step cycles are preferred. Furthermore, although some of the most efficient thermochemical cycles rely on the use of gas phase molecules, many of them are also corrosive such as HI or SO_3. For those reasons, the use of a solids' reagent provides clear operational advantages because they can be handled more easily. In fact, some of the most promising cycles are based on metal oxides that can swing from and be oxidised to a reduced state through the cycles. The characteristic of these solids will be discussed in detail in Section 4.3.

The feasibility of the thermochemical route for solar hydrogen production has been evaluated at pilot scale (up to 1 MW) at the Plataforma Solar de Almería in the framework of research projects such as the HYDROSOL (I, II and 3D with funding from the European Commission) developed in the past 10 years.[31] Also, in the field of thermochemical valorisation of CO_2 and H_2O mixtures, several demonstration projects such as SOLAR-JET and SUN-to-LIQUID have been financed by FP7 and H2020 programs of European Commissions for the production of syngas and eventually synthetic fuels such as kerosene by subsequent Fischer–Tropsch synthesis.[32] Figure 4.2 shows a picture of the solar field and the tower built in the IMDEA Energy facilities at Móstoles (Madrid, Spain) with a nominal power of 250 kW and fully dedicated to solar fuel production within the framework of the project SUN-to-LIQUID.

Figure 4.2 Image on the solar field and the tower with a nominal power of 250 kW at IMDEA Energy facilities in Móstoles (Madrid, Spain), which is fully dedicated to solar fuel production within the framework of the project SUN-to-LIQUID (see ref. 32 for further details). Reproduced with permission. Courtesy of IMDEA Energy.

These ambitious initiatives for scaling-up the production have overcome the other photonic routes for the direct solar transformations, which still remain in the lab scale.[33] The remarkable impulse given to thermochemical fuel development relies on several factors, including accumulated experience in solar engineering and the use of more efficient redox oxides, which has led to a notable improvement in energy conversion efficiency experienced in the period of just a few years (*ca.* 300% improvement in 3 years).[29] Furthermore, higher yields (in excess of 15%) are theoretically possible as estimated for the global thermodynamic analysis of the whole plant.[34]

However, before commercial production of thermochemical solar fuels becomes a reality, there are still many challenges to overcome, many of them associated with the complexity of operating at extreme temperatures. This has prompted the exploration of hybrid routes that include the utilisation of an external reducer, generally CH_4 but also H_2, to decrease the maximum operation temperature and increase the yield for CO_2 and/or H_2O splitting. Obviously, these benefits occur at the expense of feeding external fuels (*e.g.* CH_4) that can reduce the sustainability of the process. However, coupling with renewable resources such as biogas or wind-generated H_2, can compensate, at least partially, the consumption of external gas reagents.[35] Furthermore, reduced temperature for the metal oxide operation may contribute to a more straightforward implementation of combined technologies, as they can benefit not only from the synergies but also of the access to reliable and more mature engineering. The materials used for these hybridised thermochemical transformations are reviewed in Section 4.4.

In summary, the incorporation of renewable energy in fuel production is a very active and broad field that is bound to provide many exciting developments in the next few years. Different competing transformation routes are expected to play a significant role in the progressive construction of a new paradigm for fuel production, not based in extraction but in sun-powered synthesis. Thermochemical routes are currently in a privileged position for taking the lead in this race for producing sustainable fuels, boosted by the development of active redox metal oxides. However, the future relevance of thermochemical routes will depend on additional development of even more efficient materials and reactors, as well as the improvement in the engineering of systems for an optimal energy utilisation in the whole plant.

4.2 Thermodynamics and Kinetics Aspects of Redox Oxides for Thermochemical Applications

4.2.1 Reactivity of Redox Oxides at High Temperatures

Redox-based metal oxides are the most suitable oxygen carrier materials for thermochemical conversion processes forming one of the pillars of the recent boost in efficiency of the conversion of solar to chemical energy. Nevertheless, the utilisation of solids entails challenges for the reactivity in oxygen uptake and release steps, which depends not only on surface reactions, but also on the intrinsic ionic oxygen diffusion of the bulk material.

Examining carefully the metal oxides used for thermochemical fuel conversion, one can conclude that formation of oxygen vacancies on its surface is the initial step of the reduction process. In a second step, initiated for example through temperature and oxygen partial pressure change, re-oxidation to recover the initial oxygen capacity of the metal oxide forms the backward reaction. In a simplified manner, this reaction can be considered to take place according to:

$$O_O^x \leftrightarrow V_O^{\cdot\cdot} + 2e' + \tfrac{1}{2}O_2(g) \tag{4.3}$$

where $V_O^{\cdot\cdot}$ represents the oxygen vacancy and O_O^x an oxide anion. This equation has been formulated using the Kröger–Vink notation,[36] which allows one to specify the formal charges and positions of the point defects on the lattice.

According to the chemical equation (eqn (4.3)), the releasing of oxygen implies the delocalisation of two electrons in the solid, which are frequently up-taken by the redox cations, M^{n+}, of the structure, thus changing their oxidation state to $M^{(n-1)+}$. Therefore, the mixed ionic-electronic conductivity and, in particular, the type of conductor, namely n or p-type, are important characteristics of the solid because they have an influence on the capacity of its structure to promote the mobility of charge carriers and to accommodate these ionic and electronic defects.[37]

A correlation between the free energy for the formation of oxygen vacancies at the metal oxide surface and enthalpy of formation of the bulk oxide have been found to exist.[38] This implies that the generation of oxygen vacancies is energetically more costly in the oxides of which enthalpy of formation is higher, as intuitively could be expected. For thermochemical processes, transformation of the oxygen carrier can, in some cases, imply a phase transformation between well-defined stoichiometric solid compositions (*e.g.* Fe_2O_3 to Fe_3O_4), or lead to a partial deoxygenation of a non-stoichiometric material (*e.g.* CeO_2 to $CeO_{2-\delta}$).

Extensive reduction of the solid also entails vacancies' diffusion through the bulk, which influences the rate of these redox processes. Thermodynamically, the migration of the oxygen vacancy into the bulk is favoured if the free energy of the formation of a vacancy at subsurface position is lower than in the surface.[37] This is the case of many oxides relevant to thermochemical storage such as CeO_2 or $La_xSr_{1-x}MnO_3$. However, the existence of energy barriers to the diffusion of oxygen vacancies entails a kinetic control of the process. As can be observed for some redox oxide shown in Figure 4.3, when a surface vacancy moves into a subsurface site, overcoming a substantial energy barrier is usually required to reach the transition state. The magnitude of the activation energy for oxygen vacancies' migrations appears to be related to the charge density on the metal atoms in

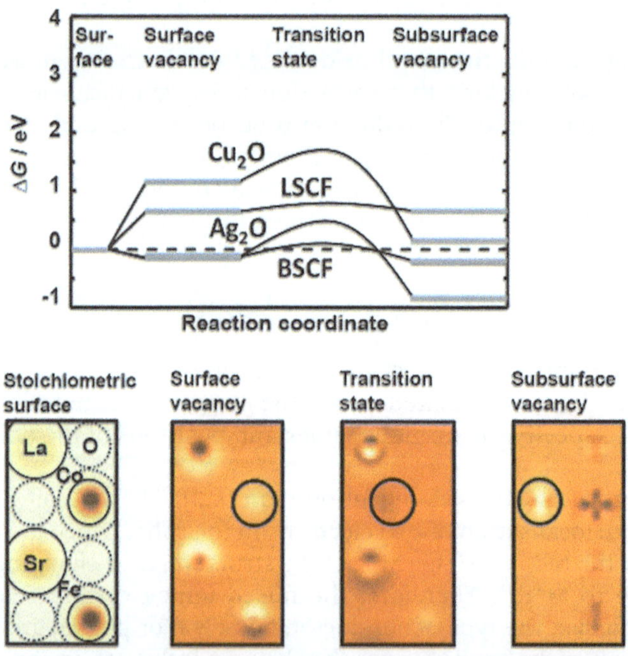

Figure 4.3 Variation of Gibbs free energy diagrams for the formation and diffusion of oxygen vacancies in different oxides (energies below the dashed line correspond to spontaneous processes). The lower panel shows variations in the charge density for the different stages of the process in the case of the LSCF perovskite. Darker colours represent higher electronic density. Adapted from ref. 38 with permission from John Wiley and Sons, © 2014 The Authors. Published by WILEY-VCH Verlag GmbH & Co. KGaA, Weinheim.

the vicinity of the vacancy. This would be determined by the nature of the cations and geometric constraints imposed by the crystalline structure.[38]

Surface reactivity of some of these redox materials have been studied theoretically using density functional theory (DFT). These quantum mechanical calculations show that, in the case of perovskites (ABO_3) and using a molecular orbital (MO) perspective, coordinately unsaturated cations of the surface, BO_5, interact through an e_g orbital with the π^* orbital of O_2 in a tilted configuration.[39] Figure 4.4 displays a scheme of the orbital implied in the interaction of an O_2 molecule with a coordinatively unsaturated cation

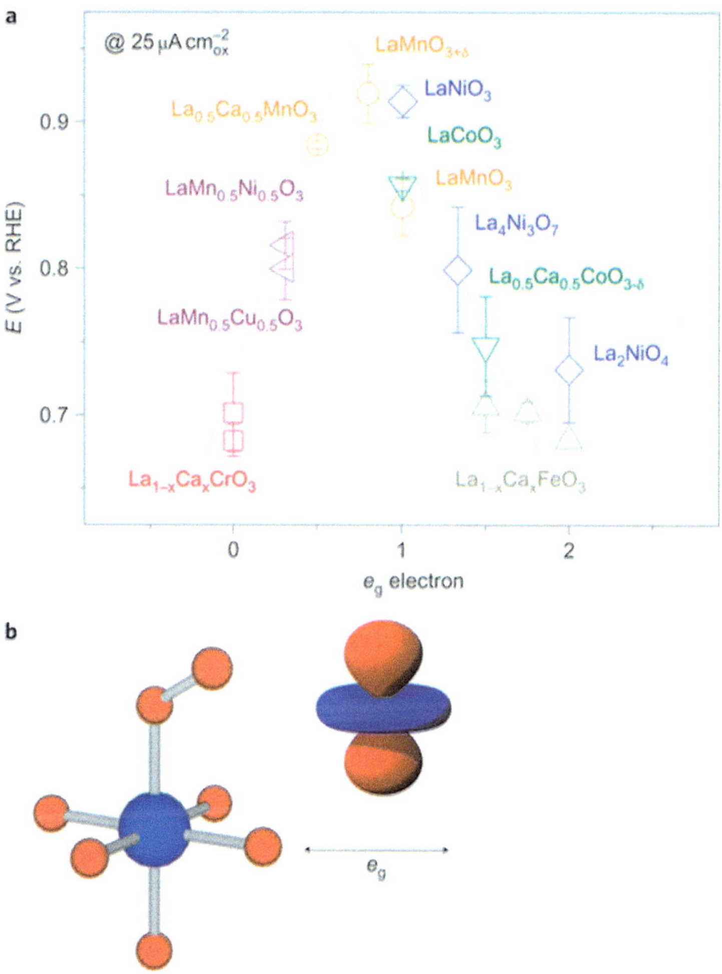

Figure 4.4 (a) Variation of the activity for the oxygen reduction reaction (ORR) with the corresponding error bars, as a function of the occupancy of the e_g of the coordinatively unsaturated metal and (b) schematic representation of the orbital, e_g, of the coordinatively unsaturated cation of the oxide surface, BO_5, which interacts with the π^* orbital of the O_2 molecule Adapted from ref. 40 with permission from Springer Nature, Copyright 2011.

on an oxide surface. Then the electron filling of the e_g orbitals of the B cation affects the strength of the binding with molecular oxygen. Thus, for low occupancy, e_g^0, the bond is strong, while for higher energy density, e_g^2, the interaction becomes weak. As a consequence, a volcano-like dependence of the activity for oxygen reduction (see Figure 4.4) in solution as a function of the e_g, occupancy is observed for these materials, as in the case of many catalytic systems.[40] In addition, DFT calculations of the band structure of these mixed oxides have shown that increasing the difference between unoccupied 3-d orbital and occupied 2p oxygen orbitals, which is equivalent to an increment in the covalent character of the M–O bonding, results in more rapid kinetics for the surface oxygen exchange reaction (*i.e.* the backward reaction described by eqn (4.3)).[32] This has been explained by considering the higher degree of covalence results in lower energy for oxygen vacancies' formation and lower energy barriers in the limiting steps of the reaction.

The atomic-scale processes described above determine the kinetics of the redox reactions, which are defined by both the surface exchange coefficient, k_{chem}, and the chemical diffusion coefficient, D_{chem}, in the solid.[37] These are important parameters because the efficiency of the fuel production is determined by the frequency of the cycles, which depends on the rate of direct and reverse reactions. In this respect, it is worth noting that although oxidation reactions are usually thermodynamically favoured at low temperature, a trade-off between that fact and the increasingly lower rate with decreasing temperature has to be established in order to select the optimal conditions for fuel production. However, for the kinetics of solids at high temperature it is frequently found that bulk diffusion of oxide anions is quick and the surface reaction is the rate-limiting step. This is usually the case in a highly porous solid, where the characteristic size of the particles is significantly smaller than the ratio between D_{chem} and k_{chem}.[41] Experimental and theoretical results have shown that the formation of oxygen vacancies can be promoted in the metal–oxide interphase and this phenomenon has been exploited in many catalytic processes.[42] In contrast, the use of specific catalysts has been still seldom used for thermochemical processes, although some studies have proven that deposition of a noble metal catalyst such as IrO_x or Rh can lead to substantial enhancements of efficiency in solar fuel production.[43,44] However, the use of such metals has a toll on material costs and a trade-off needs to be established considering the expected improvement on the energy conversion yield and the possible economic penalty. Furthermore, considering the harsh conditions of these processes, which are far away from those used in conventional catalysis, further studies are required in order to take full advantage of the catalysts' ability to speed-up reactions. In this respect, a recent work shows the potential of inexpensive CrO_x to promote CO_2 and H_2O splitting.[45]

The need for an adequate interface for favouring the reactivity of solids constitutes an important constraint when working with solid oxide materials at high temperatures. For example, sintering of particles in which the driving force is the reduction of surface energy occurs at the typical operation conditions of these processes and can cause a progressive decay of the activity. In fact, this is a usual concern for the long-term stability of materials in thermochemical cycles. In order to cope with these phenomena, open structures such as monoliths or macroporous foams have been developed for these applications because they allow a good contact with gases at scales that are not appreciably affected by thermal agglomeration. Nevertheless, since the energy storage density depends on the amount of redox oxide, preparation of these shaped materials requires a limited use of additives such as binders in the final processed solid oxide. On the other hand, it is worth mentioning that these ceramic materials must be able to withstand extreme temperature gradients that appear in solar reactors due to the inherent asymmetric heating of most of these systems.

4.2.2 Thermodynamics of Redox Cycles in Metal Oxides

Determining the thermodynamic properties of the materials used as oxygen carriers provides a crucial guide for finding suitable candidates that are able to perform CO_2 or H_2O splitting. In this respect it is important to note that although materials with high capacity for oxygen release are desirable, that fact does not guarantee that they can be re-oxidised in the presence of CO_2 and/or H_2O vapour. This point stresses the relevance of performing a global thermodynamic analysis of properties of a metal oxide to find the best candidates for thermochemical cycling and defining operation conditions (*i.e.* temperature cycle swings). As it is well known for a reaction to occur spontaneously it must be exergonic, which implies that a change in the Gibbs free energy should be negative. So, in the particular case of CO_2 splitting in thermochemical cycles, the overall free energy of the process, ΔG, considering isothermal conditions, can be obtained using the following equation:

$$\Delta G_{\mathrm{ox,CO_2}} = -\Delta G_{\mathrm{MO}} + \Delta G^{r}_{\mathrm{CO_2}} \tag{4.4}$$

where ΔG_{MO} represents the free energy of the oxygen carrier reduction and $\Delta G^{\mathrm{r}}_{\mathrm{CO_2}}$ for CO_2 splitting under the same conditions. The same consideration can be established for water splitting and an analogous equation can be obtained. On the other hand, assuming that solids activity is one and that oxygen follows an ideal gas behaviour, the Gibbs free energy for the reduction of an oxide for a fixed degree of non-stoichiometry, δ, follows the relation:

$$\Delta G_{\mathrm{MO}} = \tfrac{1}{2} RT\,ln(p_{\mathrm{O_2}}) \tag{4.5}$$

where $p_{\mathrm{O_2}}$ is the oxygen partial pressure, R the gas constant and T the temperature. This equation indicates that, as expected, the reduction of a metal oxide is favourable at high temperatures and low oxygen partial pressure.[37,38] Similarly, for the case of CO_2 splitting it can be calculated according to:

$$\Delta\bar{g}^0_{O_2}(\delta,T)=\Delta G^0_r+2RT\,ln\left(\frac{p_{CO_2}}{p_{CO}}\right) \quad (4.6)$$

where ΔG^0_r is the standard Gibbs free energy change of CO_2 dissociation, and p_{O_2} and p_{CO} are the partial pressures of CO_2 and CO, respectively.[46] According to this formula, oxidation of the solid by CO_2 is promoted at low temperatures and high CO_2 partial pressure. Again, water splitting follows an equivalent equation and, therefore, similar trends are expected for the oxidation of the redox material.

In general terms, the thermodynamic properties of many oxides, especially those with the simplest compositions, have been experimentally determined, in many cases by means of careful thermogravimetric studies. However, inconsistencies in the values reported in the literature are frequent, and these uncertainties can affect the reliability in the selection of suitable redox oxides. Due to this fact, in order to obtain reliable parameters in a wider range of temperatures and pressures, as well as for estimating the properties of materials of complex composition, computational analysis can be of high interest. In this respect, an interesting approach is the Computer Coupling of Phase Diagrams and Thermochemistry, abbreviated as CALPHAD.[47] This methodology takes advantage of all the data available in the literature on thermodynamics and defect chemistry to perform the best fit to parameters such as heat capacity, enthalpy and Gibbs free energy. Since different data sources are used in CALPHAD possible errors can be minimised, while extrapolation allows one to explore a wider range of conditions that can be relevant for thermochemical tests. Finally, this approach is able to provide information on multicomponent systems using a combination of properties such as molar Gibbs energies of the pure materials to extrapolate the properties of phases of more complex composition. That information can be very useful for estimating the behaviour of oxides with different degrees of cation doping. Nevertheless, full exploitation of the potential of this approach depends on the quality and availability of data on the defect chemistry of relevant oxides. An example of the application of CALPHAD for thermodynamic prediction for (La, Sr)MnO_3 perovskites is further detailed elsewhere.[47,48]

Another computational approach is the use of quantum DFT calculations for screening materials suitable for thermochemical cycles. An impressive example of the huge potential of theoretical methods is the high-throughput computational study of more than 5000 different compositions of cubic and distorted perovskites, which were analysed as possible candidates for water splitting in thermochemical cycles.[49]

4.2.3 Energy Efficiency of Thermochemical Cycles

A paramount aspect of thermochemical processes is solar-to-fuel yield, $\eta_{solar\text{-}to\text{-}fuel}$, of the conversion. This can be calculated considering the energy stored in the produced fuel divided by the energy input:

$$\eta_{\text{solar-to-fuel}} = \frac{HHV_{\text{Fuel}} \cdot n_{\text{Fuel}}}{Q_{\text{solar}} + E} \tag{4.7}$$

where HHV_{Fuel} refers to the higher heating value of the obtained fuel, either H_2 or syngas, n_{Fuel} is the amount produced, Q_{solar} the received solar heat, and E corresponds to the parasitic losses of the system due to the energy consumption in the auxiliary components of the system, such as pumping work.[47] In this equation, the term Q_{solar} should comprise the energy required to vaporise and heat the reagents to the operation temperature, together with the enthalpy change of the oxygen carriers required to activate the process in each cycle. On the other hand, the term E emphasises the relevance of optimising the energy utilisation in the whole plant for maximising the efficiency of fuel generation.

From another angle, it is also important to consider the maximum efficiency of the solar receiver. Taking into account the balance between absorption and radiation losses this can be determined by the following equation:[24,50]

$$\eta_{\text{rec}} = \alpha - \frac{\varepsilon\sigma\left(T_{\text{H}}^4 - T_{\text{amb}}^4\right)}{\eta_{\text{conc}}\text{CG}} \tag{4.8}$$

where α is the receiver absorptance, ε is the receiver emittance, σ is the Stefan–Boltzmann constant, T is the operation temperature of the receiver temperature, T_{amb} is the ambient temperature, η_{conc} is the concentrator efficiency, C is the concentration ratio, and I is the solar irradiance. Furthermore, considering the efficiency of the equivalent Carnot heat engine, the variation of the energy yield of a concentrating system as a function of the temperature for different values of C follows the curves, which can be appreciated in the graph of Figure 4.5. This dependence of the efficiency imposes a limit to the amount of energy that can be harvested for use in thermochemical transformations. Accordingly, although working at elevated temperatures benefits the efficiency of the Carnot cycle, radiation losses impose strict practical limits to the increment of this operation parameter.

4.3 Thermochemical Cycles for Fuel Production

4.3.1 Early Studies

As in many other topics in the field of energy, pioneering research on thermochemical cycles were carried out in the 1970s and early 1980s, prompted by the oil crisis.[51,52] One of the first two-step cycles based on oxides to be investigated was the Fe_3O_4/FeO couple.[23,42] The thermal reduction of the magnetite to wüstite occurs at extremely high temperatures, well above the melting point of this phase (1377 °C) at atmospheric pressure, while re-oxidation of FeO with water vapour to produce hydrogen is spontaneous below 700 °C. This wide temperature gap between both reactions requires an extremely fine control of the heat recovery to fulfil its potential energy

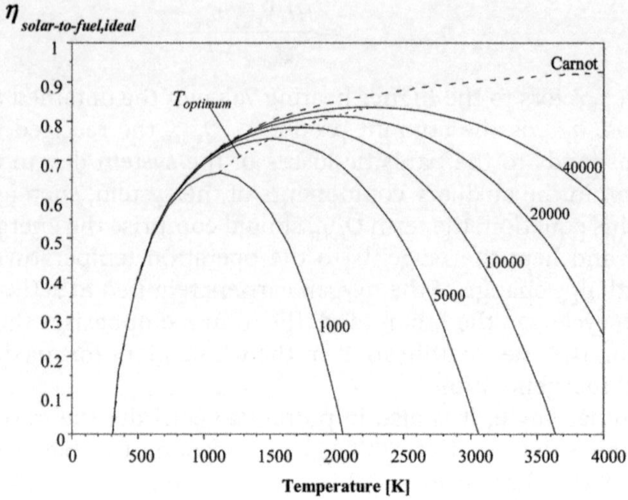

Figure 4.5 Variation of the ideal efficiency, η, of a concentrating thermosolar system as a function of the operation temperature for different concentration ratios. $T_{optimum}$ for achieving maximum yield in solar-to-fuel conversion is also indicated. Reproduced from ref. 50 with permission from the Royal Society of Chemistry.

yield and the formation of liquid phases is impractical. Furthermore, although high theoretical energy yields were initially proposed for the Fe_3O_4/FeO system, once all the relevant energy losses were accounted for the estimated overall efficiency was lower than 10%. This relatively high volatility (low melting point) is partly related to the instability of the metal oxide for long-term operation.[53] In order to lower the extreme reduction temperatures and to increase the hydrogen yield in water splitting, alternative systems based on partially substituted ferrites with a spinel structure such as MFe_2O_4 (with M being a transition metal) were proposed. Experimental lab-scale tests revealed that the best results in terms of stability and hydrogen production can be obtained with $NiFe_2O_4$,[54] although the performance of $CoFe_2O_4$ was only slightly lower.[55] During the reduction stage, phase segregation takes place according to:

$$NiFe_2O_4 \rightarrow NiO + 2FeO + \tfrac{1}{2}O_2 \tag{4.9}$$

In reality, formation of a solid solution with incorporation of some Fe in the Ni phase appears more likely.[44] However, although a minor $Ni_{1-x}Fe_xO$ contribution remains in the sample after several cycles, apparently this fact does not affect the hydrogen production. A ferrite system has been successfully tested at pilot scale in the Plataforma Solar de Almería (PSA, Spain) in a 100 kW prototype provided with a double reactor system for continuous operation and SiC honeycomb coated with Zn ferrite as active elements.[56] The use of ZrO_2 and yttria-stabilised zirconia (YSZ) as supports for the ferrites has been also investigated at lab scale with promising results due to its high melting point and consequently reduced sintering of the redox active phase.[23,57]

Interestingly, an alternative redox cycle was discovered when Co-ferrrites where supported on alumina, forming a Co–Al spinel and hercynite ($FeAl_2O_4$), after which this new redox cycle was named.[58] This process takes place according to these two stages:

$$CoFe_2O_4 + 3Al_2O_3 \rightarrow CoAl_2O_4 + 2FeAl_2O_4 + 1/2O_2 \tag{4.10}$$

$$CoAl_2O_4 + 2FeAl_2O_4 + H_2O \rightarrow CoFe_2O_4 + 3Al_2O_3 + H_2 \tag{4.11}$$

Such a redox cycle was tested isothermally at 1350 °C yielding 102 μmol g^{-1} of H_2.[59] Further DFT studies revealed that water splitting redox cycles based on such aluminum spinels operate through oxygen vacancy formation rather than stoichiometrically.[60]

The Zn/ZnO system has been also extensively investigated for many years due to the favourable theoretical yield (>29%), despite its elevated reduction temperature (*ca.* 2000 °C).[41] However, at those extreme temperatures Zn^0 is vaporised. Accordingly, condensation of the metal while avoiding its recombination with O_2 are important hurdles to achieve complete conversion of ZnO. In order to circumvent this problem, rapid temperature quenching and an efficient heat recovery are required. Some pilot plant tests have been performed with ZnO, such as the one performed in PROMES-CNRS (France) in a 100-kWth solar reactor.[61] These assays showed the possibility of achieving a solar-to-fuel yield of 16%.

In addition, more complex three-step cycles, such as MnO–NaOH,MnO–Na_2CO_3, have also received some attention due to the remarkable hydrogen yields achieved in lab conditions.[62–64] However, continuous operation of these cycles is problematic due to the low temperature regeneration stage, which requires a challenging heat recovery to efficiently save temperature differences of a few hundred degrees.

The study of these systems paved the way for future developments and contributed to widen the knowledge on the engineering of these high temperature processes driven by concentrated solar flux. Nevertheless, long-term stability, and limited hydrogen production rates constitute important barriers for the progress of these ground-breaking redox systems. Accordingly, further advances called for the development of alternative materials with improved properties, which are reviewed in the next subsections.

4.3.2 CeO_2 for Solar Fuel Production

Recent increasing interest in solar fuel production by thermochemical cycles relies on the utilisation of CeO_2, which has been revealed as an outstanding stepping-stone in the advancement of this high temperature route to generate sustainable fuels. A schematic illustrating how this process operates is shown in Figure 4.6a.

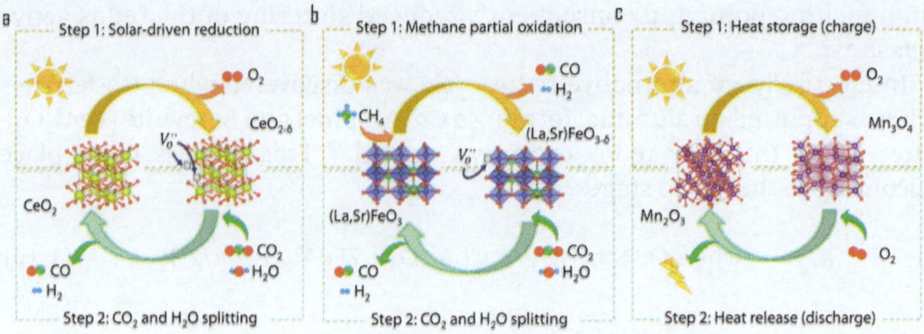

Figure 4.6 Schematic of the main types of thermochemical cycles applied to solar energy storage: (a) CO_2 and H_2O splitting over CeO_2; (b) methane chemical looping reforming coupled with CO_2 valorisation and/or H_2O splitting using (La, Sr)FeO_3; and (c) heat storage using the Mn_2O_3/Mn_3O_4 system.

Although there were some previous reports on the potential interest of thermochemical cycles based on CeO_2,[65] it was the demonstration of the stable production of hydrogen and oxygen over 500 cycles that was the event that marked the current focus on this material.[66] As a consequence of the intensive research on both the redox materials and the optimisation of the reactor and operation conditions, the reported efficiency for this system has experienced an astonishing rapid improvement, in only a few years there has been an increase in the solar-to-fuel yield from 0.7%[66] to 5.25%.[29] This represents more than a two-fold increment per year, although that progression should eventually level off when approaching the maximum theoretical efficiency (*ca.* 20%). An illustration of the solar reactor used to achieve these results, showing the arrangement of CeO_2, is displayed in Figure 4.7.

The oxygen exchange capacity of mixed oxygen ionic electronic conducting CeO_2,[67,68] which crystallises in a cubic fluorite structure (space group $Fm\bar{3}m$),[69] is moderate and under usual conditions only partial reduction to $CeO_{2-\delta}$ is achieved. Thus, $\Delta\delta$ is about 0.016 at 1500 °C and $p_{O_2} = 10^{-3}$ bar. Nevertheless, due to the oxygen ion diffusivity, which leads to rapid reduction, and its high crystallographic stability at elevated temperatures, this material appears to be a very suitable candidate for thermochemical applications.[70] Another remarkable characteristic of CeO_2 is the large variation in entropy associated with reduction. Theoretical studies using DFT indicate that whereas the electron area preferentially localised at Ce^{3+} sites, oxygen vacancies tend to repel each other. This distribution of defects leads to a large number of possible arrangements, which results in a remarkable increment in the configuration entropy as a consequence of oxygen release.[71]

From a practical point of view, cerium is relatively abundant in the Earth's crust (*ca.* 66 ppm) and accordingly is not a particularly costly

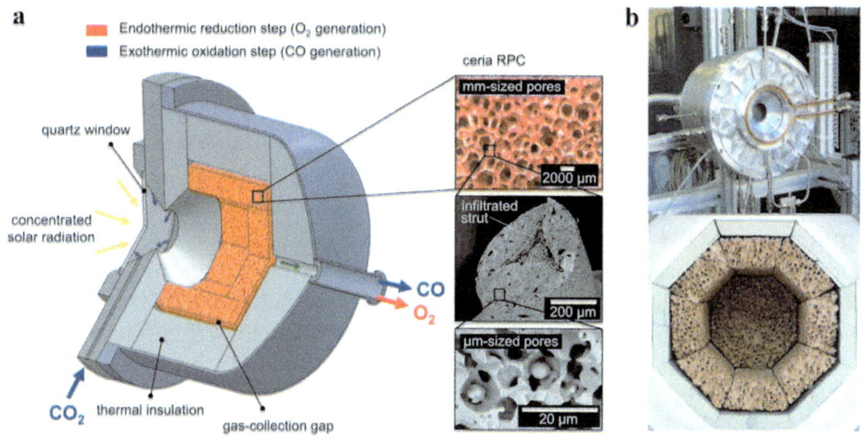

Figure 4.7 (a) Schematic of the solar reactor configuration for CO_2 splitting *via* a thermochemical redox cycle and using a porous CeO_2 as oxygen carrier, which is directly exposed to concentrating solar radiation, and (b) pictures showing details of the solar reactor. Reproduced from ref. 29 with permission from the Royal Society of Chemistry.

element, although the distribution of ores with exploitable concentration, mainly those containing the mineral bastnäsite, (Y, Ce, La)CO_3F, and monazite, (Ce, La, Nd, Th)PO_4, is more uneven. Currently the main producer is by far China, with Australia and USA in second and third place, respectively.[72]

Solid solutions through extrinsic doping of CeO_2 have been proposed as a potential way to improve the efficiency of solar fuel production. The elements proposed for doping of ceria include rare earths such as La, Sm and Gd[73] and transition metals such as Ta, Hf or Zr.[74] Incorporation of some of the latter elements allows one to increase the oxygen storage capacity, as it has been extensively investigated in the context of catalysis. In particular, $Ce_{1-x}Zr_xO_2$ materials show some of the highest reduction yields, which for Zr content below 50 mol% follows a linear dependence of the degree of doping as a full solid solution is formed.[75,76] However, enhanced stability is reported based on hydrogen production measurements for about 25 mol% of zirconia in the ceria lattice.[77] Thermodynamic studies show that for oxidation, both enthalpy and entropy decrease with Zr doping for smaller degree changes in the oxygen non-stoichiometry (δ ~0.1), which is usually attainable in thermochemical cycles.[78] At atomic scale, this behaviour has been attributed to the preference of smaller Zr^{4+} cations for a 7-fold coordination, which in the fluorite structure can be achieved by the formation of vacancies. In other words, oxygen vacancies are preferentially formed on the Zr coordination sphere. Recently, a new doping strategy based on metal cation co-doping with trivalent (La^{3+},Y^{3+}) and pentavalent cations (Nb^{5+}) that aims to balance 4+ cation substitution has been successfully developed based on DFT predictions.[79,80] Interestingly, such "paired charge-compensating" doped ceria, namely $Ce_{0.9}Y_{0.05}Nb_{0.05}O_2$, shows higher solar-to-fuel efficiency than CeO_2, requiring smaller temperature swings (the gap between reduction and oxidation temperatures) than those necessary for dopants such as Zr or Hf, resulting in a very attractive class of materials for further implementation.[81]

Morphological modifications are also crucial to achieve an adequate performance of CeO_2. In fact, as can be appreciated in Figure 4.7, a record energy yield has been obtained with CeO_2 foam with cavities in the mm range and pores in the μm range. As mentioned in Section 4.2.1, the reason behind that is the need to maintain enough active surface for efficient gas–solid exchange reactions. Experimental work carried out with both non-porous and macroporous CeO_2 have proved that the latter show a 10-fold increment of CO production.[82] This is due to the differences in surface area, which prevail despite the sintering induced by the high temperature. Furthermore, CeO_2 prepared with 3D-ordering of the pores is more stable and improves the oxidation kinetics. On the other hand, CeO_2 foams with dual distribution of pore diameters in the millimeter and micron scale, allow simultaneously an adequate irradiation due to the larger cavities and good contact with the gases.[83] Recently, CeO_2 membranes have been used in a reactor for CO_2 splitting in a continuous process.[84] This is possible because this configuration allows the separation of CO and O_2 by means of the chemical potential gradient in the oxide membrane. However, this system requires further optimisation in order to fully exploit its potential.

Direct formation of methane, instead of syngas, has been demonstrated using Rh-decorated CeO_2 to catalyse the methanation during the reoxidation stage in the presence of water and CO_2.[44] Despite the interest of this proof of concept, it is important to note that selectivity obtained under these conditions is rather limited, most likely due to thermodynamic limitations of the methanation and, accordingly, CO and H_2 are also detected along with CH_4.

Kinetic studies of CeO_2 have proven that above certain temperatures (>1200 °C for oxidation and >1500 °C for reduction) the rate is not limited by surface reaction but by the gas flow.[85] Under these conditions ionic diffusion and surface processes at those extreme temperatures are so fast that their contribution to the overall kinetics can be neglected. That behaviour has relevant implications in the optimisation not only of the oxygen carrier, but for the whole reaction system. In this way, this study has revealed that, although at relative low temperature (≤1100 °C) the addition of Rh on CeO_2 surface shows a beneficial catalytic effect in hydrogen production, differences with bare CeO_2 at higher temperatures become inappreciable because under that regime the system is not kinetically controlled.[85] Therefore, under these circumstances, the rate of transformation is only constrained by thermodynamics and the availability of reagents and products, as described by the thermo-kinetic model.

4.3.3 Perovskites

Over the last few years, perovskite oxides have emerged as an attractive alternative to ceria among the family of non-stoichiometric thermochemical redox cycles.[37] The main advantage of these mixed oxides with respect to ceria relies on the high degree of compositional tunability[37] that such a class of materials offers; this can ease the modification of the thermodynamic properties of the redox oxide. Perovskite oxides are identified by the formula ABO_3, in which A is a lanthanide or alkaline-earth cation (*e.g.* La, Ca, Sr, Ba) and B is typically a transition metal cation of the first row (*e.g.* Ti, Mn, Fe, Co, Cr, Cu). The ideal perovskite structure is a cubic lattice of space group *Pm*3*m*, in which the A and B cations are in 12- and 6-fold coordination, respectively[37,86] (Figure 4.8). Partial substitution of both the A and B-site cations ($A_{1-x}A_x'B_{1-y}B_y'O_3$) can lead, however, to deviations of the cubic phase crystallising, for instance, in rhombohedral or orthorhombic phases. Such alterations of the cubic phase can be identified by applying the Goldschmidt tolerance factor (t),[87] described in eqn (12), in which r_A, r_B and r_O represent the ionic radii of the A, B-site cations and oxide anion, respectively. Ideal cubic perovskites exhibit a Goldschmidt tolerance factor equal to one. Additional perovskite-related crystal structures such as brownmillerites ($A_2B_2O_5$), double layer perovskites or Ruddlesden–Popper phases,[37,86,88] which have attracted the

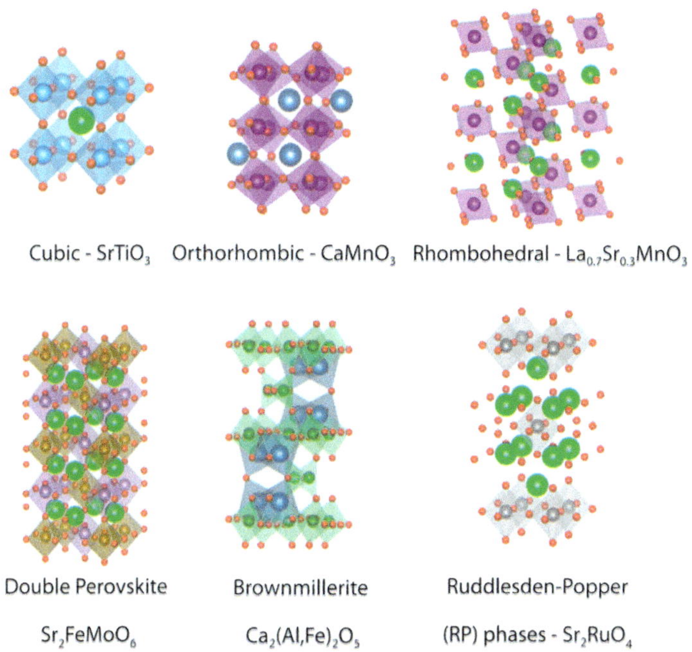

Figure 4.8 Crystal structures of perovskites and relevant perovskite related-materials for solar-driven thermochemical cycles for syngas production.

attention of the SOFC community, have been scarcely explored for solar-driven thermochemical fuel production,[89] but can represent a niche for further materials development.

$$t = \frac{r_A + r_O}{\sqrt{2}(r_B + r_O)} \tag{4.12}$$

Analogously to ceria, redox cycles for syngas production based on perovskites encompass a two-step process eqn (13–15): a first-step consisting in the solar-driven reduction of the perovskite, in which oxygen vacancies are formed ($V_O^{\bullet\bullet}$), with the consequent release of gaseous oxygen; and a second step, normally occurring at lower temperatures, in which CO_2 and H_2O re-oxidised the reduced perovskite producing syngas.

$$ABO_3 \rightarrow ABO_{3-\delta} + \delta/2O_2(g) \tag{4.13}$$

$$ABO_{3-\delta} + \delta\, H_2O(g) \rightarrow ABO_3 + \delta H_2(g) \tag{4.14}$$

$$ABO_{3-\delta} + \delta\, CO_2(g) \rightarrow ABO_3 + \delta CO(g) \tag{4.15}$$

The same thermodynamic constraints apply in this case: in order to have a high fuel production, the oxygen release (δ) during the reduction must be high, and the ΔG for the CO_2 and H_2O splitting must be negative in order to favour the production of syngas. Here is where the potential advantage of perovskite materials with respect to ceria becomes apparent. The vast compositional range that this class of oxides exhibit allows tuning their properties in order to increase the oxygen release and lower the reduction temperatures, which directly impact the higher fuel yield during the oxidation step. Additionally, lowering the working temperature during the solar-driven step presents the advantage of reducing the radiation heat losses. A successful example of perovskite development in order to achieve reduction at milder conditions is $La_{0.6}Sr_{0.4}Cr_{0.8}Co_{0.2}O_{3-\delta}$, which can be reduced at 300 °C lower but maintains similar fuel production yield when compared to ceria.[90]

However, lowering the reduction temperature also implies a decrease in the value of the reduction enthalpy, which can drastically affect the solar-to-fuel efficiency[91,92] and slow down the kinetics significantly.[48] Generally, perovskite oxides present much lower enthalpy and entropy than ceria, which on the one hand facilitates the reduction step, increasing the oxygen generation at lower temperatures. This is illustrated in Figure 4.9a, in which a comparison of oxygen release profiles between manganese-based perovskites and ceria is depicted. There it can be observed that at a given temperature (*e.g.* 1300 °C) the amount of lattice oxygen that is released follows the trend $La_{0.6}Ca_{0.4}Mn_{0.6}Al_{0.4}O_3 > La_{0.6}Sr_{0.4}Mn_{0.6}Al_{0.4}O_3 > La_{0.6}Ca_{0.4}MnO_3 > La_{0.6}Sr_{0.4}MnO_3 > CeO_2$,[93] which correlates well with CO yields obtained experimentally (Figure 4.9b),[94] evidencing the higher fuel production that perovskite exhibits. However, the favourable reduction thermodynamics comes at the price of less favourable CO_2 and H_2O splitting thermodynamics if compared to the benchmark material, ceria. This is illustrated in Figure 4.8c, in which the Ellingham diagrams of $La_{0.6}Sr_{0.4}Mn_{1-x}Al_xO_{3-\delta}$ perovskites and ceria are further compared to the H_2O and CO_2 splitting reactions. Ellingham diagrams of the Gibbs energy *vs.* temperature and oxygen partial pressure for a metal oxide indicate the ease of the metal oxide to reduce, which is a helpful tool in discussing those materials in the context of solar-to-fuel conversion.

Due to the higher enthalpy and entropy, the driving force for syngas production, reduced ceria ($CeO_{1.95}$), exhibits more favourable thermodynamics for splitting CO_2 and H_2O at higher temperatures (1000 °C) when

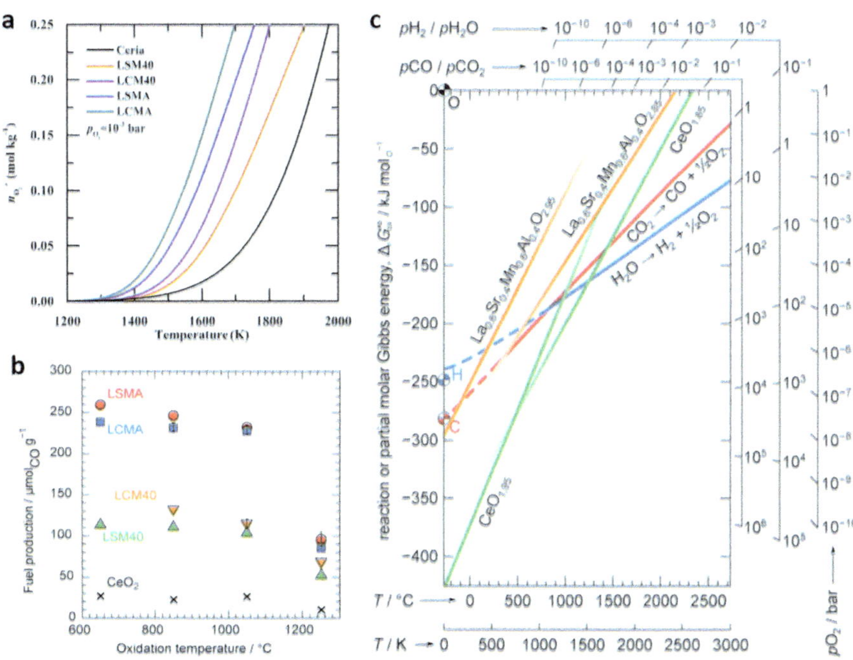

Figure 4.9 (a) Comparison of the temperature dependence of O_2 release between ceria and Mn-based perovskites (LSM40 = $La_{0.6}Sr_{0.4}MnO_3$, LCM40 = $La_{0.6}Ca_{0.4}MnO_3$, LSMA = $La_{0.6}Sr_{0.4}Mn_{0.6}Al_{0.4}O_3$, LCMA = $La_{0.6}Ca_{0.4}Mn_{0.6}Al_{0.4}O_3$). Adapted with permission from ref. 93 (b) Comparison of the CO production *versus* oxidation temperature for the same materials. (c) Ellingham plot for the oxidation reactions of $La_{0.6}Sr_{0.4}Mn_{0.6}Al_{0.4}O_3$ and CeO_2, comparing two different extents of reduction. The temperatures at which the yellow ($La_{0.6}Sr_{0.4}Mn_{0.6}Al_{0.4}O_3$) or green ($CeO_2$) lines cross the red ($CO_2$ splitting) or blue (H_2O splitting) denote the values at which oxidation under CO_2 and H_2O becomes thermodynamically favourable for each type of oxide. Adapted from ref. 94 with permission from Wiley, Copyright 2015.

compared to $La_{0.6}Sr_{0.4}Mn_{0.6}Al_{0.4}O_{2.85}$ (500 °C).[94] This fact implies that such manganites require a considerable excess of CO_2 and H_2O to produce syngas.[93,94] Namely, the amount of moles of oxidant gas can exceed 1000 times the moles of perovskite material in order to attain high fuel productions.[93] The excess of H_2O and CO_2 leads to energy penalties due to higher amounts of gas that need to be heated up to the operational temperature, and the gas separation of reactant and product gases, reflecting the impact of thermodynamic properties on the overall efficiency of solar-to-fuel conversion.[34,93]

Additionally, the overall less favourable thermodynamic values of perovskites can influence the CO_2/H_2O splitting kinetics, as observed for the Sr-doped $LaMnO_3$ system (Figure 4.10).[48,95] An increase in the amount of Sr in $La_{1-x}Sr_xMnO_3$ materials leads to higher reduction extents and lower reduction enthalpies, resulting in higher fuel yields but with slower splitting rates.[48,95] Due to the existence of a thermo-kinetic regime, slower fuel production rates can be explained from a thermodynamic point of view since the driving force for H_2O splitting decreases with lower enthalpy values and almost unvaried entropy values with Sr increase. Computed peak production values obtained from thermo-kinetic modelling reveal 20 times faster kinetics for ceria with respect to Sr-doped $LaMnO_3$ (60 and 3 mL_{H2} min^{-1} g^{-1}, respectively), evidencing the kinetic limitations of perovskites compared to ceria.[48,85]

The drawbacks exposed here suggest that Sr-doped lanthanum manganites, the first perovskite systems exhaustively explored for thermochemical fuel production,[91] might not be a real alternative to ceria. However, this recent progress has paved the route for further exploration of perovskite materials. In this respect, thermodynamic modelling is a crucial tool for the discovery of new perovskite oxides with enhanced enthalpy and entropy values. Thermodynamic assessments have made use of already existing data (sets of δ, T, p_{O_2}, for a given composition) found in the literature,[91,95] extracted from thermogravimetric analyses[93,94,96] or from thermodynamic CALPHAD libraries.[47] Additionally, as mentioned in Section 4.2.1, DFT calculations have been utilised to screen potential redox candidates over more than 5000 ABO_3-type perovskites, illustrating the potential of this technique.[49] Among them, authors show that at least 139 fulfil the requirements of stability and favourable energy for oxygen vacancy formation that make them promising redox materials for thermochemical syngas production. Interestingly, most of the perovskites with the ability to split water crystallise in orthorhombic form, as depicted in Figure 4.11. Considering further A- or B-site doping, the amount of possible compositions can be even greater. DFT analyses have been also employed to identify the free energy of oxygen vacancy formation in A- and/or B-site doped perovskites, demonstrating the capacity of expanding such type of calculations to more complex formulations.[38,46,96,97] Recently, DFT-based thermodynamic data have been developed and made publicly accessible at the Materials Project, where theoretical and experimental data can be

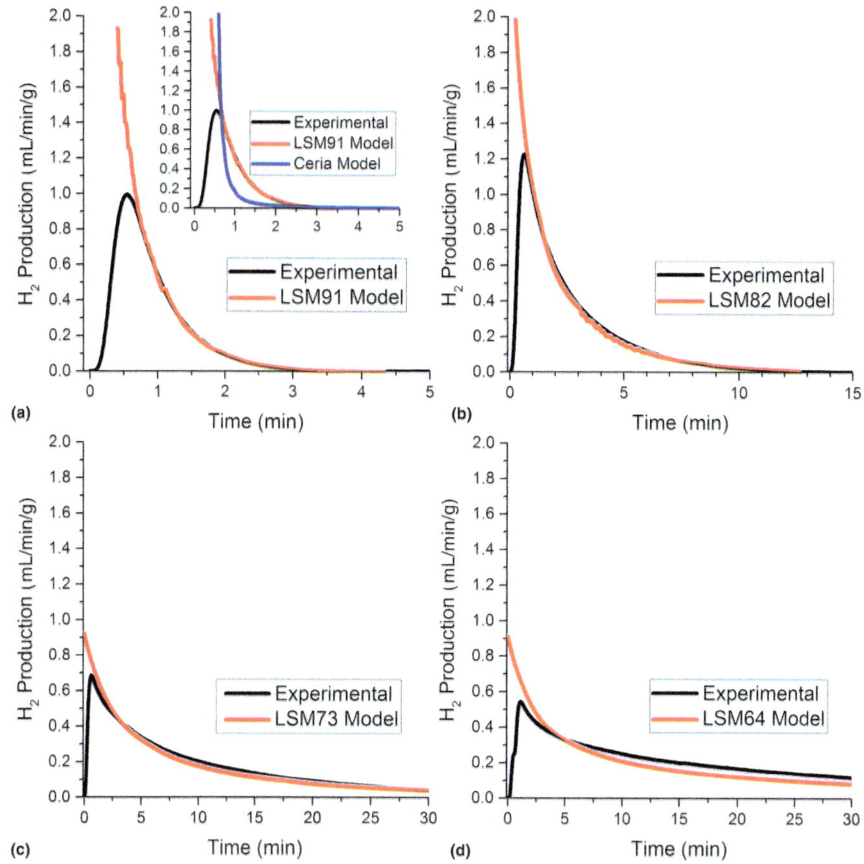

Figure 4.10 Influence of the Sr content on the H_2O splitting kinetics of $La_{1-x}Sr_xMnO_3$ perovskites (x = (a) 0.1, (b) 0.2, (c) 0.3, (d) 0.4). The black lines represent experimental data, and the red lines the data obtained from a thermo-kinetic model computed using thermodynamic data extracted from CALPHAD libraries. Reproduced with permission from ref. 48 with permission from Cambridge University Press, Copyright 2017.

readily used for studying the thermodynamics of several perovskite compositions in order to design more efficient redox materials for thermochemical water splitting.[98]

In addition, DFT can provide guidelines to elucidate the stability of cubic perovskites against carbonate formation.[46] This information is extremely useful for the design of new materials, since bivalent alkaline earth cations (Ca^{2+}, Ba^{2+} and Sr^{2+}) show a great tendency to react with CO_2 and form carbonates, which is detrimental to the production of CO. DFT findings corroborated by experimental thermogravimetric analyses point towards an easier carbonate formation for perovskites with less endothermic enthalpy of metal oxide reduction, with the following trend $SrCoO_3$ > $BaMnO_3$ > $SrTiO_3$.[99]

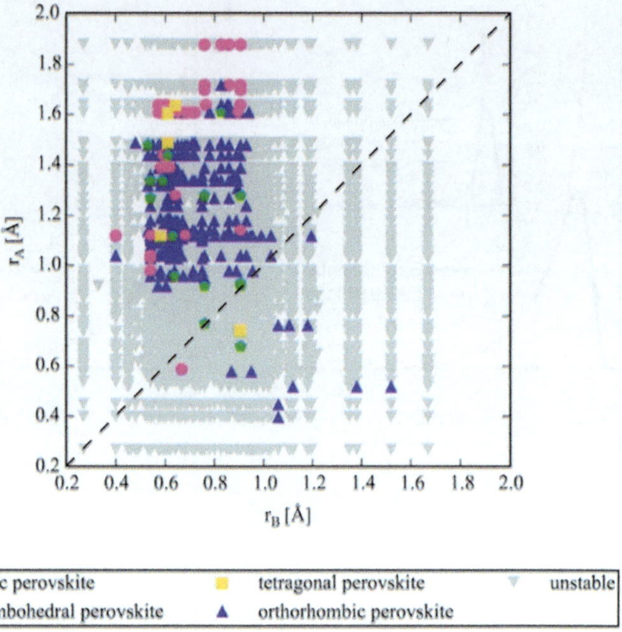

Figure 4.11 Stability plot for 5329 perovskite compositions computed by DFT. The *x*- and *y*-axis represents the ionic radii of the B- and A-site cations, respectively. Adapted from ref. 49 with permission American Chemical Society, Copyright 2016.

Carbonate formation has been also observed for Sr- and Ca-doped $LaMnO_3$ perovskites.[100] However, carbonation in such oxides can be partially counteracted by incorporation of Al^{3+} on the B-site.[100] XPS analyses of such $La_{1-x}(Ca/Sr)_xMn_{1-y}Al_yO_3$ showed a Mn enrichment of the surface of such oxides, with decreased presence of Sr, which could explain the higher resistance towards carbonate formation with Al-doping.[99]

$La_{1-x}Sr_xMn_{1-y}Al_yO_3$ perovskites have been proven to withstand long-term stability, when cycled between 1350/1000 °C without materials degradation and exhibiting constant fuel production between 125–150 µmol g^{-1}.[101] In this class of perovskites, Mn cations are the redox active cations (Mn^{4+}/Mn^{3+} or Mn^{3+}/Mn^{2+} redox trades).[99] Oxygen release is enhanced with increasing Sr-content, the Al doping having a less important role on this metric.[96,99,102] Additional doping studies showed higher oxygen release with replacement of Sr by Ca, which crystallises in the orthorhombic phase instead of the rhombohedral.[103] Replacement of La^{3+} by other trivalent cations (Sm^{3+}, Gd^{3+}, Dy^{3+}, Nd^{3+}) has also been explored – $Y_{0.5}Sr_{0.5}MnO_3$ exhibiting the best performance with the highest O_2 release and CO production (757 µmol g^{-1} after being reduced at 1400 °C).[104] Besides Al^{3+}, B-site doping with other trivalent cations such as Sc^{3+} and Ga^{3+} have been found to improve the performance of manganites.[105,106] Other perovskites with Co, Fe, Ni or Cr on the B-site and Ruddlesden–Popper related phases[89,107] have been also experimentally explored, $La_{0.6}Sr_{0.4}Cr_{0.8}Co_{0.2}O_3$ being the most attractive, which has the ability to produce similar amounts of CO when compared to ceria while being reduced at 1200 °C and exhibits stability against carbonate formation.[90] Alternatively $BaCe_xMn_{1-x}O_3$ has been proposed as a new class of perovskite oxygen carrier with promising performance.[108] The rationale behind such composition relies on mixing an easy-to-reduce perovskite, $BaMnO_3$, with $BaCeO_3$, in which the Ce–O bond is much stronger than Mn–O, resulting in a material with an optimised oxygen vacancy formation energy between the values of both end members.[108] In particular, $BaCe_{0.25}Mn_{0.75}O_3$ exhibited very stable H_2 production over the course of 50 cycles, with average H_2 yields about 140 µmol g^{-1}. Although the net H_2 production is lower than for Sr- and Al- doped $LaMnO_3$, $BaCe_{0.25}Mn_{0.75}O_3$ shows faster redox kinetics and improved performance in "high conversion" operation for H_2O/H_2 ratios between 1333 and 285, which are more relevant for practical application.[108] Operating under such "high conversion" conditions was also proved for polycation oxides with an approximated formula of $(FeMgCoNi)O_{1.2}$.[109] This novel class of oxides crystallise in a mixture of rock-salt and spinel structures and shows higher H_2 yields than Ni and Co ferrites and ceria when reduced just at 1100 °C.[109]

Another promising perovskite system for water splitting is $LaGa_{0.5}Co_{0.5}O_3$, which points out the potential interest of exploring alternative materials to Mn-based oxides. The hydrogen production obtained with this perovskite at 800 °C after reduction at 1350 °C is 478 µmol H_2 g^{-1}.[110] This value is significantly higher than for other perovskite oxides although the kinetics of oxidation are slow; however, it can be moderately increased by adding 1 wt% IrO_x as catalyst. However, long-term studies are necessary to confirm the stability of this system. In summary, perovskite materials are a relevant class of oxides for solar-driven thermochemical fuel production, with a profusion of compositional formulations that such a structure can accommodate. This fact has allowed the preparation of materials that can operate at lower temperatures than ceria while providing high yields during fuel production. This can be beneficial regarding the reduction of thermal stresses in the material.[46] However, that fact also implies lower enthalpy values, which result in limited process efficiency, since a high excess of CO_2, and H_2O is required for improving fuel production. In general, kinetics is rather sluggish and needs careful consideration when designing perovskites for solar-to-fuel conversion. Therefore, new materials design should focus on the enhancement of the enthalpy and entropy properties, and this endeavour can benefit from the recent advancement of accurate predictions based on thermodynamic modelling and DFT calculations and a close look at thermo-kinetic models.

4.3.4 Thermochemical Cycles for the Solar Upgrading of Conventional Fuels

As discussed in the previous sections, the use of sunlight as an energy source and the utilisation of abundant low energy molecules such as H_2O and CO_2 are the pillars of the notable potential for the production of sustainable fuels that provide thermochemical cycles. However, extreme operation temperature poses important constraints to the design of reactors and may limit the applicability of this route for the transformation of solar power into chemical energy. This has prompted the exploration of alternative routes that introduce an external fuel, generally CH_4 but also H_2, in the reduction step. For the case of methane this reaction takes place according to:

$$MO_x + \delta\, CH_4 \rightarrow MO_{x-\delta} + \delta\, CO + 2\,\delta\, H_2 \quad (4.16)$$

Therefore, during the reduction stage syngas is obtained, while the re-oxidation with CO_2 and/or H_2O will be entirely equivalent to the fully renewable cycles described above. An example of this kind of hybrid thermochemical cycle is shown schematically in Figure 4.6b for the particular case of (La, Sr)FeO_3 perovskites. This approach allows much milder operation temperatures, while increasing the extent of the reduction and change in oxygen non-stoichiometry, δ. This has the potential of boosting the production of fuels. Obviously, these advantages do not come without cost, and the sustainability of the process relies now on the origin, generally fossil, of the reducing agent. In this respect, the coupling of this thermochemical cycle with biogas as a source of both CH_4 and CO_2 could be an interesting opportunity for the hybridisation of biofuels with solar energy. On the other hand, these hybrid processes can open up new avenues for reducing the carbon footprint in the production of chemicals. In this respect, a recent evaluation has estimated that acetic acid can be obtained with an 84% reduction of CO_2 emission by combining a solar-driven thermochemical cycle fed with methane with the subsequent cascade of reactions: first the production of methanol from syngas and then acetic acid by the Cativa process.[111]

Theoretical thermodynamic consideration indicates that by using CeO_2 as the oxygen carrier and methane for the activation it is possible to achieve a solar-to-fuel efficiency that is 39% for CO_2 splitting at 950 °C. The feasibility of this process has been recently tested experimentally using CeO_2 of different morphologies under isothermal conditions at 1000 °C.[112] This study reveals the higher reactivity of hydrothermally prepared CeO_2, which presumably presents a larger surface area, although the best syngas yields are obtained with CeO_2 modified with MgO.[113] Furthermore, a CeO_2 aerosol was tested in a 2 kW solar reactor with methane reduction and the analysis of the results showed a maximum energy upgrade of 24% in the outlet stream.[114] These assays achieved a maximum reduction of $\delta = 0.25$ of the CeO_2 particles and they obtained a solar-to-fuel energy conversion yield of 12%. Recently, Ce-modified Mn_3O_4 has been successfully used for syngas production in a hybrid cycle fed with methane, achieving a CO yield for CO_2 splitting of 17.72

mmol g^{-1}, which is appreciably higher than the rate obtained in conventional thermochemical cycles. In this process, formation of Mn carbide takes place during methane reduction with the segregation of CeO_2. However, in the presence of CO_2 and H_2O the carbide is transformed back to Ce-doped MnO with the simultaneous release of CO and H_2. Despite this profound chemical transformation of the solid, the system is very stable as was confirmed by performing 100 consecutive cycles.[115] Similarly, the system CeO_2–TiO_2/$Ce_2Ti_2O_7$ modified with Ni has been revealed to be very active for methane-activated CO_2 and H_2O splitting.[116] Although, in the absence of NiO_x productivities are remarkably lower, with the incorporation of this component as catalyst 1.6 mmol g^{-1} of CO and 1.8 mmol g^{-1} of H_2 are obtained at 900 °C. Formation of the mixed phase with a pyrochlore structure takes place during reduction in CH_4, while re-oxidation under CO_2 and H_2O yields to the segregation of CeO_2 and TiO_2.

Another potentially interesting oxygen carrier for this kind of cycle is Fe_2O_3 because it can be deeply reduced to metallic Fe^0, which can be subsequently re-oxidised by H_2O producing a large yield of hydrogen. However, significant carbon deposition takes place during the reduction stage due to methane decomposition:

$$CH_4 \rightarrow C + 2H_2 \quad (4.17)$$

This side reaction is catalysed by Fe^0 and results in a decrease of the CO : H_2 ratio. In order to improve the performance of this system, Fe_2O_3 has been modified with Cu, yielding the spinel $CuFe_2O_4$, and dispersed on $MgAl_2O_4$ with good productivity of hydrogen and low carbon deposition.[117] A more complex composite containing a reforming catalyst, Ni/$MgAl_2O_4$, an oxygen carrier component, Fe_2O_3/$MgAl_2O_4$, and a CO_2 adsorbent, CaO/Al_2O_3, has been used to boost the efficiency of CO production in a two-step dry reforming of methane.[118] With this system, a CO production rate of 7.5 μmol g_{Fe}^{-1} s^{-1} at 850 °C has been achieved.

Perovskites have also been used in redox cycles coupled with methane partial oxidation to syngas. In particular, $La_{1-x}Sr_xFeO_{3-\delta}$ has successfully been applied to hydrogen generation upon reduction with methane.[119,120] The highest hydrogen yield was obtained with $La_{0.7}Sr_{0.3}FeO_3$ and no significant formation of carbon was observed operating at 1000 °C. The addition of Ni as catalyst increments remarkably the activity of this perovskite, but it also enhances carbon deposition. In another recent study, CO_2 splitting after activation in methane was investigated using an oxygen carrier on phases of the Sr–Fe–O system-related perovskite structure including brownmillerite oxide $Sr_2Fe_2O_5$ and the Ruddlesden–Popper phase $Sr_3Fe_2O_6$.[111] Remarkably, results in terms of stability and CO production were obtained with a composite of $SrFeO_{3-\delta}$ and CaO, which is used to prevent deactivation by sintering. Employing this heterogeneous system and operating at 980 °C, the CO generation rate, 9.13 μmol g^{-1} s^{-1}, is about eight times more than the rate obtained with conventional thermochemical cycles with CeO_2 at around 1000 °C.[24] Higher rates have been obtained recently using $La_{0.9}Sr_{0.1}FeO_3$, which exhibits overall syngas yields considering the two steps of 12.8 mmol H_2 g^{-1} and 15.9 mmol CO g^{-1} per cycle. This high productivity is attributed to the contribution of both methane cracking in the first step and the Boudouard reaction:[120]

$$CO_2 + C \rightarrow 2CO \tag{4.18}$$

This implies that carbon deposits originated by methane decomposition can be removed continuously, avoiding (at least partly) encapsulation of the active solids. Furthermore, similarly to the case of Ce-doped MnO, the formation of metal carbides, in this case Fe_3C, may play a significant role on the reactivity of this perovskite system.

Activation with hydrogen of redox oxides, in a process that is globally equivalent to reverse water gas shift reaction:

$$CO_2 + H_2 \rightarrow CO + H_2O \tag{4.19}$$

has received less attention. Nevertheless, a recent theoretical and experimental study of a number of perovskites of Earth-abundant elements has been evaluated in this process. Results obtained with $La_{0.6}Ca_{0.4}MnO_3$ and $La_{0.6}Ca_{0.4}Fe_{0.4}Mn_{0.6}O_3$ perovskites were particularly promising, with a CO production rate of about 2.3 μmol g^{-1} s^{-1} at 450–500 °C, which are remarkably low temperatures for thermochemical processes.[121]

In summary, the research on hybrid thermochemical cycles assisted by external reducing agents is gathering a growing interest despite their lower capacity to reduce the carbon footprint, due to the very attractive rates for CO_2 and H_2O splitting. Accordingly, these routes will probably receive further attention in the coming years.

4.4 Heat Storage in Redox Oxides

In addition to fuel production, the heat exchange events (endo/exothermic) that take place during reduction/oxidation cycles of metal oxides can be utilised efficiently for thermal energy storage purposes. In this case, no other product than O_2 is exchanged by the solids but, as will be discussed below, the apparent simplicity of the chemical process helps to reveal the underlying complexity of these redox systems, showing the influence of a number of parameters in their performance.

4.4.1 General Considerations of Thermochemical Heat Storage

Thermal energy storage (TES) systems are one of the four main components of concentrated solar power (CSP) plants, together with the optical concentrator, the receiver and the heat engine.[122] These systems play a crucial role since their implementation allows the production of electricity on demand, with the possibility of generating renewable energy around-the-clock.[27,50] The working principle is simple: the excess of energy collected from the sun during daily hours is stored thermally (charging period) and then utilised to supply heat (discharging period) during off-sun hours. Thermal energy storage systems can be classified in three types based on the physicochemical mechanism employed to store the excess heat: sensible heat storage, latent heat storage and thermochemical heat storage.[123] Sensible thermal energy storage makes use of changes in temperature, of a liquid or solid media, to store thermal energy. It is the most developed technology, already implemented in commercial CSP plants,[24] which use a mixture of melted K and Na nitrate salts. However, this technology presents the lowest gravimetric energy storage density and problems of heat losses over time, requiring additional insulation. Latent TES utilises the heat involved in phase changes, normally solid to liquid, to store thermal energy.[124] These materials are normally referred to as *phase change materials* (PCM), which can be metals, alloys, organic compounds or salts.[24] In thermochemical TES, heat is stored and released using reversible chemical reactions. In these systems, the enthalpy of reaction dictates the energy storage density, although it should be also taken into account that there is an additional sensible heat storage contribution, especially when gas–solid reactions are utilised. Regardless of the type, ideal TES systems should present the following characteristics:

- High energy storage density;
- Low heat losses high exergy efficiency;
- Inexpensive components, preferentially based on Earth-abundant materials;
- High mechanical and chemical stability;
- Low interaction with confinement media: low corrosion, no side reactions.

The energy storage density establishes the amount of energy, collected from the sun, per mass (or volume) that a material can capture. High values of this parameter will imply TES units of less size, which at the same time implies lower operational and maintenance costs. Attending to the energy storage density, thermochemical TES is the most promising, *ca.* 0.1–1.0 kW h kg^{-1} (360–3600 kJ kg^{-1}) against 0.02–0.03 kW h kg^{-1} (~70–100 kJ kg^{-1}) of sensible TES[125] and presents very low heat losses. However, it also entails a higher level of complexity since it involves chemical reactions, which requires a detailed knowledge of reaction thermodynamics and kinetics. Finally, optimum reactor design, which is strongly linked with the nature of the process, is also crucial for achieving high-energy yields. Of the three types, thermochemical energy storage is the less mature technology, although it has experienced remarkable progress during the past few years, with some developments already reaching pilot plant scale.[126]

As mentioned above, thermochemical heat storage utilises reversible reactions for the discharge and charge step. During discharge, the heat collected from the sun drives the thermal decomposition of a generic compound A into B and C (eqn (4.20)). Through this endothermic process, solar energy is stored chemically in the bonds of the products. Such compounds, B and C, are generally kept in separate tanks to avoid recombination. This process is facilitated because at least one of these components is in the gas phase. During the off-sun period, B and C are mixed again resulting in an exothermic reaction, releasing the heat collected in the endothermic step, which will be utilised to drive the power block and to produce electricity during peaks of high-energy demand.

$$A + \Delta H_r \leftrightarrow B + C \tag{4.20}$$

The pressure and temperature at which both reactions (direct and reverse) take place might require re-adjustment between both operations in order to guarantee high process efficiencies, *i.e.* full reaction conversion and selectivity, absence of side reactions and fast kinetics. Differences in the temperature between the charge and discharge processes can affect the exergy efficiency.[24] Exergy efficiency (η_{ex}) of TES systems is defined as

$$\eta_{ex} = \frac{Q_{out}\left(1 - \frac{T_c}{T_{out}}\right)}{Q_{in}\left(1 - \frac{T_c}{T_{in}}\right)} \tag{4.21}$$

where Q_{out} and Q_{in} are the heat recovered from and supplied to the TES system, respectively, and T_{out} and T_{in} are the temperatures at which discharge and charge take place. T_c is the cold reservoir temperature.[24] Heat losses and temperature difference between charging and discharging are the main factors affecting the exergy efficiency. According to eqn (4.21), maximum exergy efficiency is achieved by reducing heat losses and for systems with temperature gaps between heat storage and release as small as possible. Additionally, high-temperature can lead to higher exergy efficiencies in the CSP plant.[127] Thus, ideal heat storage processes should operate at conditions that match these requirements.

In general, high-temperature thermochemical heat storage for concentrating solar energy can be performed by using either gas–gas or gas–solid reactions, the latter being the focus of this chapter. Regarding gas–gas reactions, ammonia dissociation is the most explored process, benefiting from decades of research experience. However, this a challenging process due to the use of corrosive chemicals, high pressures and different catalysts.[128] Additionally, methane reforming reactions have also been suggested, although issues with side reactions and high costs derived from gas storage can be detrimental for future implementation.[129]

Gas–solid reactions are classified depending on the solid that is thermally decomposed, the most prominent being: hydrides, hydroxides, carbonates and oxides (redox) systems.[125,130]

Hydrides store thermal energy by decomposing into a metal and hydrogen (eqn (4.22)). The hydrogen generated in the charge step requires storage. This can be done in a high-pressure tank or by utilising a second metal hydride/metal system working at lower temperatures, although this approach has higher complexity.[131] The temperature required for the charging step varies considerably depending on the hydride used, ranging from 280 °C (equilibrium temperature at 1 bar) for MgH_2, the most studied hydride, to 950 °C for CaH_2, providing a high flexibility depending on the application. This class of materials exhibit excellent values of energy storage density, *e.g.* 2814 kJ kg^{-1} for MgH_2 or 8397 kJ kg^{-1} for LiH.[132] However, hydrides still present challenges associated with slow reaction kinetics and implementation of the hydrogen storage subsystem.[131]

$$MH(s) + \Delta H_r \leftrightarrow M(s) + \tfrac{1}{2}H_2(g) \qquad (4.22)$$

Hydroxide systems work on dehydration/hydration cycles. The endothermic step consists of the decomposition of a hydroxide into a metal oxide and water (steam), which both recombine in the backward step releasing the stored heat (eqn (4.23)). $Ca(OH)_2$[133–137] and $Mg(OH)_2$[138,139] have been proposed and tested for thermochemical heat storage. A $Ca(OH)_2$/CaO system has been thoroughly studied reaching the pilot plant level.[140] It operates in the 400–600 °C range, exhibiting a 1406 kJ kg^{-1} energy density storage and low cost. In addition, its stability has been proven at lab scale by performing over 100 cycles.[135] However, its low thermal conductivity[141] and the poor mechanical properties of the powder,[142] mainly associated with changes in volume during dehydration/hydration and particle agglomeration, still present issues for reactor up-scaling. In addition, the mid-range temperatures might not suit the next generation of CSP plants working at temperatures close to 1000 °C.

$$M(OH)_2(s) + \Delta H_r \leftrightarrow MO(s) + H_2O(g) \qquad (4.23)$$

Alkaline-earth carbonates can fulfil such requirements since $CaCO_3$ and $SrCO_3$ operate as a thermochemical heat storage medium at 850 and 1200 °C, respectively.[143,144] Besides being inexpensive and abundant materials, carbonates show high energy storage density values (1790 and 1585 kJ kg^{-1}). As drawbacks, calcination/carbonation cycles exhibit cycle-to-cycle conversion decay due to materials deactivation caused by particle sintering and agglomeration. Chemical and structural modifications by incorporation of dopants (Mg, Al), addition of supports (ZrO_2) or creation of porous structures (*e.g.* by a sacrificial template method) were successful in partially counteracting the multicycle deactivation.

$$MCO_3(s) + \Delta H \leftrightarrow MO(s) + CO_2(g) \tag{4.24}$$

Several of the systems described above are promising for high-temperature thermochemical energy storage. Carbonates ($CaCO_3$ and $SrCO_3$) and hydrides (LiH or CaH_2) exhibit high gravimetric energy storage at high temperatures. In addition, due to their low price, carbonates are an attractive option if deactivation problems associated with sintering can be overcome, ensuring long-term stability. However, hydroxides, carbonates and hydrides operate in a *closed system* configuration, requiring the incorporation of a gas storage subsystem, which can be costly, especially in the case of hydrogen due to safety considerations.

In this sense, redox-systems present the advantage of working, generally, as an *open system*. They run in a reduction–oxidation cycle using multivalent metal oxides. The endothermic step in which energy is stored is the thermal reduction of an oxide, resulting in a reduced phase with a lower oxidation state and the release of gaseous oxygen (eqn (4.24)). The exothermic step consists of the oxidation, normally using air, of the reduced form of the metal oxide, recovering the initial oxidation state.

$$MO_x(s) + \Delta H_r \leftrightarrow MO_{x-y}(s) + y/2O_2(g) \tag{4.25}$$

The great advantage of redox systems is the possibility of utilising air simultaneously as a heat transfer fluid and oxidant for the reduction and oxidation steps, respectively, for most of the oxides tested for such application. This fact allows the use of atmospheric air as the operating fluid, avoiding the need of gas storage (*open system* operation), as is the case for carbonates or hydrides, which makes redox storage systems ideal candidates for implementation in CSP plants working with volumetric air receivers. This new generation of concentrated solar receivers allows for electricity production at temperatures around 1000 °C, which is the range of most of the prominent redox pairs that will be discussed in the next section.

4.4.2 Redox Systems: Materials, Physicochemical Aspects and Challenges

Redox systems were already suggested in the late 1970s to 1980s,[145–147] but it was 30 years later in 2010 that Wong *et al.* carried out the first exhaustive screening of potential oxide systems for thermochemical heat storage.[148] The way that these systems operate is shown in the drawing of Figure 4.6c

Table 4.1 Characteristics of stoichiometric redox couples for thermochemical heat storage.[161, 162, 167]

Redox couple	ΔH_r (kJ kg^{-1})	T_{red} (°C)	T_{ox} (°C)	ΔT_{hyst} (°C)	Advantages/ Disadvantages
BaO_2/BaO	474	714	704	10	+Low T_{red}, small hysteresis −Reactivity with residual CO_2 and H_2O
Co_3O_4/CoO	844	915	880	35	+Outstanding cyclability and energy storage density −High price and toxicity
Mn_2O_3/ Mn_3O_4	202	940	715	225	+Inexpensive, abundant, non-toxic −Slow oxidation, broad hysteresis gap, low energy storage density
CuO/Cu_2O	811	1042	1021	21	+High energy storage density −T_{red} close to melting point of Cu_2O
Fe_2O_3/Fe_3O_4	496	1392	1374	18	+Inexpensive, abundant, non-toxic −High working temperature

for the case of the Mn_2O_3/Mn_3O_4 couple. Initial thermodynamic and experimental studies allowed one to exclude oxides presenting low (MgO_2) or extremely high (V_2O_5) temperature transitions or poor re-oxidation kinetics (MnO_2). In addition, highly toxic (U_2O_3, PbO_2) and expensive oxides (PtO_2, Rh_2O_3) were not further considered. This left the redox pairs BaO_2/BaO, CuO/Cu_2O, Co_3O_4/CoO, Mn_2O_3/Mn_3O_4 and Fe_2O_3/Fe_3O_4 as promising candidates (Table 4.1). Figure 4.12a depicts an Ellingham diagram for the reduction reaction in which the transition temperature for each of these oxides can be observed. These values can accurately predict the temperature range at which the thermal decomposition of the metal oxide into the reduced oxide and oxygen takes place. However, the oxidation reaction occurs at lower temperatures than the reduction for each redox couple. This interesting phenomenon, the so-called *thermal hysteresis,* exemplified in Figure 4.12c–d for Mn and Co oxides, respectively,[149] causes heat to be released at lower temperatures than it is collected, implying a loss of exergy in the process. According to Wokon *et al.*, hysteresis is caused by overheating and undercooling (*viz.* heating up or cooling down to temperatures away from the equilibrium value) events that enhance the driving force for triggering the phase transitions of stoichiometric metal oxides during reduction and oxidation, respectively.[149] Both effects thermally decouple the charge and discharge steps, displacing reduction and oxidation working temperatures from the thermodynamic equilibrium value.[149] This effect is especially

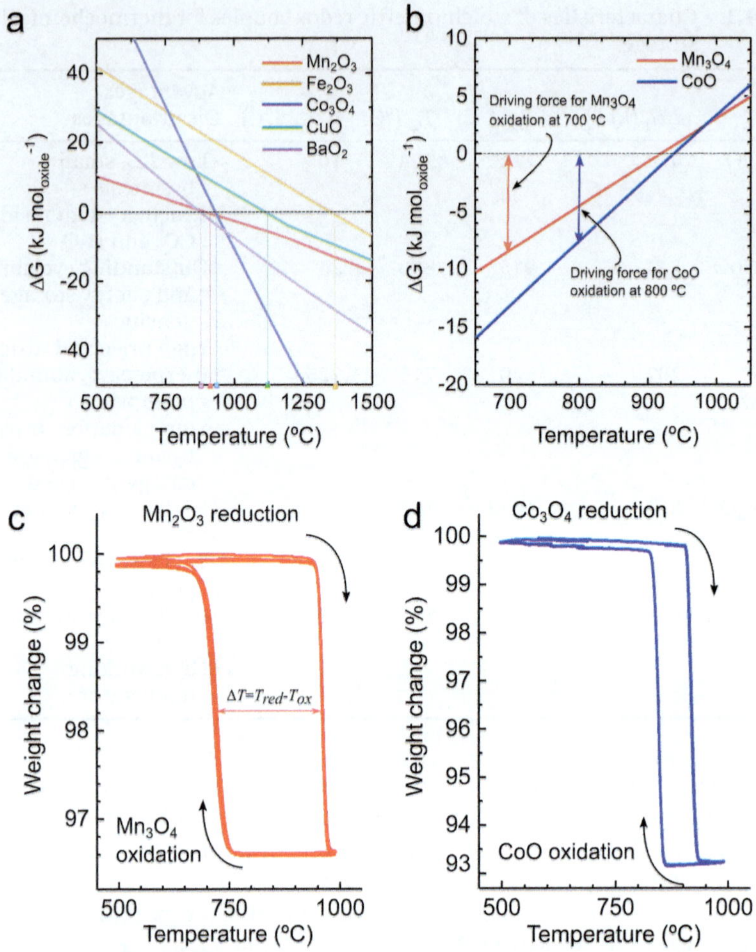

Figure 4.12 (a) Ellingham plot of the reduction of the most promising oxides for thermochemical heat storage, depicting the theoretical reduction temperature. Adapted from ref. 27 with permission from American Chemical Society, Copyright 2019. (b) Ellingham plot for the oxidation of Mn_3O_4 and CoO. Weight change *vs.* temperature for the redox reactions between Mn_2O_3/Mn_3O_4 (c), and Co_3O_4/CoO (d), recorded by thermogravimetric analysis (TGA) showing the thermal hysteresis resulting from redox transitions. TGA data were extracted from ref. 167 and the weight change was normalised to the start of the reduction reaction. Adapted from ref. 167 with permission from Elsevier, Copyright 2014.

pronounced for oxidation (Table 4.1), since the undercooling provides the driving force for the oxidation reaction (Figure. 4.12c).[149] While the small overheating required during reduction facilitates a faster reaction due to the temperature increase, oxidation kinetics are slowed down due to undercooling.[149] Thus, redox systems exhibiting higher oxidation driving force (*viz.* more negative ΔG in Figure 4.12b) are sought. This can be exemplified for the cases of Mn and Co oxides. According to the Ellingham plot for the oxidation reaction of Mn_3O_4 and CoO (see Figure 4.12b), this latter solid presents a higher driving force for oxidation, that is, it has more negative ΔG values than Mn_3O_4. This results in a steeper slope and, consequently, the gain in driving force with respect to the transition temperature (ΔG = 0) is higher with narrower temperature gaps. This points to an important influence of entropy increment, ΔS, for CoO in this process. Conversely, Mn_3O_4 requires cooling down to lower temperatures to reach similar levels of driving force, which, in turn, leads to a slower oxidation reaction according to the expectations of the Arrhenius law. This is a common characteristic of redox cycles involving stoichiometric phase transitions, in which reduction is generally a much faster process than oxidation. For instance, reduction of Mn_2O_3 to form Mn_3O_4 takes about 6 min, whereas the backward process around 70 min.[150] On the other hand, redox cycles involving non-stoichiometric oxides (as described in detail in Section 4.3) generally exhibit oxidation steps faster than the reduction ones.

The most investigated non-stoichiometric materials for thermochemical heat storage are perovskites.[151–153] Contrary to the redox couples described above, the thermal reduction of perovskites involves the creation of oxygen vacancies ($V_O^{\bullet\bullet}$) in the crystal lattice with the consequent release of gaseous oxygen, which are re-filled in the backward step releasing heat. In this case, the energy stored is not a fixed value, since ΔH_r varies with the extent of reduction, δ, which depends on the temperature and p_{O_2} at which the reduction is performed. For a given composition, δ increases with high temperatures and low p_{O_2}. This fact implies that the use of perovskites normally requires performing the reduction under an inert atmosphere to guarantee high δ, which in turn accounts for a high energy storage density, although it has been demonstrated that perovskite oxides can be also reduced under air atmosphere, which will be more practical for TCS applications.[154] For instance, as depicted in Figure 4.13, increasing δ from 0.15 to 0.25 enhances the energy storage density from 200 to 350 kJ kg^{-1}. Although reduction under vacuum and/or under a sweep gas flow could be also beneficial for stoichiometric redox pairs if the reduction temperature needs to be lowered, those oxides do not require such operational constraint, which eventually leads to energy penalties in the form of gas separation or gas compressing units.

Among the advantages, the perovskite crystal structure offers a broad compositional range, since a huge variety of different possible materials can be generated by doping either the A-site, B-site or both. This allows an easy tuning of the perovskite's physicochemical properties. To date, the most prominent perovskites for heat storage are calcium manganites.[153,155] In particular, $Ca_{0.95}Sr_{0.05}MnO_3$ stores 555 kJ kg^{-1} if reduced at 1000 °C and p_{O_2} of 10^{-4} atm. The total energy storage density augments to 985 kJ kg^{-1} if the sensible heat storage capacity of such ceramic oxides is taken into account.[153] On the other hand, Fe-doping is an interesting strategy to widen

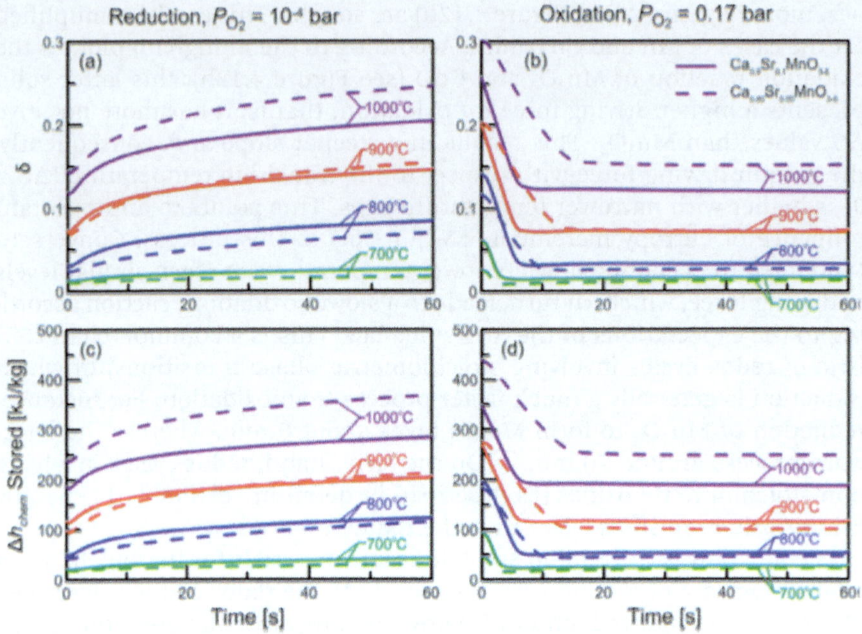

Figure 4.13 Influence of the reduction extent (δ) (plotted in a and c) on the energy storage density (b,d) for Sr-doped calcium manganites as a function of time for low (a,c) and high (b,d) p_{O_2}. The higher the temperature used for reduction, at low p_{O_2}, of the perovskites, the higher the energy is stored during the process. Reproduced from ref. 153 with permission from Elsevier, Copyright 2017.

the temperature operation window of $CaMnO_3$, which is limited by decomposition at 1200 °C. In this respect, $CaMn_{0.9}Fe_{0.1}O_{3-\delta}$ has proven to have enhanced thermal stability and an attractive heat storage capacity of 298 kJ kg^{-1}.[156] Besides exhibiting fast oxidation kinetics, which can be less than one minute for some materials,[153] perovskites present high chemical and mechanical stability over prolonged cycling. For instance, $Ca_{0.95}Sr_{0.05}MnO_3$ cyclability was proven for 1000 redox loops, with stable values of multicycle reaction conversion and without the presence of phase segregations or severe microstructural deterioration.[153]

Particle sintering is one of the main causes of material deactivation in redox TCS. The high temperatures attained during redox cycling can lead to severe particle coarsening and agglomeration, decreasing the oxide active surface. In addition, the remarkable increase of particle size caused by sintering enlarges the diffusion path length within the bulk of the solid. Both effects have a negative impact on the cycle stability of these redox materials, slowing down the oxidation reaction. This normally leads to a cycle-to-cycle deactivation since oxidation does not reach full conversion, limiting the amount of materials that can be actively reduced on the heating step of

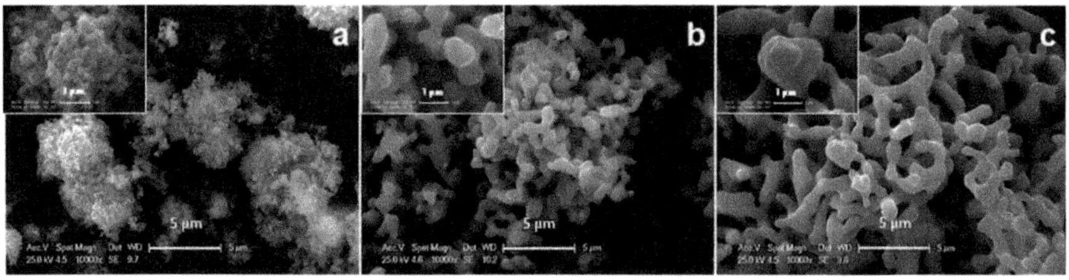

Figure 4.14 Morphology evolution of a Mn_2O_3 sample prepared by precipitation with ammonia of Mn^{2+} solutions: (a) before redox cycling, (b) after 5 redox cycles, and (c) after 30 redox cycles. Reproduced from ref. 157 with permission of the Royal Society of Chemistry.

the next cycle.[157] The point at which sintering starts is known as the Tamman temperature (T_T), which is equal to half of the melting point (T_{mp}, in K) value at which the bulk-to-surface migration becomes fast.[158] Additional factors that increase the sintering effect can be particle size, cycling time, the gas composition and the presence of dopants (*e.g.* Cu in Mn oxides[159]). The Mn_2O_3/Mn_3O_4 system is a good example of loss of cyclability due to sintering.[157] Both oxides exhibit mild Tamman temperature T_T values[160] that trigger a remarkable sintering effect transforming the initial particles, normally aggregates of particles of sub-micron size, into open macro-porous structures with a coral-like shape (Figure 4.14). It was found that smaller particles (<88 nm) with higher surface areas tend to densify in a more pronounced manner, resulting in rapid loss of activity if compared to particles with larger size (>127 nm).[157] Additionally, it was observed that the morphology change was more marked during the first five cycles, where the sintering effect stabilised between cycles 5 and 30 (Figure 4.14).[157]

An additional source of deactivation is the possible interaction of the oxides with residual CO_2 or moisture present in the air stream. This is the case of the BaO_2/BaO system, which was the first redox pair suggested for thermochemical heat storage in the late 1970s.[145] This redox cycle, also utilised for oxygen separation (the Brin process), has been proven to withstand 30 cycles if initial carbonate impurities were removed by a thermal pretreatment.[161] However, *ex situ* physicochemical characterisation of the BaO phase reveals that this oxide reacts easily with ambient CO_2 and H_2O[161,162] at room temperature. Prevention of such side reactions is crucial for the barium oxide system, since barium hydroxides melt easily (T_{mp} of $Ba(OH)_2$ $8H_2O$ is 78 °C) causing a complete degradation of the heat storage media.[161] In this respect, proximity to the melting point is one of the main drawbacks of the CuO/Cu_2O system, which exhibits a eutectic point at 1091 °C,[163] very close to the reduction temperature (1042 °C).[150]

In summary, in order to guarantee the feasibility of thermochemical storage, materials require high-energy storage density, low thermal hysteresis, long-term cycling stability without side reactions and phase segregations and resistance to severe morphological degradation. Once the main challenges that redox materials face from a thermodynamic and kinetic point of view have being identified, the approaches employed for optimising the heat storage performance of these redox oxides are described in the next section.

4.4.3 Strategies to Improve the Redox Performance

(a) Chemical modifications

Partial substitution of the cations in the oxide crystal lattice by other elements has been widely employed as a tool for improving the thermochemical heat storage properties, either by partial doping or by the formation of a different crystal structure with redox activity, such as bimetallic spinels. In this respect, several oxide crystal structures can allocate a certain amount of dopants without phase segregations. The presence of an additional solid can be detrimental for performance if it does not show redox activity in the desired temperature range, since part of the oxide mass is not acting in the thermochemical storage process, leading to a loss in the gravimetric energy storage density. In this respect, phase diagrams of binary oxides serve as a useful tool for predicting the formation of segregated phases and phase transitions as a function of temperature and molar ratio.[150,164,165]

The limited redox characteristics of Mn oxides have motivated the study of the incorporation of several metal cations (Al^{3+}, Co^{3+}, Cr^{3+} or Cu^{2+}),[150,159,166,167] the most promising being Li, in the form of Li–Mn spinels,[168] and the incorporation of Fe^{3+}.[149,169–171] In the latter case, the identical cationic radii of Mn^{3+} and Fe^{3+} in six-fold coordination allows the introduction of Fe cations to almost 50% of molar ratio without the formation of the hematite phase (Fe_2O_3).[172] The progressive insertion of Fe atoms in the bixbyite structure led to the increase of both the reduction (Figure 4.15) and oxidation temperatures, inducing a remarkable narrowing of the hysteresis gap.[169] Namely, with the incorporation of 40% Fe, ΔT_{hyst} could be decreased to *ca.* 100 °C.[169] In addition it was observed that the addition of Fe increases the energy storage density[149,150,169] to 271 kJ kg^{-1} for the composition $(Mn_{0.75}Fe_{0.25})_2O_3$. Besides thermodynamic effects, cationic doping can also benefit the reaction kinetics as observed for Cu-doped Co oxides[150] and Fe-doped Mn oxides.[169,173] In the latter case, the improvement is remarkable considering the poor oxidation kinetics of Mn_3O_4. With the incorporation of 20 mol% Fe, oxidation full conversion can be reached in 10 min,[173] whereas it takes *ca.* 70 min for an un-doped Mn oxide.[150] Interestingly, both Mn_3O_4 and $(Mn_{0.8}Fe_{0.2})_3O_4$ oxidation are characterised by Johnson–Mehl–Avrami (or Avrami–Erofeev) kinetic models,[149,173] which describe nucleation and growth reaction mechanisms. In this type of process, a reaction starts by the formation of initial nuclei on an active nucleation site

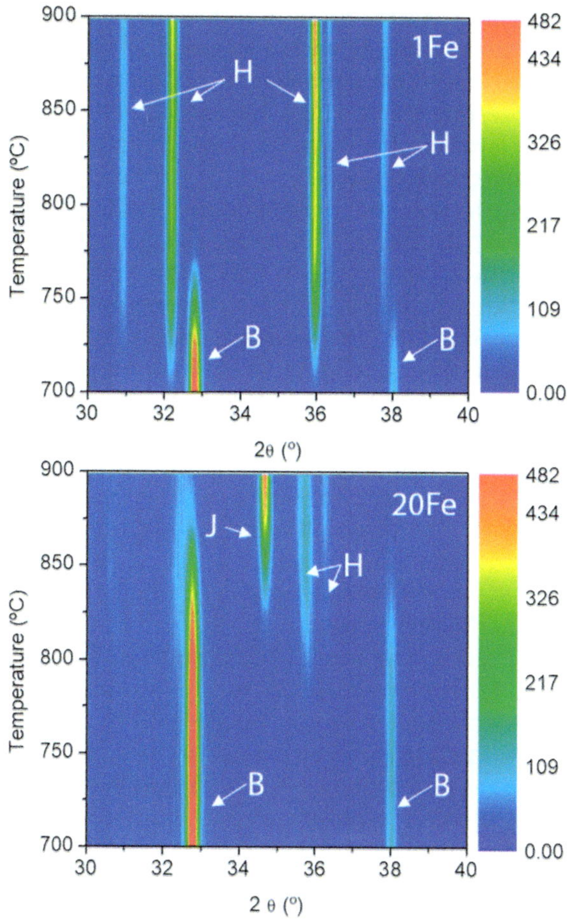

Figure 4.15 *In situ* XRD analyses of the reduction of Fe-doped Mn oxides with 1 (top panel) and 20% mol of Fe (bottom panel). Reaction was carried out under Ar flow of 100 ml min^{-1} (H = tetragonal (Mn, Fe)$_3$O$_4$; B = Mn$_2$O$_3$ and J = cubic (Mn, Fe)$_3$O$_4$). The right bar indicates the peak intensity. Reproduced from ref. 173 with permission from American Chemical Society, Copyright 2017.

of the oxide particle (induction period), followed by an increase of the population (nucleation) and growth of the nuclei until the reaction completes with the formation of the new phase.

At the atomic level, the transformation of Mn_3O_4 into Mn_2O_3 involves mainly cationic diffusions. Mn_3O_4 is a tetragonal spinel (hausmannite) with Mn^{2+} and Mn^{3+} occupying the tetrahedral and octahedral sites, respectively, while Mn_2O_3 crystallises in the bixbyite cubic structure, with Mn^{3+} occupying the octahedral sites. During such crystallographic transformation a cationic diffusion occurs when Mn^{2+} transforms into Mn^{3+}, the unique cation that is present in Mn_2O_3, migration that can occur through hopping within octahedral sites, hopping within tetrahedral sites, or between both sites.[173,174] Such processes require overcoming an activation energy that defines how fast the migration of the relevant species (ions, electrons...) can happen. Here is where the cationic doping can play a crucial role in improving the total oxidation rate, since the incorporation of cations inside the lattice is going to alter metal–oxygen bond lengths, which can facilitate the cation migration path through the reduced state to the transition state and from the transition state to the oxidised one. In this respect, application of experimental techniques such as synchrotron-based X-ray absorption or Raman spectroscopy can provide useful insights on oxidation states, coordination and bond lengths.[175] For instance, a shift to lower wavenumbers is observed for the A_{1g} mode of Mn_3O_4 when doped with 20% Fe, a composition that exhibits faster kinetics.[173] That particular Raman mode defines Mn–O breathing vibration of Mn^{2+} cations in tetrahedral positions, and a shift to lower values could be related to the enlargement of the Me–O bonds, indicating that incorporation of Fe^{3+} modifies Mn–O bond lengths. This can facilitate cation diffusion and, in turn, it could help to increase the oxidation rate.[173] Another good example of efficient chemical modification in Mn oxides is the incorporation of Mg,[176,177] but in this case applied to the Mn_3O_4–MnO redox transition, which occurs at higher temperatures, *viz.* 1200–1500 °C. Such elevated temperatures cause sintering problems that were efficiently alleviated by Mg incorporation, enabling less densification and higher specific surface areas after 30 high-temperature redox cycles. Additionally, Mg incorporation, which induces cation vacancies in the oxide lattice, resulted in superior energy storage densities, namely 1070 kJ kg^{-1}, for a Mn : Mg molar ratio of 2 : 3.[177]

(b) Morphological modifications

Morphological modifications encompass alterations of textural properties, *e.g.* porosity, specific surface area, particle size and shape of the redox materials. The main goal of this type of modification relates to the improvement of the solid–gas (oxide-air) phase contact for enhancing the oxidation rate. Here it should be pointed out that the type of reactor selected for the thermochemical storage unit has direct implications on the degree of aggregation of the redox material. For the Mn_2O_3/Mn_3O_4 system, it was found that the particle size and specific surface area could be tuned by adjusting the synthesis parameters. Namely, materials prepared by precipitation with ammonia using diluted concentrations of Mn precursor exhibit smaller particle size and higher specific surface area. Interestingly, the materials with higher specific surface area showed a faster deactivation caused by increased marked sintering.[157] The specific surface area can show a decrease of one order of magnitude after a few cycles,[169] since the principal driving force of sintering is the elimination of the surface area. Additionally, it can cause the oxide porous network to collapse after several cycles at high temperatures.[82]

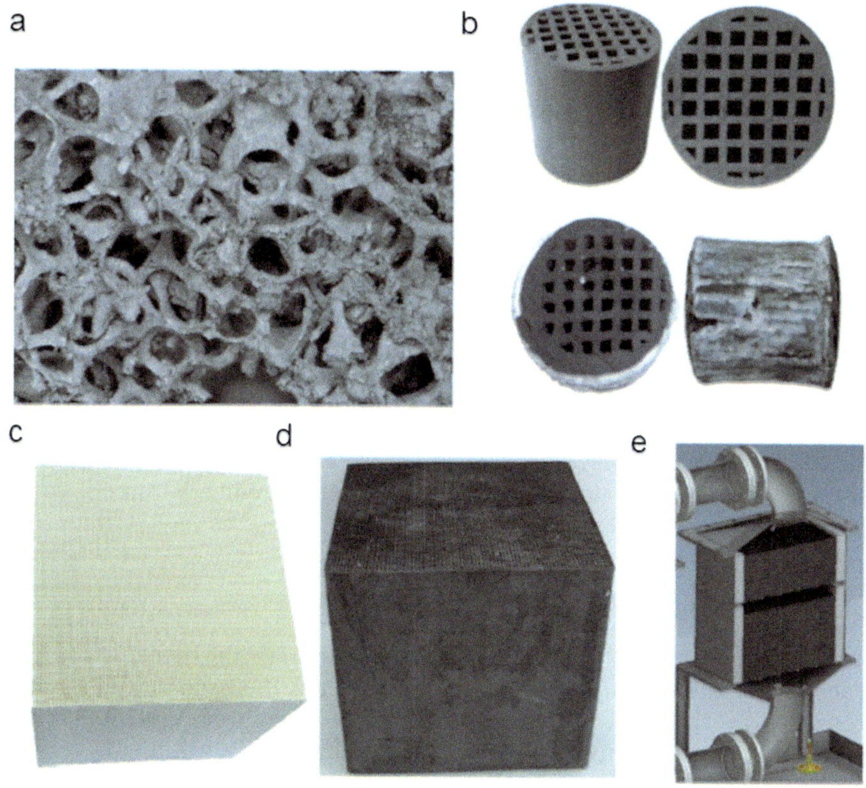

Figure 4.16 (a) Co_3O_4-based porous foams. Reproduced from ref. 184 with permission from Elsevier, Copyright 2015. (b) Pictures of Co_3O_4/Al_2O_3-based honeycombs (top: freshly made, bottom: after 104 cycles). Reproduced from ref. 186 with permission from Elsevier, Copyright 2016. Uncoated (c) and Co oxide-coated cordierite honeycomb, (d) used for thermochemical heat storage at pilot plant. (e) 3D model of a TCS unit utilising Co oxide coated honeycombs. Reproduced from ref. 126 with permission from Elsevier, Copyright 2017.

Experimental tests with oxide granulates of different particle sizes involved the use of fluidised bed reactors,[178] packed bed reactors[170] and rotary kilns.[179,180] Depending on the reactor configuration and its implementation inside the CSP plant, the granule size and shape can strongly affect the heat storage/release operation. For instance, Ströhle *et al.* discovered that for the Mn_2O_3/Mn_3O_4 system, a fluidised bed reactor with spheres of 100 μm size performed better in a *serial* configuration. However, a packed bed of granules of 5 mm better suits a *parallel* configuration, pointing out the remarkable influence of reactor type and particle properties on the TCS unit performance.[181] An alternative reactor scheme consists of the incorporation or extrusion of redox oxides into ceramic honeycombs, see Figure 4.16. This heat exchanger/reactor concept was successfully tested for Co_3O_4 coated onto porous monoliths system[162,182–184] reaching implementation on a pilot plant scale at the Solar Tower Jülich (Germany), Figure 4.16c–e.[126] Several porous ceramic structures have been experimentally tested from which we can separate two arrangements of the composite: monolithic structures in which cobalt oxide is directly coated onto a ceramic (Al_2O_3, ZrO_2, SiC, Cordierite) foam or a honeycomb,[183] and porous structures entirely made of Co_3O_4, as shown in Figure 4.16.[184–186] For this last approach, techniques such as the replica method[184] or ceramic extrusion[186] were assessed. However, in most of the cases, oxide-based porous structures deteriorate upon cycling, due to thermal and chemical stresses caused by the high-temperature cycling and the phase changes occurring in the redox transition. On the other hand, coated monoliths exhibit better multicycle performance, but they also present a decrease in the total volumetric heat storage density, since normally the ceramic support is not redox active. In addition, these structures can facilitate the heat transference if the substrate presents higher thermal conductivity, as in the case of SiC. Nevertheless, the remarkable results and knowledge gained in cobalt oxide-coated honeycombs, already tested at pilot plant scale, represent a noteworthy milestone in the development of redox-thermochemical heat storage for concentrated solar power plants.

4.4.4 Concluding Remarks on TCS

Redox oxides represent a promising class of materials for thermochemical heat storage. The relatively easy tunability of the thermodynamic and kinetic properties by chemical modifications enables finding compositions that greatly improve the performance of the pure redox pair. Additionally, recent research advances have shown that such oxides can be shaped as porous structures, which have been successfully tested at pilot plant scale.

In summary, we have shown how materials research plays an important role in improving the oxide properties and reaction kinetics, leading to optimised ceramic as heat storage media. These experimental modifications could greatly benefit from theoretical approaches including DFT calculations and thermodynamic modelling, which can provide guidelines for selecting redox compositions with increased energy storage density and narrow hysteresis gaps using affordable elements. Besides, gaining a better understanding of the rate-limiting factor at atomic scale by means of *in situ* or *operando* techniques can also contribute to selecting the most adequate strategies to enhance the performance of redox oxides for these applications, by means of faster reaction rates and more stable multicycle stability. Finally, long-term stability, not only in terms of redox reversibility but also concerning mechanical integrity, is also a crucial aspect for the applicability of these systems, which need to withstand the severe conditions of cycling operation at high temperature. Since the replacement of the material is determined by its durability, this aspect can have an important economic impact on the plant operation.

4.5 Future Outlook of Thermochemical Cycles with Redox Oxides

Recent achievements in the thermochemical storage of solar energy using redox oxides have prompted the current research interest in this field. The huge chemical, structural and morphological variety of solid oxides allows the forecasting of new exciting developments in a short–medium term. In the search for materials, even more active theoretical studies, either by classic thermodynamics or using systematic quantum mechanical calculations such as those based on DFT, would play a key role in narrowing the range of suitable structures and compositions. Nevertheless, experimental work will still be crucial to confirm the adequacy of the selected materials because complex aspects of the processes such as kinetics and deactivation are more difficult to implement in theoretical studies than energetic aspects. Similarly, modulation of the activity of the relevant redox solids by morphological and textural modifications or, if necessary, by the incorporation of suitable catalytic components, will require most likely extensive and systematic experimental tests. In any case, considering the probed flexibility of solid redox oxides and the relatively small number of compositions so far assayed, expectations for a rapid advance are high.

The possible success of future developments at the commercial scale of thermochemical processes for solar fuel production based on redox oxides will depend not only on improved performance and cost reduction, but also on the evolution of competing photo- and electro-activated transformation chains. Reduction of the operation temperatures and concentration factors may simplify design and allow for more modular systems that can be more viable for a broader deployment of these technologies. In this respect, a hybrid system activated by methane, despite providing a more modest sustainability gain, can accelerate the adoption of thermochemical technologies by taking advantage of existing infrastructures and industrial experience.

In contrast, thermal storage, conceptually simpler but challenging to implement, most likely will find a number of application opportunities. Nevertheless, improving the energy density, heat transference and long-term stability in economically affordable materials are still important goals to be reached.

Acknowledgements

Financial support from the projects SOLARKITE of the '*Ramón Areces*' Foundation and ACES2030 from "Comunidad de Madrid" and European Structural Funds to (P2018/EMT-4319) is fully acknowledged.

References

1. http://unfccc.int/meetings/paris_nov_2015/meeting/8926.php.
2. International Energy Agency, Key World Energy Statistics 2017, https://www.iea.org/publications/freepublications/publication/Key-World2017.pdf.
3. BP ENergy Outlook, 2017 Edition, https://www.bp.com/content/dam/bp/pdf/energy-economics/energy-outlook-2017/bp-energy-outlook-2017.pdf.
4. S. Y. Searle and C. J. Malins, *Biomass Bioenergy*, 2014, **65**, 3–12.
5. E. A. Olivetti, G. Ceder, G. G. Gaustad and X. Fu, *Joule*, 2017, **1**, 229–243.
6. I. Garbayo, M. Struzik, W. J. Bowman, R. Pfenninger, E. Stilp and J. L. M. Rupp, *Adv. Energy Mater.*, 2018, **1702265**, 1–14.
7. J. van den Broek, S. Afyon and J. L. M. Rupp, *Adv. Energy Mater.*, 2016, **6**, 1–11.
8. S. B. Walker, U. Mukherjee, M. Fowler and A. Elkamel, *Int. J. Hydrogen Energy*, 2016, **41**, 7717–7731.
9. T. Schaaf, J. Groenig, M. R. Schuster, T. Rothenfluh and A. Orth, *Energy Sustain. Soc.*, 2014, **4**, 1–14.
10. K. Ghaib and F. Z. Ben-Fares, *Renewable Sustainable Energy Rev.*, 2018, **81**, 433–446.
11. L. K. Rihko-Struckmann, A. Peschel, R. Hanke-Rauschenbach and K. Sundmacher, *Ind. Eng. Chem. Res.*, 2010, **49**, 11073–11078.
12. F. Studt, I. Sharafutdinov, F. Abild-Pedersen, C. F. Elkjær, J. S. Hummelshøj, S. Dahl, I. Chorkendorff and J. K. Nørskov, *Nat. Chem.*, 2014, **6**, 320–324.
13. M. Bjørgen, F. Joensen, M. Spangsberg Holm, U. Olsbye, K. P. Lillerud and S. Svelle, *Appl. Catal., A*, 2008, **345**, 43–50.
14. C. Wulf, J. Linßen and P. Zapp, *Energy Procedia*, 2018, **155**, 367–378.
15. S. Rönsch, J. Schneider, S. Matthischke, M. Schlüter, M. Götz, J. Lefebvre, P. Prabhakaran and S. Bajohr, *Fuel*, 2016, **166**, 276–296.
16. http://www.ceops-project.eu/sites/default/files/documents/ceops_ws_5._rieke.pdf.
17. L. Chen, F. Chen and C. Xia, *Energy Environ. Sci.*, 2014, **7**, 4018–4022.
18. Z. Han, R. Kortlever, H.-Y. Chen, J. C. Peters and T. Agapie, *ACS Cent. Sci.*, 2017, **3**, 853–859.
19. J. A. Herron, J. Kim, A. A. Upadhye, G. W. Huber and C. T. Maravelias, *Energy Environ. Sci.*, 2015, **8**, 126–157.
20. F. Fresno, R. Portela, S. Suárez and J. M. Coronado, *J. Mater. Chem. A*, 2014, **2**, 2863–2884.
21. K. Maeda, K. Teramura, D. Lu, T. Takata, N. Saito, Y. Inoue and K. Domen, *Nature*, 2006, **440**, 295.
22. Q. Wang, T. Hisatomi, Y. Suzuki, Z. Pan, J. Seo, M. Katayama, T. Minegishi, H. Nishiyama, T. Takata, K. Seki, A. Kudo, T. Yamada and K. Domen, *J. Am. Chem. Soc.*, 2017, **139**, 1675–1683.

23. S. Y. Reece, J. A. Hamel, S. Sung, T. D. Jarvi, A. J. Esswein, J. J. H. Pijpers and D. G. Nocera, *Science*, 2011, **1**, 3.
24. L. A. Weinstein, J. Loomis, B. Bhatia, D. M. Bierman, E. N. Wang and G. Chen, *Chem. Rev.*, 2015, **115**, 12797–12838.
25. W. F. Sawin J., K. Seyboth, *Renewables 2017 Global Status Report*, 2017. www.ren21.net/gsr-2017.
26. J. Lilliestam, M. Labordena, A. Patt and S. Pfenninger, *Nat. Energy*, 2017, **2**, 17094.
27. A. J. Carrillo, J. González-Aguilar, M. Romero and J. M. Coronado, *Chem. Rev.*, 2019, **119**, 4777–4816.
28. T. Kodama and N. Gokon, *Chem. Rev.*, 2007, **107**, 4048–4077.
29. D. Marxer, P. Furler, M. Takacs and A. Steinfeld, *Energy Environ. Sci.*, 2017, **10**, 1142–1149.
30. S. Abanades, P. Charvin, G. Flamant and P. Neveu, *Energy*, 2006, **31**, 2469–2486.
31. http://www.psa.es/en/areas/ussc/grupocsspi/projects/hydrosol-3d.php.
32. http://www.sun-to-liquid.eu/.
33. D. M. Fabian, S. Hu, N. Singh, F. A. Houle, T. Hisatomi, K. Domen, F. E. Osterloh and S. Ardo, *Energy Environ. Sci.*, 2015, **8**, 2825–2850.
34. C. L. Muhich, S. Blaser, M. C. Hoes and A. Steinfeld, *Int. J. Hydrogen Energy*, 2018, **43**, 18814–18831.
35. M. Kathe, A. Empfield, P. Sandvik, C. Fryer, Y. Zhang, E. Blair and L.-S. Fan, *Energy Environ. Sci.*, 2017, **10**, 1345–1349.
36. J. Maier, *Physical Chemistry of Ionic Materials: Ions and Electrons in Solids No Title*, John Wiley and Sons, 2004.
37. M. Kubicek, A. H. Bork and J. L. M. Rupp, *J. Mater. Chem. A*, 2017, **5**, 11983–12000.
38. R. Michalsky, V. Botu, C. M. Hargus, A. A. Peterson and A. Steinfeld, *Adv. Energy Mater.*, 2015, **5**, 1401082.
39. J. Hwang, R. R. Rao, L. Giordano, Y. Katayama, Y. Yu and Y. Shao-Horn, *Science*, 2017, **358**, 751–756.
40. J. Suntivich, H. A. Gasteiger, N. Yabuuchi, H. Nakanishi, J. B. Goodenough and Y. Shao-Horn, *Nat. Chem.*, 2011, **3**, 546–550.
41. S. F. Bychkov, I. I. Gainutdinov, S. A. Chizhik and A. P. Nemudry, *Solid State Ionics*, 2018, **320**, 297–304.
42. A. Ruiz Puigdollers, P. Schlexer, S. Tosoni and G. Pacchioni, *ACS Catal.*, 2017, **7**, 6493–6513.
43. Q. Jiang, Z. Chen, J. Tong, M. Yang, Z. Jiang and C. Li, *ACS Catal.*, 2016, **6**, 1172–1180.
44. F. Lin, M. Rothensteiner, I. Alxneit, J. A. van Bokhoven and A. Wokaun, *Energy Environ. Sci.*, 2016, **9**, 2400–2409.
45. S. Mostrou, R. Büchel, S. E. Pratsinis, J. A. Van Bokhoven and J. A. van Bokhoven, *Appl. Catal., A*, 2017, **537**, 40–49.
46. M. Ezbiri, M. Takacs, B. Stolz, J. Lungthok, A. Steinfeld and R. Michalsky, *J. Mater. Chem. A*, 2017, **5**, 15105–15115.

47. A. H. Bork, E. Povoden-Karadeniz and J. L. M. Rupp, *Adv. Energy Mater.*, 2017, **7**, 1601086.
48. M. J. Ignatowich, A. H. Bork, T. C. Davenport, J. L. M. Rupp, C. Yang, Y. Yamazaki and S. M. Haile, *MRS Commun.*, 2017, **7**, 873–878.
49. A. A. Emery, J. E. Saal, S. Kirklin, V. I. Hegde and C. Wolverton, *Chem. Mater.*, 2016, **28**, 5621–5634.
50. M. Romero and A. Steinfeld, *Energy Environ. Sci.*, 2012, **5**, 9234–9245.
51. T. Nakamura, *Sol. Energy*, 1977, **19**, 467–475.
52. E. Bilgen, M. Ducarroir, M. Foex, F. Sibieude and F. Trombe, *Int. J. Hydrogen Energy*, 1977, **2**, 251–257.
53. A. Steinfeld, S. Sanders and R. Palumbo, *Sol. Energy*, 1999, **65**, 43–53.
54. F. Fresno, R. Fernández-Saavedra, M. Belén Gómez-Mancebo, A. Vidal, M. Sánchez, M. Isabel Rucandio, A. J. Quejido and M. Romero, *Int. J. Hydrogen Energy*, 2009, **34**, 2918–2924.
55. N. B. Goikoetxea, M. B. Gómez-Mancebo, R. Fernández-Saavedra, F. García-Pérez, J. A. Jiménez, J. Rodríguez, I. Rucandio and A. J. Quejido, *Int. J. Hydrogen Energy*, 2016, **41**, 16696–16704.
56. M. Roeb, J. P. Säck, P. Rietbrock, C. Prahl, H. Schreiber, M. Neises, L. de Oliveira, D. Graf, M. Ebert, W. Reinalter, M. Meyer-Grünefeldt, C. Sattler, A. Lopez, A. Vidal, A. Elsberg, P. Stobbe, D. Jones, A. Steele, S. Lorentzou, C. Pagkoura, A. Zygogianni, C. Agrafiotis and A. G. Konstandopoulos, *Sol. Energy*, 2011, **85**, 634–644.
57. F. Fresno, T. Yoshida, N. Gokon, R. Fernández-Saavedra and T. Kodama, *Int. J. Hydrogen Energy*, 2010, **35**, 8503–8510.
58. J. R. Scheffe, J. Li and A. W. Weimer, *Int. J. Hydrogen Energy*, 2010, **35**, 3333–3340.
59. C. L. Muhich, B. W. Evanko, K. C. Weston, P. Lichty, X. Liang, J. Martinek, C. B. Musgrave and A. W. Weimer, *Science*, 2013, **341**, 540–542.
60. C. L. Muhich, B. D. Ehrhart, V. A. Witte, S. L. Miller, E. N. Coker, C. B. Musgrave and A. W. Weimer, *Energy Environ. Sci.*, 2015, **8**, 3687–3699.
61. W. Villasmil, T. Cooper, E. Koepf, A. Meier and A. Steinfeld, *J. Sol. Energy Eng.*, 2016, **139**, 021015.
62. B. Xu, Y. Bhawe and M. E. Davis, *Chem. Mater.*, 2013, **25**, 1564–1571.
63. A. Bayón, V. A. de la Peña O'Shea, J. M. Coronado and D. P. Serrano, *Int. J. Hydrogen Energy*, 2016, **41**, 113–122.
64. C. Herradón, R. Molina, J. Marugán and J. Á. Botas, *Int. J. Hydrogen Energy*, 2019, **44**, 91–100.
65. S. Abanades and G. Flamant, *Sol. Energy*, 2006, **80**, 1611–1623.
66. W. Chueh, C. Falter, M. Abbott, D. Scipio, P. Furler, S. M. Haile and A. Steinfeld, *Science*, 2010, **330**, 1797–1801.
67. J. L. M. Rupp, E. Fabbri, D. Marrocchelli, J. W. Han, D. Chen, E. Traversa, H. L. Tuller and B. Yildiz, *Adv. Funct. Mater.*, 2014, **24**, 1562–1574.
68. J. L. M. Rupp, A. Infortuna and L. J. Gauckler, *J. Am. Ceram. Soc.*, 2007, **90**, 1792–1797.
69. M. Mogensen, N. M. Sammes and G. A. Tompsett, *Solid State Ionics*, 2000, **129**, 63–94.

70. S. Ackermann, J. R. Scheffe and A. Steinfeld, *J. Phys. Chem. C*, 2014, **118**, 5216–5225.
71. G. E. Murgida, V. Ferrari, M. V. Ganduglia-Pirovano and A. M. Llois, *Phys. Rev. B: Condens. Matter Mater. Phys.*, 2014, **90**, 1–12.
72. Mineral Commodities Summaries 2017, US Geological Survey. www.usgs.gov/centers/nmic.
73. A. Le Gal and S. Abanades, *J. Phys. Chem. C*, 2012, **116**, 13516–13523.
74. J. R. Scheffe, R. Jacot, G. R. Patzke and A. Steinfeld, *J. Phys. Chem. C*, 2013, **117**, 24104–24114.
75. A. Le Gal, S. Abanades and G. Flamant, *Energy Fuels*, 2011, **25**, 4836–4845.
76. M. Kuhn, S. R. Bishop, J. L. M. Rupp and H. L. Tuller, *Acta Mater.*, 2013, **61**, 4277–4288.
77. A. Le Gal and S. Abanades, *Int. J. Hydrogen Energy*, 2011, **36**, 4739–4748.
78. Y. Hao, C.-K. Yang and S. M. Haile, *Chem. Mater.*, 2014, **26**, 6073–6082.
79. C. Muhich and A. Steinfeld, *J. Mater. Chem. A*, 2017, **5**, 15578–15590.
80. C. Muhich, M. Hoes and A. Steinfeld, *Acta Mater.*, 2018, **144**, 728–737.
81. M. Hoes, C. L. Muhich, R. Jacot, G. R. Patzke and A. Steinfeld, *J. Mater. Chem. A*, 2017, **5**, 19476–19484.
82. S. G. Rudisill, L. J. Venstrom, N. D. Petkovich, T. Quan, N. Hein, D. B. Boman, J. H. Davidson and A. Stein, *J. Phys. Chem. C*, 2013, **117**, 1692–1700.
83. P. Furler, J. R. Scheffe, D. Marxer, M. Gorbar, A. Bonk, U. Vogt and A. Steinfeld, *Phys. Chem. Chem. Phys.*, 2014, **16**, 10503–10511.
84. M. Tou, R. Michalsky and A. Steinfeld, *Joule*, 2017, **1**, 146–154.
85. T. C. Davenport, M. Kemei, M. J. Ignatowich and S. M. Haile, *Int. J. Hydrogen Energy*, 2017, **42**, 16932–16945.
86. M. A. Peña and J. L. G. Fierro, *Chem. Rev.*, 2001, **101**, 1981–2018.
87. V. M. Goldschmidt, *Naturwissenschaften*, 1926, **14**, 477–485.
88. J. T. S. Irvine, D. Neagu, M. C. Verbraeken, C. Chatzichristodoulou, C. Graves and M. B. Mogensen, *Nat. Energy*, 2016, **1**, 15014.
89. M. M. Nair and S. Abanades, *Sustainable Energy Fuels*, 2018, **2**, 843–854.
90. A. H. Bork, M. Kubicek, M. Struzik and J. L. M. Rupp, *J. Mater. Chem. A*, 2015, **3**, 15546–15557.
91. J. R. Scheffe, D. Weibel and A. Steinfeld, *Energy Fuels*, 2013, **27**, 4250–4257.
92. A. H. McDaniel, *Curr. Opin. Green Sustain. Chem.*, 2017, **4**, 37–43.
93. M. Takacs, M. Hoes, M. Caduff, T. Cooper, J. R. Scheffe and A. Steinfeld, *Acta Mater.*, 2016, **103**, 700–710.
94. T. Cooper, J. R. Scheffe, M. E. Galvez, R. Jacot, G. Patzke and A. Steinfeld, *Energy Technol.*, 2015, **3**, 1130–1142.
95. C.-K. Yang, Y. Yamazaki, A. Aydin and S. M. Haile, *J. Mater. Chem. A*, 2014, **2**, 13612–13623.
96. M. Ezbiri, M. Takacs, D. Theiler, R. Michalsky and A. Steinfeld, *J. Mater. Chem. A*, 2017, **5**, 4172–4182.

97. A. M. Deml, V. Stevanović, A. M. Holder, M. Sanders, R. Ohayre and C. B. Musgrave, *Chem. Mater.*, 2014, **26**, 6595–6602.
98. J. Vieten, B. Bulfin, P. Huck, M. Horton, D. Gubán, L. Zhu, Y. Lu, K. Persson, M. Roeb and C. Sattler, *Energy Environ. Sci.*, 2019, **12**, 1369–1384.
99. M. Ezbiri, V. Becattini, M. Hoes, R. Michalsky and A. Steinfeld, *ChemSusChem*, 2017, **10**, 1517–1525.
100. M. E. Gálvez, R. Jacot, J. Scheffe, T. Cooper, G. Patzke and A. Steinfeld, *Phys. Chem. Chem. Phys.*, 2015, **17**, 6629–6634.
101. A. H. McDaniel, E. C. Miller, D. Arifin, A. Ambrosini, E. N. Coker, R. O'Hayre, W. C. Chueh and J. Tong, *Energy Environ. Sci.*, 2013, **6**, 2424.
102. D. Sastre, A. J. Carrillo, D. P. Serrano, P. Pizarro and J. M. Coronado, *Top. Catal.*, 2017, **60**, 1108–1118.
103. S. Dey, B. S. Naidu, A. Govindaraj and C. N. R. Rao, *Phys. Chem. Chem. Phys.*, 2015, **17**, 122–125.
104. S. Dey, B. S. Naibu and C. N. R. Rao, *Chem.–Eur. J.*, 2015, **21**, 7077–7081.
105. S. Dey, B. S. Naidu and C. N. R. Rao, *Dalton Trans.*, 2016, **45**, 2430–2435.
106. L. Wang, M. Al-Mamun, Y. L. Zhong, L. Jiang, P. Liu, Y. Wang, H. G. Yang and H. Zhao, *Sustainable Energy Fuels*, 2017, **1**, 1013–1017.
107. A. Demont, S. Abanades and E. Beche, *J. Phys. Chem. C*, 2014, **118**, 12682–12692.
108. D. Romero Barcellos, M. D. Sanders, J. Tong, A. H. McDaniel and R. O'Hayre, *Energy Environ. Sci.*, 2018, **11**, 3256–3265.
109. S. Zhai, J. Rojas, N. Ahlborg, K. Lim, M. F. Toney, H. Jin, W. C. Chueh and A. Majumdar, *Energy Environ. Sci.*, 2018, **11**, 2172–2178.
110. Z. Chen, Q. Jiang, F. Cheng, J. Tong, M. Yang, Z. Jiang and C. Li, *J. Mater. Chem. A*, 2019, **7**, 6099–6112.
111. J. Zhang, V. Haribal and F. Li, *Sci. Adv.*, 2017, **3**, e1701184.
112. P. T. Krenzke and J. H. Davidson, *Energy Fuels*, 2014, **28**, 4088–4095.
113. M. M. Nair and S. Abanades, *Energy Fuels*, 2016, **30**, 6050–6058.
114. M. Welte, K. Warren, J. R. Scheffe and A. Steinfeld, *Ind. Eng. Chem. Res.*, 2017, **56**, 10300–10308.
115. X. Gao, G. Liu, Y. Zhu, P. Kreider, A. Bayon, T. Gengenbach, T. Lu, Y. Liu, J. Hinkley, W. Lipiński and A. Tricoli, *Nano Energy*, 2018, **50**, 347–358.
116. L. Li, J. Lin, X. Wang, C.-R. Chang, C. Ruan, Z.-Q. Huang, X. Liu, M. Tian, J. Li and C. Huang, *Energy Environ. Sci.*, 2019, **12**, 767–779.
117. Q. Imtiaz, N. S. Yüzbasi, P. M. Abdala, A. M. Kierzkowska, W. van Beek, M. Broda and C. R. Müller, *J. Mater. Chem. A*, 2016, **4**, 113–123.
118. L. C. Buelens, L. C. Buelens, V. V. Galvita, H. Poelman, C. Detavernier and G. B. Marin, *Science*, 2016, **354**, 449–452.
119. A. Evdou, V. Zaspalis and L. Nalbandian, *Fuel*, 2010, **89**, 1265–1273.
120. D. Sastre, D. P. Serrano, P. Pizarro and J. M. Coronado, *J. CO2 Util.*, 2019, **31**, 16–26.

121. D. Maiti, B. J. Hare, Y. A. Daza, A. E. Ramos, J. N. Kuhn and V. R. Bhethanabotla, *Energy Environ. Sci.*, 2018, **11**, 648–659.
122. M. Romero and J. González-Aguilar, *Wiley Interdiscip. Rev.: Energy Environ.*, 2014, **3**, 42–59.
123. S. Kuravi, J. Trahan, D. Y. Goswami, M. M. Rahman and E. K. Stefanakos, *Prog. Energy Combust. Sci.*, 2013, **39**, 285–319.
124. A. Gil, M. Medrano, I. Martorell, A. Lázaro, P. Dolado, B. Zalba and L. F. Cabeza, *Renewable Sustainable Energy Rev.*, 2010, **14**, 31–55.
125. L. André, S. Abanades and G. Flamant, *Renewable Sustainable Energy Rev.*, 2016, **64**, 703–715.
126. S. Tescari, A. Singh, C. Agrafiotis, L. de Oliveira, S. Breuer, B. Schlögl-Knothe, M. Roeb and C. Sattler, *Appl. Energy*, 2017, **189**, 66–75.
127. M. Paskevicius, D. A. Sheppard, K. Williamson and C. E. Buckley, *Energy*, 2015, **88**, 469–477.
128. R. Dunn, K. Lovegrove and G. Burgess, *Proc. IEEE*, 2012, **100**, 391–400.
129. X. Peng, T. W. Root and C. T. Maravelias, *Green Chem.*, 2017, **19**, 2427–2438.
130. P. Pardo, A. Deydier, Z. Anxionnaz-Minvielle, S. Rougé, M. Cabassud and P. Cognet, *Renewable Sustainable Energy Rev.*, 2014, **32**, 591–610.
131. D. N. Harries, M. Paskevicius, D. A. Sheppard, T. E. C. Price and C. E. Buckley, *Proc. IEEE*, 2012, **100**, 539–549.
132. M. Fellet, C. E. Buckley, M. Paskevicius and D. A. Sheppard, *MRS Bull.*, 2013, **38**, 1012–1013.
133. Y. A. Criado, A. Huille, S. Rouge and J. C. Abanades, *Chem. Eng. J.*, 2017, **313**, 1194–1205.
134. Y. A. Criado, M. Alonso and J. C. Abanades, *Ind. Eng. Chem. Res.*, 2014, **53**, 12594–12601.
135. F. Schaube, L. Koch, A. Wörner and H. Müller-Steinhagen, *Thermochim. Acta*, 2012, **538**, 9–20.
136. F. Schaube, A. Kohzer, J. Schütz, A. Wörner and H. Müller-Steinhagen, *Chem. Eng. Res. Des.*, 2013, **91**, 856–864.
137. M. Schmidt and M. Linder, *Appl. Energy*, 2017, **203**, 594–607.
138. E. Mastronardo, L. Bonaccorsi, Y. Kato, E. Piperopoulos, M. Lanza and C. Milone, *Appl. Therm. Eng.*, 2017, **120**, 626–634.
139. E. Mastronardo, Y. Kato, L. Bonaccorsi, E. Piperopoulos and C. Milone, *Energies*, 2017, **10**, 70.
140. M. Schmidt, M. Gollsch, F. Giger, M. Grün and M. Linder, Solar PACES, *AIP Proceedings*, 2016, **1734**, 050041.
141. M. Schmidt, A. Gutierrez and M. Linder, *Appl. Energy*, 2017, **188**, 672–681.
142. C. Roßkopf, M. Haas, A. Faik, M. Linder and A. Wörner, *Energy Convers. Manage.*, 2014, **86**, 93–98.
143. M. Benitez-Guerrero, J. M. Valverde, P. E. Sanchez-Jimenez, A. Perejon and L. A. Perez-Maqueda, *Sol. Energy*, 2017, **153**, 188–199.
144. N. R. Rhodes, A. Barde, K. Randhir, L. Li, D. W. Hahn, R. Mei, J. F. Klausner and N. AuYeung, *ChemSusChem*, 2015, **8**, 3793–3798.

145. R. G. Bowrey and J. Jutsen, *Sol. Energy*, 1978, **21**, 523–525.
146. M. A. Fahim and J. D. Ford, *Chem. Eng. J.*, 1983, **27**, 21–28.
147. D. Chadda, J. D. Ford and M. A. Fahim, *Int. J. Energy Res.*, 1989, **13**, 63–73.
148. B. Wong, L. Brown, F. Schaube, R. Tamme and C. Sattler in *16th Solar PACES International Symposium*, Perpignan, France, 2010, pp. 1–8.
149. M. Wokon, T. Block, S. Nicolai, M. Linder and M. Schmücker, *Sol. Energy*, 2017, **153**, 471–485.
150. T. Block and M. Schmücker, *Sol. Energy*, 2016, **126**, 195–207.
151. S. M. Babiniec, E. N. Coker, J. E. Miller and A. Ambrosini, *Sol. Energy*, 2015, **118**, 451–459.
152. S. M. Babiniec, E. N. Coker, A. Ambrosini and J. E. Miller, *AIP Conf. Proc.*, 2016, **1734**, 050006.
153. L. Imponenti, K. J. Albrecht, J. W. Wands, M. D. Sanders and G. S. Jackson, *Sol. Energy*, 2017, **151**, 1–13.
154. L. Imponenti, K. J. Albrecht, R. Kharait, M. D. Sanders and G. S. Jackson, *Appl. Energy*, 2018, **230**, 1–18.
155. S. M. Babiniec, E. N. Coker, J. E. Miller and A. Ambrosini, *Int. J. Energy Res.*, 2016, **40**, 280–284.
156. E. Mastronardo, X. Qian, J. M. Coronado and S. M. Haile, Solar PACES, *AIP Conf. Proc*, 2019, **2126**, 210005.
157. A. J. Carrillo, D. P. Serrano, P. Pizarro and J. M. Coronado, *J. Mater. Chem. A*, 2014, **2**, 19435–19443.
158. J. E. Miller, A. H. McDaniel and M. D. Allendorf, *Adv. Energy Mater.*, 2014, **4**, 1300469.
159. A. J. Carrillo, D. P. Serrano, P. Pizarro and J. M. Coronado, *J. Energy Storage*, 2016, **5**, 169–176.
160. D. Hosseini, Q. Imtiaz, P. M. Abdala, S. Yoon, A. Kierzkoska, A. Weidenkaff, C. R. Müller, A. M. Kierzkowska, A. Weidenkaff and C. R. Müller, *J. Mater. Chem. A*, 2015, **3**, 10545–10550.
161. A. J. Carrillo, D. Sastre, D. P. Serrano, P. Pizarro and J. M. Coronado, *Phys. Chem. Chem. Phys.*, 2016, **18**, 8039–8048.
162. C. Agrafiotis, M. Roeb and C. Sattler, *Sol. Energy*, 2016, **139**, 695–710.
163. M. Deutsch, F. Horvath, C. Knoll, D. Lager, C. Gierl-Mayer, P. Weinberger and F. Winter, *Energy Fuels*, 2017, **31**, 2324–2334.
164. L. Andre, S. Abanades and L. Cassayre, *J. Solid State Chem.*, 2017, **253**, 6–14.
165. T. Block, N. Knoblauch and M. Schmücker, *Thermochim. Acta*, 2013, **577**, 25–32.
166. A. J. Carrillo, D. P. Serrano, P. Pizarro and J. M. Coronado, *AIP Conf. Proc.*, 2016, **2**, 050009.
167. A. J. Carrillo, J. Moya, A. Bayón, P. Jana, V. A. de la Peña O'Shea, M. Romero, J. Gonzalez-Aguilar, D. P. Serrano, P. Pizarro and J. M. Coronado, *Sol. Energy Mater. Sol. Cells*, 2014, **123**, 47–57.
168. F. Varsano, C. Alvani, A. La Barbera, A. Masi and F. Padella, *Thermochim. Acta*, 2016, **640**, 26–35.

169. A. J. Carrillo, D. P. Serrano, P. Pizarro and J. M. Coronado, *ChemSusChem*, 2015, **8**, 1947–1954.
170. M. Wokon, A. Kohzer and M. Linder, *Sol. Energy*, 2017, **153**, 200–214.
171. C. Agrafiotis, T. Block, M. Senholdt, S. Tescari, M. Roeb and C. Sattler, *Sol. Energy*, 2017, **149**, 227–244.
172. J. V. Crum, B. J. Riley and J. D. Vienna, *J. Am. Ceram. Soc.*, 2009, **92**, 2378–2384.
173. A. J. Carrillo, D. P. Serrano, P. Pizarro and J. M. Coronado, *J. Phys. Chem. C*, 2016, **120**, 27800–27812.
174. *Diffusion in Condensed Matter*, ed. P. Heitjans and J. Kärger, Springer-Verlag, Berlin, Heidelberg, 2005.
175. K. Lee and J. R. Scheffe, *J. Sol. Energy Eng.*, 2018, **141**, 021013.
176. K. King, K. Randhir and J. Klausner, *Thermochim. Acta*, 2019, **673**, 105–118.
177. K. Randhir, K. King, N. Rhodes, L. Li, D. Hahn, R. Mei, N. Auyeung, J. Klausner and E. Lansing, *J. Energy Storage*, 2019, **21**, 599–610.
178. S. Á. de Miguel, S. Bellan, J. M. G. de María, J. González-Aguilar and M. Romero, *AIP Conf. Proc.*, 2016, **1743**, 050005.
179. E. Alonso, C. Pérez-Rábago, J. Licurgo, E. Fuentealba and C. A. Estrada, *Sol. Energy*, 2015, **115**, 297–305.
180. M. Neises, S. Tescari, L. de Oliveira, M. Roeb, C. Sattler and B. Wong, *Sol. Energy*, 2012, **86**, 3040–3048.
181. S. Ströhle, A. Haselbacher, Z. R. Jovanovic and A. Steinfeld, *Energy Environ. Sci.*, 2016, **9**, 1375–1389.
182. C. Agrafiotis, A. Becker, M. Roeb and C. Sattler, *Sol. Energy*, 2016, **114**, 440–458.
183. C. Agrafiotis, M. Roeb, M. Schmücker and C. Sattler, *Sol. Energy*, 2015, **114**, 440–458.
184. C. Agrafiotis, S. Tescari, M. Roeb, M. Schmücker and C. Sattler, *Sol. Energy*, 2015, **114**, 440–458.
185. C. Pagkoura, G. Karagiannakis, A. Zygogianni, S. Lorentzou, M. Kostoglou, A. G. Konstandopoulos, M. Rattenburry and J. W. Woodhead, *Sol. Energy*, 2014, **108**, 146–163.
186. G. Karagiannakis, C. Pagkoura, E. Halevas, P. Baltzopoulou and A. G. Konstandopoulos, *Sol. Energy*, 2016, **133**, 394–407.

CHAPTER 5

Thermoelectric Oxide Materials for Energy Conversion

DURSUN EKREN, FERIDOON AZOUGH AND ROBERT FREER*

Department of Materials, University of Manchester, Manchester M13 9PL, UK
*E-mail: Robert.Freer@manchester.ac.uk

5.1 Introduction

The thermoelectric effect, by which electric charges move in response to a temperature gradient (thereby developing an electromotive force, or voltage), has been known for almost 200 years,[1] but only exploited for power generation applications for the past 60 years and mainly in niche applications.[2] Once a thermoelectric is located in a temperature gradient and connected to an external load or circuit, a current will flow and the power generated can be harvested; it is more beneficial if both p-type and n-type materials are employed (Figure 5.1).

The efficiency of a thermoelectric generator (TEG) in converting thermal energy to electrical energy is given by η_{max}, and detailed in eqn (5.1) where T_M is the mean temperature and ZT is the dimensionless thermoelectric figure of merit. The latter is given by eqn (5.2) where σ, S, κ and T are electrical

Inorganic Materials Series No. 5
Energy Storage and Conversion Materials
Edited by Stephen Skinner

Published by the Royal Society of Chemistry, www.rsc.org

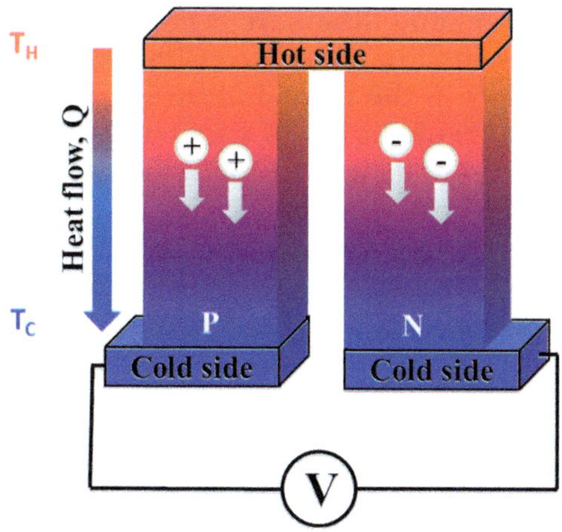

Figure 5.1 Schematic diagram showing charge flow in p-type and n-type thermoelectrics in a thermal gradient giving rise to external potential.

conductivity, the Seebeck coefficient (which is the voltage generated per degree of temperature difference), thermal conductivity and absolute temperature, respectively.

$$\eta_{\max} = \frac{T_H - T_C}{T_H}\left[\frac{\sqrt{1+ZT_M}-1}{\sqrt{1+ZT_M}+\left(T_C/T_H\right)}\right] \tag{5.1}$$

$$ZT = \frac{\sigma S^2}{\kappa}T \tag{5.2}$$

For maximum ZT, materials need to exhibit a high Seebeck coefficient, high electrical conductivity and low thermal conductivity. To achieve the optimum combination of properties, most thermoelectrics need to be semiconductors, or exhibit semiconductor characteristics. As many oxides are inherently insulators, doping is required to achieve the required conductivities. The fact that both σ and κ are directly linked together presents challenges for increasing thermoelectric performance. Furthermore, since the mean free path of phonons in oxides is typically 0.2–2 nm (much smaller than for traditional thermoelectric semiconductors), the reduction of κ by nanostructuring is more demanding and therefore, atomic-scale structural features are generally used for introducing mass and size fluctuations in oxides.[3–5]

Thermoelectric modules used to harvest energy comprise multiple pairs of n-type and p-type elements (as in Figure 5.1) connected in parallel thermally and in series electrically. The main advantages of TEGs are direct energy conversion, no moving parts, scalability, long lifespan (TEGs have been used continuously for several decades[2]) and no vibration or noise during operation.[6] The underlying principles of TEGs are well established and modules employing traditional thermoelectric materials such as Bi_2Te_3 and SiGe have been used in deep space probes by NASA for over 40 years.[2] In spite of considerable opportunities to use TEGs to harvest energy in the many industrial, transport and domestic environments,[6,7] their exploitation to date has been limited because of the modest conversion efficiency, the limited temperature range at peak performance, plus the cost and stability of the traditional thermoelectric materials.[8] To overcome these problems there have been considerable efforts to develop materials with higher thermoelectric performance and higher temperature stability. This has included a range of metallic materials including skutterudites, half Heuslers and clathrates,[9] but most tend to degrade at high temperatures and still require some form of complex, cascade-type module system to be effective in wide temperature gradients.[10] These limitations, combined with increasing concern over the use of toxic raw materials focused attention on more environmentally friendly, lighter weight alternatives and this led to the development of the first oxide thermoelectrics in the 1990s.[11] Since oxides offer considerable design flexibility, because of their chemical versatility and structural complexity, a wide variety of oxide thermoelectrics has been developed. Today, these can be broadly divided into two main groups:[12] (a) p-type oxides, which are mainly layered cobaltites ($Ca_3Co_4O_9$, $Bi_2Sr_2Co_2O_y$ and Na_xCoO_2) and (b) n-type oxides including $SrTiO_3$, $CaMnO_3$ and ZnO. For the purpose of this chapter we will concentrate on the materials that have attracted the most interest; these will be considered in six families, namely: (i) layered cobaltites, (ii) ZnO-based materials, (iii) tungsten bronze structured oxides, (iv) $CaMnO_3$, (v) TiO_2/Magnéli phases and (vi) A-site deficient perovskites, including $SrTiO_3$. As strontium titanate-based materials have attracted considerable attention, more details will be provided about them as the approach is generic and relevant to many families of thermoelectric oxides.

Numerous publications are available for oxide thermoelectric materials and there are a number of earlier reviews.[8,10,12,13] The objectives of this chapter are to provide an overview of the main p-type and n-type oxide materials, their current status, devices based on oxide materials and the prospects for future developments.

5.2 Layered Cobaltites

The most promising p-type thermoelectric oxides are families of layered cobaltites; (i) sodium cobaltite (Na_xCoO_2),[11,14–17] (ii) calcium cobaltite ($Ca_3Co_4O_9$)[18–25] and (iii) bismuth strontium cobaltite ($Bi_2Sr_2Co_2O_y$).[24,26–37] These materials are formed by stacking CoO_2 and rock-salt (RS) type layers[38] (Figure 5.2). The common feature of these cobaltites is the conductive nature of the CoO_2 layers, leading to the desired high electronic charge transport, while the non-conductive RS layers are responsible for the low thermal conductivity;[16] the stacking misfit between the CoO_2 and RS layers enhances

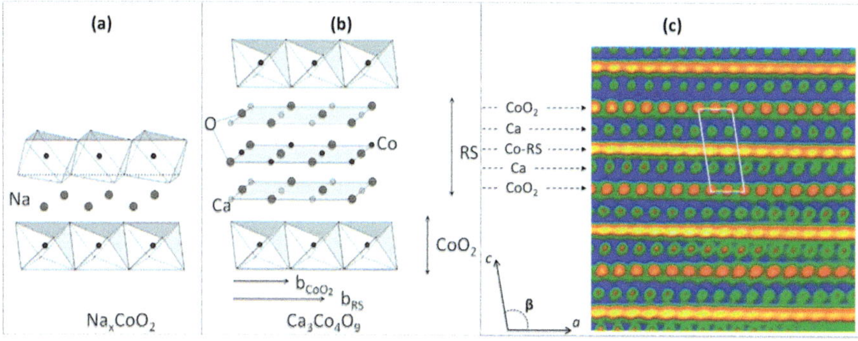

Figure 5.2 Crystal structures of thermoelectric misfit-layered cobalt oxides. (a) Na-based Na_xCoO_2, (b) Ca-based $Ca_3Co_4O_9$,[39] and (c) high angle annular dark field (HAADF) image of $Ca_3Co_4O_9$ showing the Ca, Co-rock salt (RS) and CoO_2 layers.[24] Parts a and b reproduced from ref. 39 with permission. Part c reproduced from ref. 24 with permission from American Chemical Society, Copyright 2015.

phonon scattering and further reduces thermal conductivity. The thermoelectric properties of these materials exhibit strong anisotropy perpendicular and parallel to these layers.[24,27]

5.2.1 Sodium Cobaltite

The investigation of oxide thermoelectrics started with the discovery of a high power factor, σS^2 (5000 μW m^{-1} K^{-2}) in Na_xCoO_2 single crystals by Terasaki *et al.*[11] These materials have a simple layered structure (Figure 5.2a) formed by the stacking of Na layers (which are 50% occupied) and CoO_2 layers along the *c*-axis. They exhibit high charge mobility μ (13 cm^2 V^{-1} s^{-1} at 300 K) and also high S (~100 μV K^{-1} at 300 K), indicating that S is not determined by the number of charge carriers but their degeneracy and strong correlation of 3d electrons.[15] The CoO_2 layers enable high electrical conductivity[16] while low κ results from the insulating rock salt layer. Yakabe *et al.*[14] showed that the crystallinity and densification of ceramic $NaCo_2O_4$ was improved by hot-pressing processing techniques leading to increased electrical conductivity and lower thermal conductivity. However, the maximum ZT was 0.11 at 473 K, much lower than the values reported for singe crystalline material. Adopting a different approach, Ito and Furumoto[17] prepared $Na_xCo_2O_4$/Ag composites by a polymerised complex (PC) route. Their results are summarised in Figure 5.3. The addition of Ag improved the electrical conductivity (shown as a reduction in electrical resistivity) whilst having an adverse effect on thermal conductivity. Thus, similar ZT values were achieved with and without Ag addition; the maximum ZT of 0.92 at 960 K was achieved for the composite of $Na_xCo_2O_4$ + 10% Ag. Although Na_xCoO_2 shows promise for thermoelectric applications, the preparation of dense polycrystalline samples with phase stability and optimised grain morphology is challenging.[40]

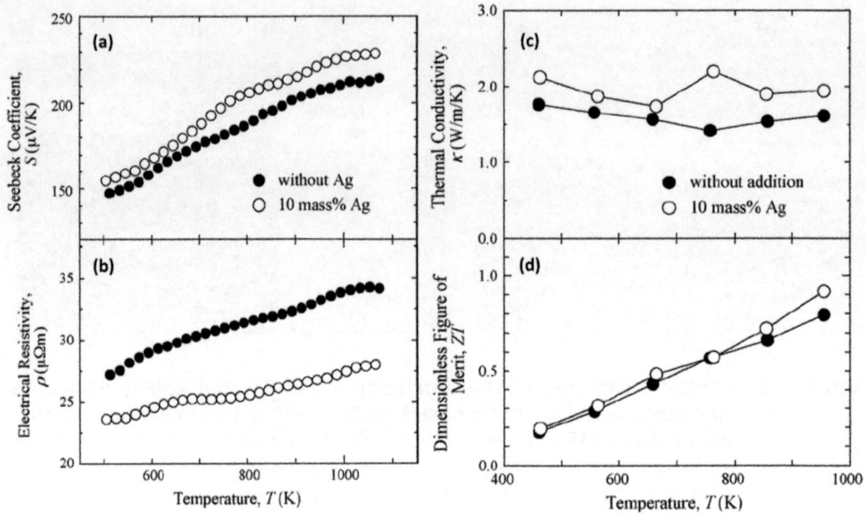

Figure 5.3 Temperature dependence of (a) the Seebeck coefficients, (b) the electrical resistivity, (c) the thermal conductivity, and (d) the dimensionless figure of merit of the sintered $Na_xCo_2O_4$ with and without 10% Ag.[17] Reproduced from ref. 17 with permission from Elsevier, Copyright 2008.

5.2.2 Calcium Cobaltite

The second member of the family of misfit-layered cobalt oxides is $Ca_3Co_4O_9$. Its structure (Figure 5.2b and c) consists of hexagonal CoO_2 layers connected to the planar CaO and CoO layers of rock-salt type, which results in disproportionate mismatch along the *b*-axis.[24,41] Moser *et al.*[23] investigated the effects of oxygen content on the thermoelectric properties of $Ca_3Co_4O_{9+\delta}$ and showed that samples prepared in a reducing atmosphere exhibited lower σ and increased S (compared to standard processing conditions). This was due to charge compensation by the formation of Co^{4+} in the Co^{3+} matrix and increased entropy of charge carriers. However, there was no significant change in κ due to the limited contribution of oxygen vacancy formation to phonon scattering. Shikano *et al.*[18] showed that single crystals of calcium cobaltite could exhibit high S (~240 μV K^{-1}), low ρ (2.3×10^{-5} S^{-1} m) and a reasonable κ (3 W m^{-1} K^{-1}) at 973 K. Therefore, a very high ZT value (~0.87 at 973 K) could be achieved (Figure 5.4). The studies on the polycrystalline samples[20,25] showed that *via* the substitution of Ba or Bi for Ca it was possible to obtain a reasonably high ZT value (~0.43 at 1073 K). The enhanced ZT values resulted from the simultaneous reduction of ρ and κ upon doping; however, optimisation of the carrier concentration was necessary to maximise the improvement in the properties. As well as anisotropy in the crystal structure (Figure 5.2b) the ceramic grains tend to be plate-like in shape; together this encourages anisotropic thermoelectric properties. Hao *et al.*[22] investigated the effect of texturing on the transport properties of undoped and Bi-doped $Ca_3Co_4O_9$. Their results highlighted the effectiveness of Bi additions for

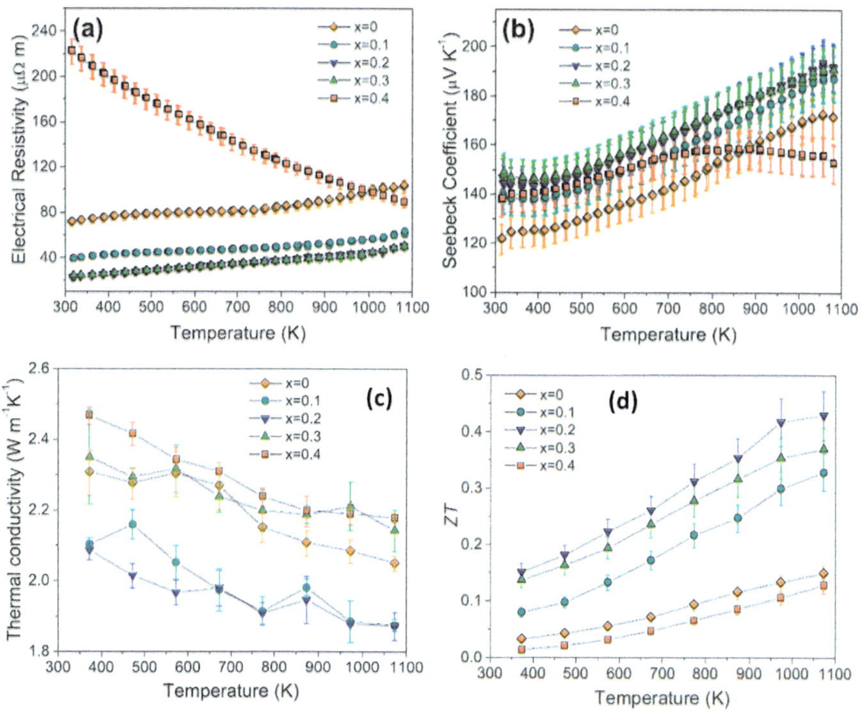

Figure 5.4 Temperature dependence of (a) electrical resistivity, (b) Seebeck coefficient, (c) thermal conductivity and (d) dimensionless figure of merit of $Ca_{3-x}Bi_xCo_4O_9$ with $x = 0$, 0.1, 0.2, 0.3 and 0.4.[25] Reproduced from ref. 25 with permission from Elsevier, Copyright 2016.

improving the growth kinetics and enhancing texturing (Lotgering factor[42] = 0.86); in contrast, no texturing was observed for undoped $Ca_3Co_4O_9$. This led to an improvement in the power factor (σS^2) in the *ab*-plane from 8200 $\mu W\ m^{-1}\ K^{-2}$ for the undoped samples to 27 700 $\mu W\ m^{-1}\ K^{-2}$ for the Bi-doped samples at 1000 K.

Donor and acceptor doping of $Ca_3Co_4O_9$ by Na, Bi and Na–Bi led to increased electrical conductivity regardless of the dopant.[19] Upon Bi doping of Ca sites a reduction in conductivity might be anticipated because of the expected reduction in hole concentration. However, Xu *et al.*[19] showed that Bi doping led to an increase in carrier mobility, and that co-doping with Bi and Na on Ca sites was beneficial for reducing thermal conductivity, which in turn increased the *ZT* value at 1000 K from 0.24 for $Ca_{2.7}Bi_{0.3}Co_4O_9$ to 0.35 for $Ca_{2.4}Bi_{0.3}Na_{0.3}Co_4O_9$. These experimental results[18,19,22] combined with modelling studies[24] suggest that the properties of $Ca_3Co_4O_9$ can be optimised by dual doping of heavy and light alkaline earth elements on the Ca sites, enabling good thermoelectric oxides; the materials have found use as the p-type legs in thermoelectric modules/devices for practical applications (Section 5.9).

5.2.3 Bismuth Strontium Cobaltite

Finally, $Bi_2Sr_2Co_2O_y$ is one of the most studied misfit-layered cobaltites. This material has a distorted CdI_2-type structure which consists of two sub-systems; the first is formed by RS type BiO and SrO layers and the second is formed by CoO_2 layers.[38] Funahashi *et al.*[26] showed that at 973 K the material could have relatively high S (~100 μV K^{-1}), low ρ (~1×10^{-6} S^{-1} m), and low κ (~1.3 W m^{-1} K^{-1}). They[30] later demonstrated that $Bi_2Sr_2Co_2O_y$ whiskers could exhibit high S and low κ, giving an exceptionally high ZT value of 1.1 at 973 K. Recent studies on polycrystalline samples[24,27] have shown that such high ZT values could not be achieved due to the anisotropic nature of the properties; textured materials are essential for very high ZT values. For $Bi_{2-x}Pb_xSr_2Co_2O_y$ single crystals, Fujii *et al.*[31] showed that the anisotropy in the electrical conductivity for undoped samples was reduced upon Pb doping, without significantly altering the Seebeck coefficients. Different doping strategies have also been explored for $Bi_2Sr_2Co_2O_y$ ceramics, including substituting Sr by Ag, Y and Ca, or substituting Co by Zr, Al and Mo.[35] It was demonstrated that doping on Sr sites was more effective for improving the electronic transport properties compared to substitutions on Co sites; Ca was regarded as one of the best dopants for overall improvement of the performance. Recent computational studies of the misfit-layered cobaltites[24,28] have demonstrated that co-doping of the rock-salt layers by isovalent cations could lead to charge transfer and induced strain, which in turn modifies the thermoelectric properties. Moreover, even modest Bi deficiency could lead to enhanced electronic transport properties, while Bi vacancies would increase phonon scattering, thereby reducing thermal conductivity.[24,28] Additionally, the use of a composite approach was investigated by Wang *et al.*;[36] data from their study are summarised in Figure 5.5. It was demonstrated that Ag addition considerably improved the power factor as a result of significant reduction of electrical resistivity while not affecting the Seebeck coefficient. It was also beneficial for the reduction of thermal conductivity, leading to a maximum ZT of 0.26 at 973 K with 3 wt% Ag addition.

Combe *et al.*[37] explored the effects of different processing routes on the microstructure and thermoelectric properties of $Bi_2Sr_2Co_2O_y$. The final density of the ceramics was highly dependent on the processing route and conditions; hot-pressing gave ceramics with high density and small grain size. It was demonstrated that partial melting was the best route for achieving higher conductivity ceramics, while having a limited effect on the Seebeck coefficients and thermal conductivity. Processing *via* partial melting led to a significant improvement in the ZT in comparison to the other techniques and a maximum ZT value of 0.25 at 973 K.

Wang *et al.*[36] investigated $Bi_2Sr_2Co_2O_y$/Ag composites and showed that electrical conductivity was substantially increased by the addition of Ag, whilst Seebeck coefficients were not affected. The net effect was a

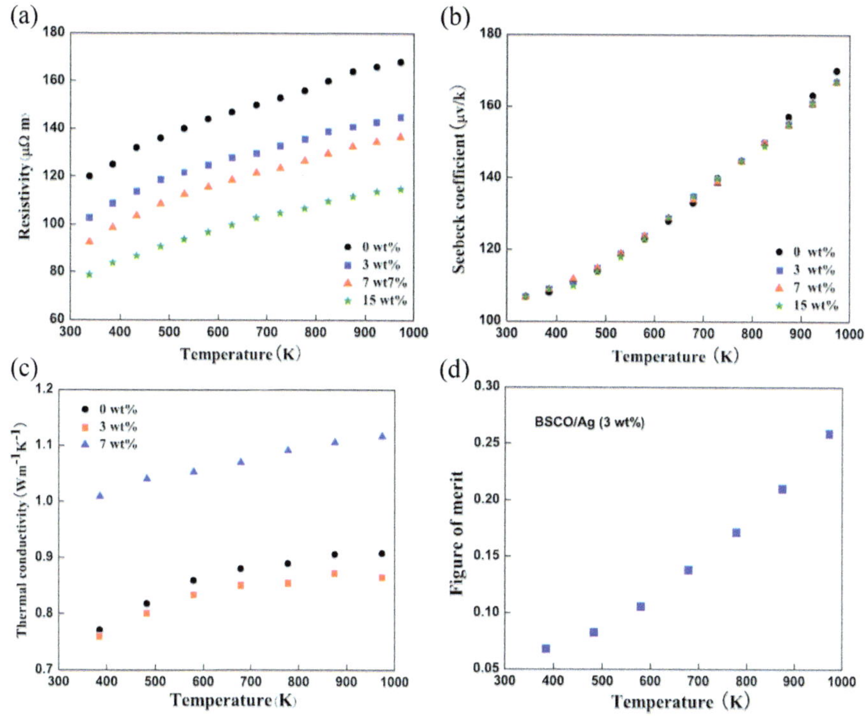

Figure 5.5 Variation of (a) electrical resistivity, (b) Seebeck coefficients, (c) thermal conductivity and (d) the dimensionless figure of merit values as a function of temperature for the Ag-added $Bi_2Sr_2Co_2O_y$.[36] Reproduced from ref. 36 with permission from Elsevier, Copyright 2013.

significant improvement in the power factor (by ~41%) to 240 μW m^{-1} K^{-2} at 973 with 15 wt% Ag addition. However, at this level, the thermal conductivity was also increased. Overall, the addition of 3 wt% Ag was regarded as optimum for both improving the power factor and reducing thermal conductivity, leading to a maximum *ZT* of 0.26 at 973 K. In a different approach, Diez *et al.*[32,33] prepared textured bismuth strontium cobaltite ceramics using a laser floating zone technique. The resulting ceramics exhibited enhanced electrical conductivity due to improved grain alignment and modification of microstructure. This had only a minor effect on the Seebeck coefficients, but distinct improvements in power factor (~100 μW m^{-1} K^{-2} at 300 K) with texturing. Sotelo *et al.*[34] studied the effect of Ag addition on the properties of textured $Bi_2Sr_2Co_{1.8}O_x$ ceramics. Similar to the ceramics without texturing,[36] the addition of Ag improved the electrical conductivity while having a minor effect on the Seebeck coefficient. Thus, the power factor at 923 K was almost doubled with only 1 wt% Ag addition.

5.3 ZnO and Homologous Compounds

5.3.1 ZnO

ZnO has a wurzite-type hexagonal crystal structure and is a conductive, wide band gap semiconductor.[43] The covalent nature of the Zn–O bonds promotes high carrier mobility (μ). There have been many investigations of the thermoelectric properties of ZnO.[43–51] Tsubota *et al.*[43] showed that pure ZnO has a good power factor (σS^2) at high temperature (~368 μW m^{-1} K^{-2} at 1273 K) but also a high κ (~5 W m^{-1} K^{-1}) due to the simple crystal structure, which limits the *ZT* value. They also examined the effect of Al doping and showed that the maximum *ZT* of 0.30 at 1273 K was achieved with 2% Al doping.[43] Later, Ohtaki *et al.*[44] studied the dual doping of ZnO with Al and Ga in detail; they showed that Ga additions prevented the formation of the $ZnAl_2O_4$ phase, enabling the preparation of near-single phase materials. However, another secondary phase, which was later determined to be a homologous $Ga_2O_3(ZnO)_m$ compound,[52] was observed in the ceramics. The presence of this phase allowed significant reduction of thermal conductivity (Figure 5.6) while not degrading the electronic transport properties. As a result, there was a significant improvement in the thermoelectric performance; for $Zn_{0.96}Al_{0.02}Ga_{0.02}O$ a maximum *ZT* of 0.65 was achieved at 1273 K.[44]

Although these (Ga, Al)-doped ZnO materials have exhibited some of the highest *ZT* values recorded among n-type thermoelectric oxides (Figure 5.6),[44] replicating the results *via* different processing routes was unsuccessful. Later work by Choi *et al.*[50] established that conventional sintering was effective for the preparation of high performance ZnO thermoelectrics. The main reason for the earlier, exceptionally high *ZT* value was the dramatic reduction of κ, with only limited reduction in σS^2. Bérardan *et al.*[47] showed that the preparation of samples under a N_2 atmosphere generated oxygen vacancies, in turn

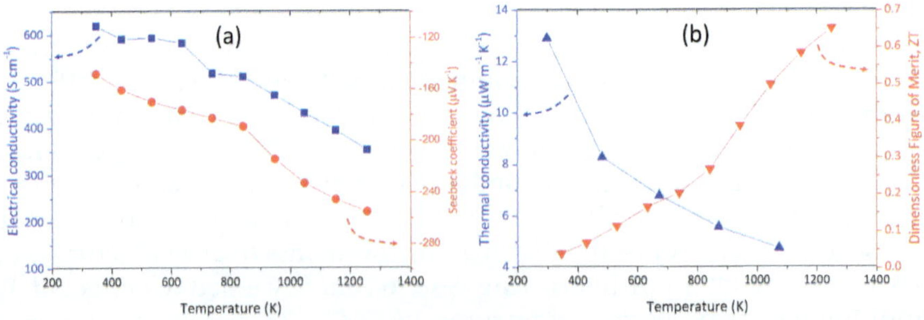

Figure 5.6 Temperature dependence of (a) electrical conductivity and Seebeck coefficients, and (b) thermal conductivity and dimensionless figure of merit for $Zn_{0.96}Al_{0.02}Ga_{0.02}O$.[44]

increasing the concentration of carriers, and that significantly increased σS^2. However, it was believed that the presence of a $Ga_2O_3(ZnO)_9$ secondary phase could also cause the reduction of thermal conductivity.[50] These encouraging results showed that ZnO can be employed as the n-type legs for thermoelectric generators (Section 5.9); however, there is still a need to understand the exact role of the $Ga_2O_3(ZnO)_9$ homologous compound secondary phases on the performance of the material.

5.3.2 Homologous Compounds

$In_2O_3(ZnO)_m$ and $Ga_2O_3(ZnO)_m$ (m = integer) type compounds are known as homologous compounds. These materials have a layered structure along the *c*-axis where *m* is the number of zinc oxide layers separated by indium oxide or gallium oxide layers, respectively. This type of structure encourages low thermal conductivity, and the electronic properties can be modified by optimising *m*.

Early studies on the thermoelectric properties of homologous compounds mainly focused on $In_2O_3(ZnO)_m$. Ohta *et al.*[53] showed that optimum *m* values between 9 and ∞ were required to achieve maximum thermoelectric performance. However, their findings also highlighted densification problems when *m* was greater than nine. Kaga *et al.*[54] studied the effect of the number of ZnO layers between each In-containing layer and found that *m* values in the range of 3 to 5 were optimum for maximising thermoelectric performance (Figure 5.7); a maximum *ZT* of 0.23 at 1053 K was obtained for formulations based on $m = 3.5$, doped with 2.5% Ca. Related studies of the effect of doping by transition metals[55,56] showed that *ZT* could be enhanced by such species because of the resulting lattice distortion. More recently, Liang *et al.*[57] showed that in order to achieve maximum σS^2 and *ZT*, the optimum value for *m* was 4. Further increase of the *m* value resulted in the formation of In-based secondary phases, which tend to degrade material properties.

In the past decade more investigations have focused on the Ga analogue, the $Ga_2O_3(ZnO)_m$ compounds.[52,58–60] They have shown that controlling the size of the superlattice and nano-twins within the structures, by changing *m*, is an effective way to reduce thermal conductivity, but may also degrade electrical conductivity. Michiue *et al.*[58] studied the thermoelectric properties of $Ga_2O_3(ZnO)_9$ and found that cold isostatic pressing (CIP) was beneficial for the preparation of high-quality ceramics. Additionally, post-annealing the ceramics in a reducing atmosphere did not affect the transport properties. Even though their best samples were of relatively low density (73% theoretical), and low stability at high temperatures, they exhibited a maximum *ZT* of 0.066 at 660 K. Subsequently, Michiue *et al.*[52] examined the pseudo-binary ZnO-Ga_2O_3 system and showed that $Ga_2O_3(ZnO)_m$ was formed for $m = 9–38$, whilst higher *m* resulted in a two-phase material (doped ZnO + $Ga_2O_3(ZnO)_{m\approx 38}$), where the solubility limit for Ga in ZnO was 0.5%. They demonstrated that the presence of a homologous phase was beneficial for

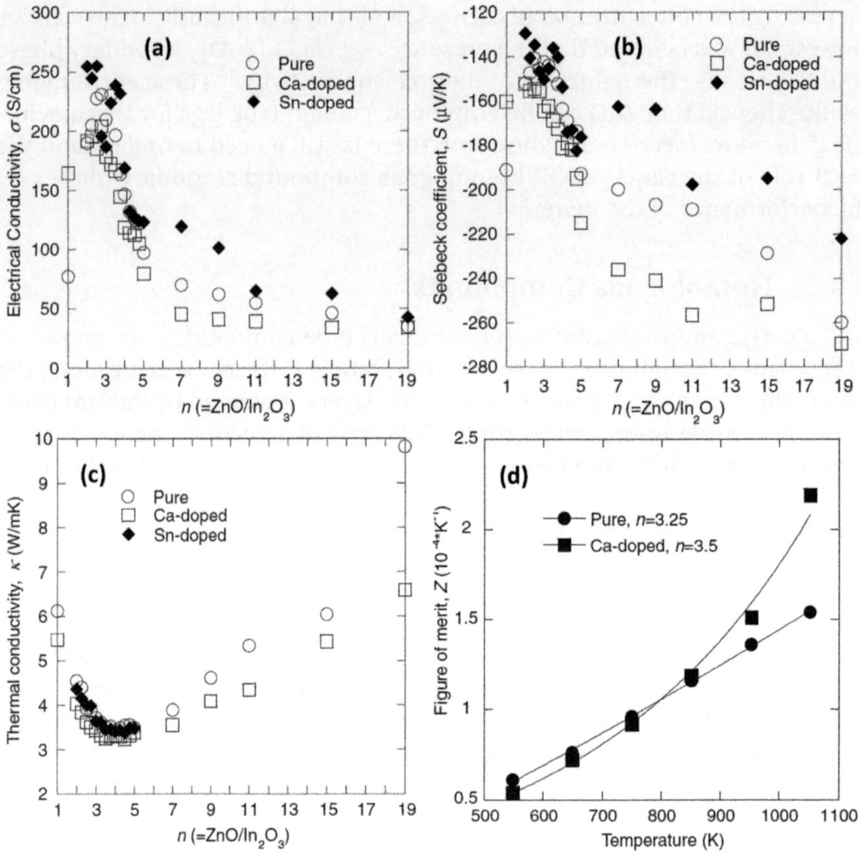

Figure 5.7 Effect of molar ratio on (a) electrical conductivity, (b) Seebeck coefficients and (c) thermal conductivity of the pure (○), Ca-doped (□), and Sn-doped (◆) $In_2O_3(ZnO)_m$ at 300 K. (d) Temperature dependence of the thermoelectric figure of merit (Z) for the pure and Ca-doped specimens.[54] Reproduced from ref. 54 with permission from IOP Publishing, Copyright (2004) The Japan Society of Applied Physics.

reducing thermal conductivity, but it had an adverse effect on electronic transport, reducing the power factor. As a result of the trade-off between electronic and thermal transport, a similar maximum ZT value (0.07 at 973 K) was achieved for compositions based on $m = 98$ and $m = 398$. It was shown that the point defect and the nanoscale interfaces in the $Ga_2O_3(ZnO)_m$ system played an important role for electronic and thermal transport. Increasing m for the material formulations led to significantly increased electrical conductivity while reducing the thermal conductivity, indicating that nanoscale interfaces were responsible for increased scattering of both charge carriers and phonons.[59,61]

Later, Alvarez-Ruiz *et al.*[61] demonstrated that the junction of twin boundaries and inversion boundaries (Figure 5.8) were occupied by Ga

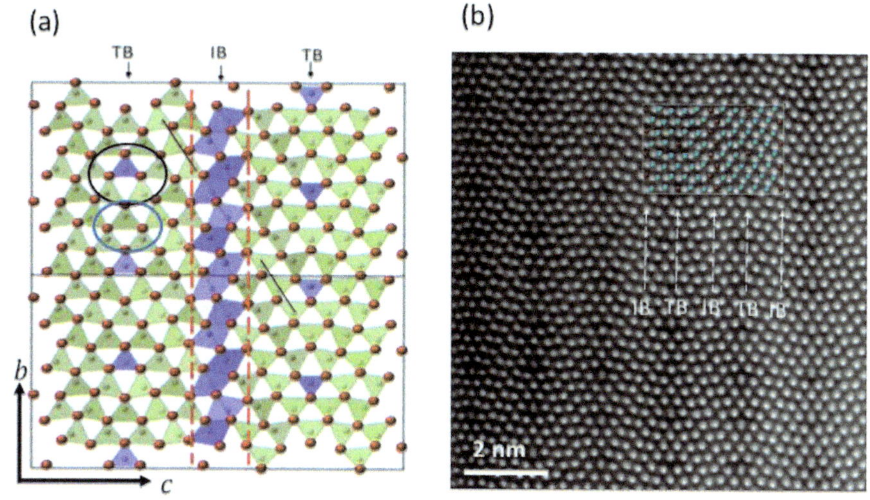

Figure 5.8 (a) [100] schematic structural representation of $Ga_2O_3(ZnO)_9$ showing twin boundaries (TB) and inversion boundaries (IB); the position of Ga and its oxygen environment is shown inside the black ellipsoid. (b) The superimposed [100] projection of the final crystal structure on the [100] HAADF image.[61] Reproduced from ref. 61, https://doi.org/10.1016/j.jallcom.2018.05.260, under the terms of the CC BY 4.0 licence, https://creativecommons.org/licenses/by/4.0/.

ions and optimisation of processing conditions could improve *ZT* values; a maximum *ZT* of 0.06 at 900 K was reported. For the same system, Guilmeau *et al.*[60] adjusted the Ga content and achieved a maximum *ZT* of 0.09 at 1000 K. Increasing Ga content led to an increase in the electrical conductivity and a reduction in thermal conductivity due to formation of interface defects, which enhanced *ZT*. Alvarez-Ruiz *et al.*[62] also used B_2O_3 and Nd_2O_3 as sintering aids and studied their effects on the thermoelectric properties of the $Ga_2O_3(ZnO)_9$–$In_2O_3(ZnO)_9$ solid solutions. Incorporation of In modified the crystal structure, which affected thermoelectric properties. $(Ga_{0.8}In_{0.2})_2O_3(ZnO)_9$ provided the best overall thermoelectric performance (Figure 5.9); a low thermal conductivity of 2 W m^{-1} K, a high power factor of 150 μW m^{-1} K^{-2} and a *ZT* value of 0.07 at 900 K. The high Seebeck coefficient and low thermal conductivity make them attractive candidates for thermoelectric applications, and provides guidance to engineer other material structures.

5.4 Tungsten Bronze Structured Oxides

Tungsten bronze structured materials have attracted attention as possible thermoelectrics because of their large complex crystal structures, containing heavy ions which could offer inherent low thermal conductivity. The (Sr, Ba)Nb_2O_6 tungsten bronze materials (SBN) have a tetragonal *P4mm* crystal structure at room temperature,[63] with two A-sites; A1

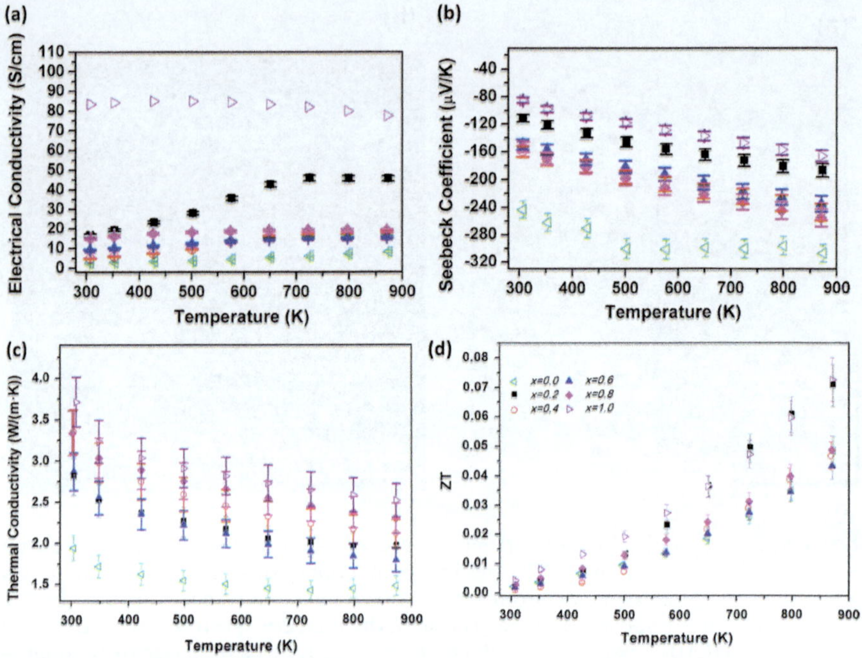

Figure 5.9 (a) Electrical conductivity, (b) Seebeck coefficient, (c) thermal conductivity and (d) dimensionless figure of merit of $(Ga_{1-x}In_x)_2O_3(ZnO)_9$ ceramics.[62] Reproduced from ref. 62 with permission from Springer Nature, Copyright 2018.

is occupied by Sr and A2 is occupied by Sr/Ba atoms. The B-site is occupied by Nb atoms (where there is usually reduction from Nb^{5+})[64] and the vacant C-site is surrounded by a distorted trigonal prism of O atoms.[64–66] The material is polarised along the *c*-axis,[67] giving rise to anisotropic properties, with the highest value of S along the *c*-axis. Early studies of single crystal SBN revealed a high power factor (σS^2) along the *c*-axis.[68] This observation, along with the anticipated low thermal conductivity, gave hope for high ZT values.[69,70] Lee *et al.*[69] determined the thermoelectric properties of single crystal $Sr_{0.61}Ba_{0.39}Nb_2O_6$ (SBN61) and confirmed a high power factor (2×10^{-3} W m^{-1} K^{-2} at 516 K) along the *c*-direction. However, the anisotropic behaviour suggested that textured ceramics would be necessary to maximise thermoelectric performance. They also concluded that the conduction mechanism for highly reduced samples was polaron hopping. Lee *et al.*[71] also prepared textured polycrystalline SBN61 ceramics using a templated grain growth approach and showed that both electrical conductivity and Seebeck coefficients were higher along the *c*-axis. However, the properties of non-textured ceramics were dominated by characteristics along the *a*-axis. The Seebeck coefficients were independent of temperature due to a polaron hopping mechanism. The electrical conductivity of the textured ceramics was higher than that of the non-textured counterparts due to anisotropy and differing numbers of grain boundaries; it was demonstrated that grain boundaries control the electrical conductivity in SBN.

Dandeneau *et al.*[72] examined the thermoelectric properties of SBN ceramics with lower Sr content, $Sr_{0.5}Ba_{0.5}Nb_2O_6$ (SBN50) and found that air-sintered materials exhibited exceptionally low electrical conductivity. However, additional heat treatment in a reducing atmosphere promoted oxygen vacancy formation and hence an increase in carrier concentration; this material also had a high Seebeck coefficient (−281 μV K^{-1}), a high power factor (1780 μW m^{-1} K^{-2}) at 930 K and a very low thermal conductivity (<1.9 W m^{-1} K^{-1}). As a result, a maximum *ZT* of 0.09 was obtained at 930 K for SBN50. In view of the low density of these SBN50 ceramics further improvement in performance was anticipated. By introduction of fluorine at oxygen sites of SBN61 Li *et al.*[73] were able to significantly increase the carrier concentration and also modified the band structure, leading to improved electrical conductivity and reduction in Seebeck coefficient. Fluorine doping had the additional benefit of reducing thermal conductivity. The ceramics with a formulation of $Sr_{0.61}Ba_{0.39}Nb_2O_{5.95}F_{0.05}$ exhibited the best thermoelectric performance due to a combination of the high power factor and relatively low thermal conductivity, leading to a maximum *ZT* of 0.21 at 1073 K.

Li *et al.*[74] also studied $Sr_{0.7}Ba_{0.3}Nb_2O_{6-\delta}$ (SBN70) and found the substitution of La for Sr was beneficial for improving the power factor (~433 μW m^{-1} K^{-2} at 1073 K) due to increased carrier concentration and reduced thermal conductivity (2.1 W m^{-1} K^{-1} at 1073 K). The net result was a maximum *ZT* of 0.20 at 1073 K for $Sr_{0.5}La_{0.2}Ba_{0.3}Nb_2O_{6-\delta}$. Later, Chan *et al.*[75] investigated filled $Sr_xBa_{6-x}Nb_{10}O_{30-\delta}$ materials and showed that those with lower Sr content exhibited higher electrical conductivity and power factor due to the enhanced carrier mobility and concentration. More recently, Ottini *et al.*[76,77] examined the effects of substituting K and Y for Sr, and Zr and Mo for Nb in $Sr_{0.5}Ba_{0.5}Nb_2O_6$ (SBN50) ceramics. The introduction of Y into SBN50 was particularly effective, increasing the power factor more than 30 times in comparison to the undoped material; there were increases in both electrical conductivity and Seebeck coefficients.[76] These improvements were attributed to changes in the spacing between NbO_6 structural layers and the weakening of metal–oxygen bonds. Their later work[77] showed that insulating samples could be obtained under oxidising conditions even for heavily doped samples; post-annealing in a reducing atmosphere was necessary to obtain the highest performance samples. For reducing thermal conductivity, Mo was the most effective dopant, but overall, for optimum electronic and thermal transport, Y proved to be the best dopant for SBN50.

Interstitial doping of SBN ceramics has also been explored as a way to improve thermoelectric performance.[78,79] Li *et al.* doped SBN70 ceramics with Yb and achieved improvements in electrical conductivity through a donor mechanism, leading to an improved power factor (426 μW m^{-1}

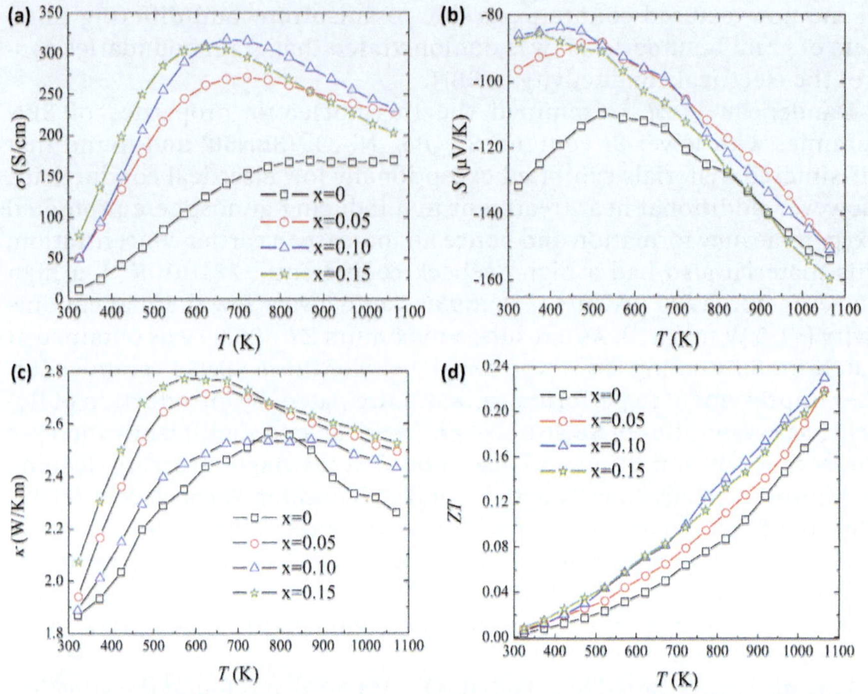

Figure 5.10 Temperature dependence of (a) the electrical conductivity, (b) the Seebeck coefficient, (c) the thermal conductivity, and (d) dimensionless figure of merit of the $Sr_{0.7}Ba_{0.3}K_xNb2O_{6-\delta}$ samples.[78] Reproduced from ref. 78 with permission from Springer Nature, Copyright 2018.

K^{-2}). At the same time interstitial doping by Yb increased the distortion of the SBN lattice, causing a reduction in lattice thermal conductivity. As a result, a maximum *ZT* of 0.21 at 1073 K was obtained for $Sr_{0.7}Ba_{0.3}Yb_{0.1}Nb_2O_{6-\delta}$. Similarly, Zhang *et al.*[78] found that K interstitial doping of SBN70 ceramics was beneficial, increasing carrier concentration and modifying the electronic structure; their results are summarised in Figure 5.10. Interstitial doping by K resulted in a significant improvement in the electrical conductivity while the Seebeck coefficients were maintained, leading to a significant improvement in the power factor (520 μW m^{-1} K^{-2} at 1073 K) but an increase in thermal conductivity. The overall effect of K doping was improvement in thermoelectric performance giving a maximum *ZT* of 0.23 at 1073 K for $Sr_{0.7}Ba_{0.3}K_{0.1}Nb_2O_{6-\delta}$. This is among the highest values reported for SBN ceramics.

Azough *et al.*[80] investigated a complementary ferroelectric ceramic with a tungsten bronze structure, $Ba_{6-x}Nd_{8+2x}Ti_{18}O_{54}$, and evaluated the effect of changing the Ba/Nd ratio on the thermoelectric properties (Figure 5.11). They demonstrated that $Ba_{5.19}Nd_{8.54}Ti_{18}O_{54}$ exhibited a high Seebeck coefficient (−210 μV K^{-1}) and an acceptable electrical conductivity

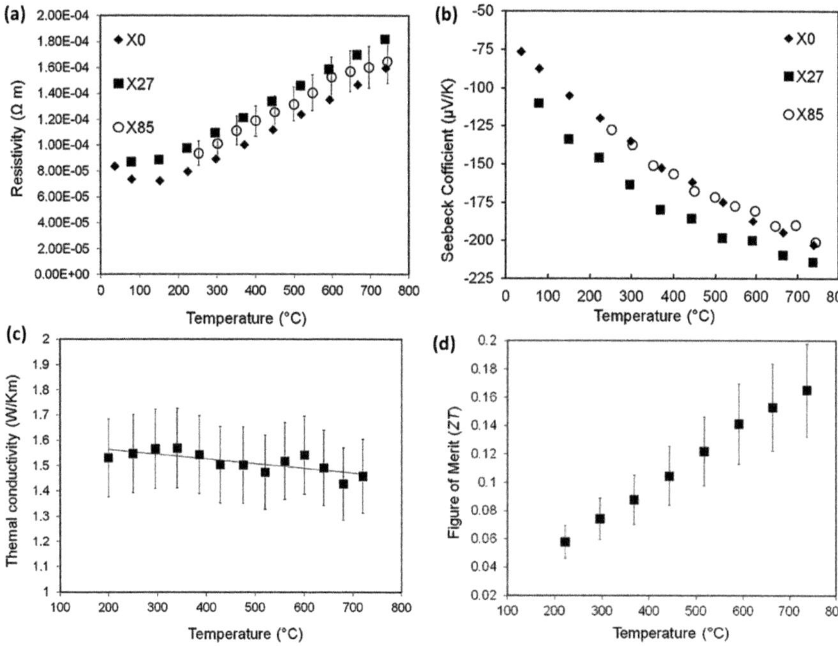

Figure 5.11 Thermoelectric data for $Ba_{6-3x}Nd_{8+2x}Ti_{18}O_{54}$ ceramics as a function of temperature: (a) electrical resistivity, (b) Seebeck coefficient, (c) thermal conductivity and (d) dimensionless figure of merit. Here, x = 0.0, 0.27 and 0.85 compositions are denoted as X0, X27 and X85.[80] Reproduced from ref. 80, https://doi.org/10.1007/s11664-015-4275-6, under the terms of the CC BY 4.0 licence, https://creativecommons.org/licenses/by/4.0/.

(60 S cm^{-1}), which led to maximum power factor of 250 μW m^{-1} K^{-2} at 1000 K. Furthermore, this material exhibited a low and stable thermal conductivity over the full measurement range, being only 1.45 W m^{-1} K^{-1} at 1000 K, which is among the lowest values reported for thermoelectric oxides. A maximum ZT of 0.16 was obtained at 1000 K. Molecular dynamics modelling of $Ba_{5.19}Nd_{8.54}Ti_{18}O_{54}$ showed that the thermal properties were anisotropic, with thermal conductivity lowest along the a-axis. These results for $Ba_{6-x}Nd_{8+2x}Ti_{18}O_{54}$ ceramics are promising; texturing and different doping strategies could further improve the properties of tungsten bronze materials.

5.5 Calcium Manganate – $CaMnO_3$

$CaMnO_3$ has an orthorhombic perovskite structure (Space group *Pnma*) and shows n-type semiconducting behaviour at room temperature. It goes through a metal-insulator phase transition at high temperature due to the changes in the spin state of Mn ions within the crystal.[81] In one of the earliest investigations Kobayashi *et al.*[81] showed that with appropriate donor doping $CaMnO_3$ has a large Seebeck coefficient and high electrical conductivity. For undoped $CaMnO_3$ the modest electrical conductivity (~50 S m^{-1}) and relatively high S (−350 μV K^{-1}) of the material led to a low σS^2 (~7.4 μW m^{-1} K^{-2}).[82] However, the high κ (~3.5 W m^{-1} K^{-1}) of the material limits the overall thermoelectric performance.[83] Two main strategies have been employed to tune the thermoelectric properties of $CaMnO_3$: (i) doping on Ca sites[81,82,84–87] and (ii) doping on Mn sites.[86,88–91] The effects of oxygen stoichiometry on thermoelectric performance have also been evaluated.[90,92] Doping on the Ca (or the perovskite A) sites has mainly focused on rare earth elements and demonstrated that the size and weight of the dopant have measurable effects on the transport properties.[82,84,85] Flahaut *et al.*[82] showed that the use of dopants of smaller ionic radius decreased the Mn–O–Mn bond angle, modifying the electronic band structure, leading to improved electrical conductivity.[82] It was suggested by Funahashi *et al.*[84] that the decrease in the average valence of Mn with doping was responsible for the enhanced electrical conductivity. Additionally, the change in the octahedral distortion with doping and increased mass difference between the host and dopant were responsible for a reduction in thermal conductivity.[82] By this approach, Wang *et al.*[85] obtained a maximum ZT of 0.20 at 1000 K for both $Ca_{0.9}Yb_{0.1}MnO_3$ and $Ca_{0.9}Dy_{0.1}MnO_3$ ceramics. From dynamical mean field theory calculations, they predicted that the maximum ZT of $CaMnO_3$ could not exceed 1.0.

Doping on the Mn (or perovskite B) sites has also been effective for enhancing the thermoelectric performance of $CaMnO_3$.[88–91] Xu *et al.*[88] showed that Nb doping was more effective than Ta doping for improving the electrical conductivity due to the differences in the electronic configurations of these elements; Nb is directly involved in the conduction process while Ta acts as an impurity ion. In contrast, Ta doping gave rise to higher Seebeck coefficients than Nb doping, but donor doping by either 5+ species led to an increase in thermal conductivity from the increased electronic contribution. These results indicated that an optimised doping concentration is crucial for maximising the thermoelectric performance. Later, Bocher *et al.*[89] examined the effect of the processing route on the thermoelectric properties of Nb-doped $CaMnO_3$ ceramics prepared by soft chemistry (SC) and solid state reaction (SSR) methods; their findings are summarised in Figure 5.12. They demonstrated the SC route enabled the preparation of sub-micron grain sizes while the SSR route gave much larger grain sizes. The Seebeck coefficients were not affected by grain size differences, whereas use of the SC route gave higher electrical conductivity; this was attributed to increased surface contact between crystallites. Moreover, the nanosized twinned domains introduced by the SC route led to lower thermal conductivity. The combination of higher power factor (190–200 μW m^{-1} K^{-2}) and lower thermal conductivity (<1 W m^{-1} K^{-1}) achieved by the SC route for $CaMn_{0.98}Nb_{0.02}O_3$ resulted in a maximum ZT value of 0.32 at 1060 K.[89]

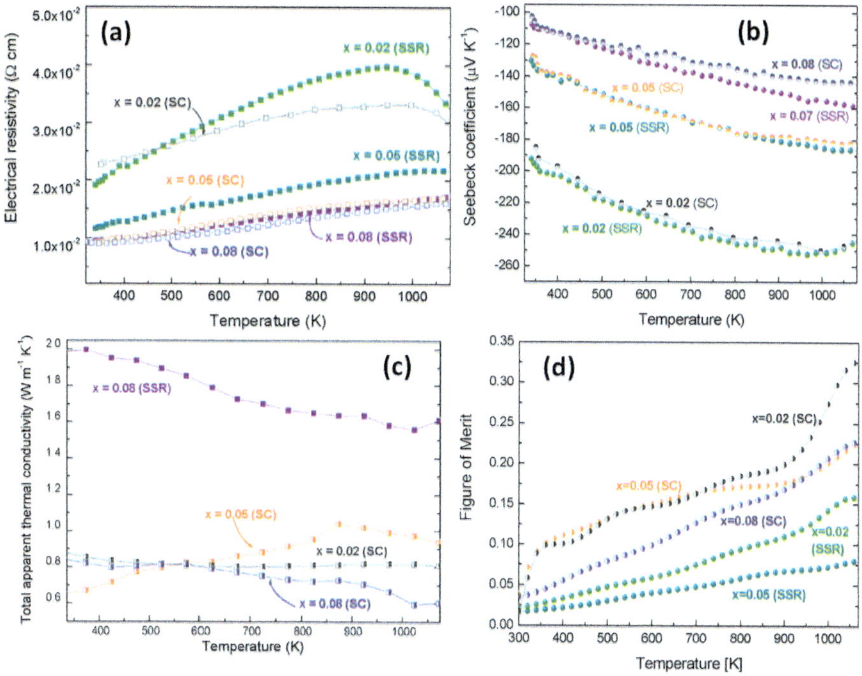

Figure 5.12 Temperature dependence of (a) electrical resistivity, (b) Seebeck coefficient, (c) thermal conductivity and (d) dimensionless figure of merit for $CaMn_{1-x}Nb_xO_3$ (for x = 0.02, 0.05 and 0.08) synthesised by SSR (closed symbols) and SC methods (open symbols).[89] Reproduced from ref. 89 with permission from American Chemical Society, Copyright 2008.

In a different approach, Kabir *et al.*[91] studied the effect of W doping on the Mn site; once the doping concentration was above 1%, there was an increase in the activation energy for the hopping mechanism, suggesting that W atoms act as scattering centres above the 1% threshold. On the other hand, the substitution of Mn by W led to a reduction in thermal conductivity as a result of the distortion of MnO_6 octahedra and reduction in the phonon mean free path. Compared to the undoped samples these changes led to an improvement in *ZT* of $CaMn_{0.99}W_{0.01}O_3$ to a maximum of 0.15 at 973 K. More recently, Wang *et al.*[87] investigated the effect of processing conditions on the thermoelectric properties of donor doped $CaMnO_3$. They demonstrated that carrier concentration could be significantly improved by multiple calcination steps, giving a microstructure with coherent interfaces and inhomogeneous distribution of dopants. This led to an increase in the electrical conductivity of $Ca_{0.9}Yb_{0.1}MnO_{3-\delta}$, and net improvement in the power factor (339 μW m^{-1} K^{-2} at 678 K). The heterogeneous structure also contributed to a reduction of thermal conductivity. Thus, the use of multiple calcination steps for $Ca_{0.9}Yb_{0.1}MnO_{3-\delta}$ enabled 30% improvement in the dimensionless

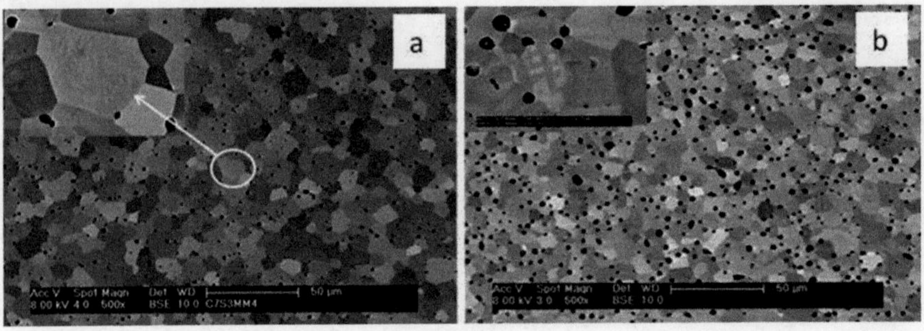

Figure 5.13 Back scattered SEM images for (a) $Ca_{0.7}Sr_{0.3}Mn_{0.96}Mo_{0.04}O_3$ and (b) $Ca_{0.4}Sr_{0.6}Mn_{0.96}Mo_{0.04}O_3$; the insets show detail of the sub-grain boundaries.[86] Reproduced from ref. 86 with permission from the Royal Society of Chemistry.

figure of merit to 0.14 at 670 K, showing that optimisation of processing parameters is crucial for maximising the thermoelectric performance of $CaMnO_3$ ceramics.

The studies of undoped[92] and donor doped[90] $CaMnO_3$ ceramics demonstrated that oxygen deficiency can be beneficial for thermoelectric performance. Schrade *et al.*[92] examined the effect of oxygen partial pressure on the electronic transport properties of $CaMnO_{3-\delta}$, employing reducing atmospheres up to 1000 °C; it was proposed that $CaMnO_{3-\delta}$ could accommodate a high degree of oxygen deficiency (δ = 0.15). The thermal treatment in a reducing atmosphere modified the Mn^{3+}/Mn^{4+} concentrations, increasing electrical conductivity and reducing the Seebeck coefficients. However, there is concern that $CaMnO_3$ may decompose in a reducing environment at elevated temperatures;[93] electrical conductivity enhancement in non-deficient $CaMnO_3$ can still be achieved by Mn on the A-sites and B-sites having different valence states.[93]

Srivastava *et al.*[86] carried out the first dual doping study of $CaMnO_3$ (using Sr and Mo) employing both experimental and computational techniques. The resulting microstructures (examples shown in Figure 5.13) exhibit twin and sub-grain boundaries arising from Sr doping on Ca sites and the structural transition from the high temperature cubic form to the low temperature orthorhombic or tetragonal forms. The twin boundaries resulted in an increase in phonon scattering, leading to a reduction of thermal conductivity; an interpretation supported by molecular dynamics calculations. The Sr and Mo doping also improved densification; Mo doping generates additional charge carriers, increasing electrical conductivity. Thus, the power factor was improved while the thermal conductivity was reduced. As a result, a maximum *ZT* of 0.11 at 1000 K was obtained for $Ca_{0.7}Sr_{0.3}Mn_{0.96}Mo_{0.04}O_3$ ceramics. Calcium manganite ceramics have good thermal stability in air at high temperatures, making them useful candidates as n-type thermoelectrics, but the transport properties need to be further improved before they can be exploited to any degree.

5.6 TiO_2-based Oxides and Magnéli Phases

5.6.1 TiO_2

TiO_2 crystallises in three types of structure; (i) brookite, (ii) anatase and (iii) rutile. Backhaus-Ricoult *et al.*[3,94] investigated the thermoelectric properties of the rutile and anatase phases of TiO_2. They demonstrated that the processing conditions affected the electronic transport properties and that the anatase phase had higher electrical conductivity and lower Seebeck coefficients ($-400\ \mu V\ K^{-1}$) than the rutile phase material ($-600\ \mu V\ K^{-1}$). It was also shown that the thermal conductivity was anisotropic for single crystals, being higher along the *c*-axis ($5\ W\ m^{-1}\ K^{-1}$ at 1050 K). As expected, the thermal conductivity was lower for polycrystalline samples and anatase exhibited a lower thermal conductivity ($3\ W\ m^{-1}\ K^{-1}$ at 1000 K) than the rutile phase. Due to the low electrical conductivity and relatively high thermal conductivity of TiO_2, the maximum *ZT* of undoped TiO_2 was limited (Figure 5.14).[3] First principles calculations by Bayerl *et al.*[95] showed that Seebeck coefficients were not dependent on the crystal orientation while the electrical conductivity was the highest along the *a*-axis for rutile and the anatase phases, and along the *b*-axis for the brookite phase. It was also demonstrated that rutile exhibited superior thermoelectric properties to other polymorphs when the carrier concentration was optimised. Their calculations agreed with earlier results and they predicted a maximum *ZT* of 0.93 at 2000 K for rutile.[95]

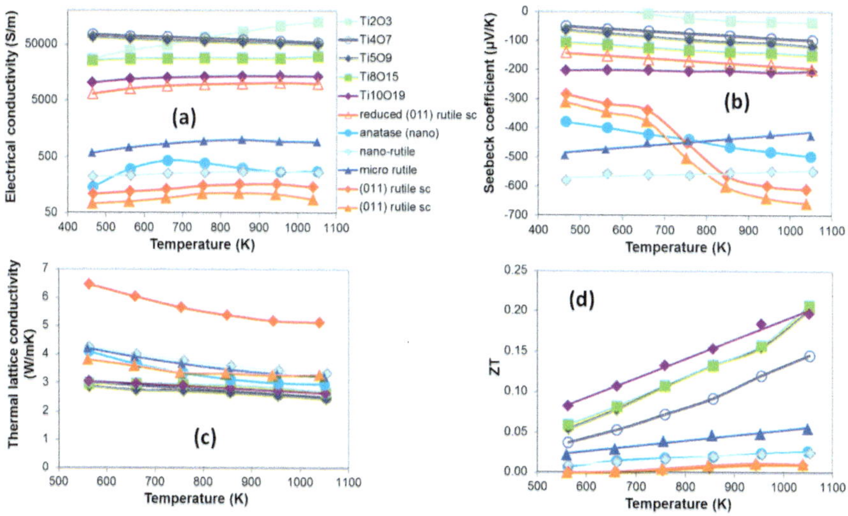

Figure 5.14 (a) Electrical conductivity, (b) Seebeck coefficient, (c) thermal lattice conductivity and (d) dimensionless figure of merit as a function of temperature for as-processed and reduced single-crystalline rutile, micro- and nano-rutile, anatase, various Magnéli phases and Ti_2O_3.[3] Reproduced from ref. 3 with permission from Springer Nature, Copyright 2014.

5.6.2 Magnéli Phases

Although the Magnéli phases of titanium dioxide (Ti_nO_{2n-1}; $n = 2, 3, \ldots$) have been the most studied for thermoelectric applications,[96] the Nb, and W based Magnéli phases have also been explored. The Ti Magnéli phases are formed by the reduction of TiO_2. Once the oxygen vacancy concentration reaches a certain level, the structure collapses to form planar shear defects, as shown in Figure 5.15.[3] This occurs with the replacement of corner-sharing TiO_6 octahedra by edge- and face-sharing octahedra. The formation of such defects makes Ti_nO_{2n-1} Magnéli phases a promising thermoelectric material because the *d*-spacing of the planar shear defects (0.2–2.0 nm) is well suited for the phonon scattering as the phonon mean free length in TiO_2 is ~1 nm. Therefore, a low κ can be expected. It was shown by Harada *et al.*[97] that these crystallographic shear planes do not cause significant electron scattering, thereby enabling simultaneous tuning of both electronic and thermal transport properties of the material.

Studies of thermoelectric properties of Ti_nO_{2n-1} ($n = 2, 3, \ldots$) have shown that preparation of single phase materials is difficult; most investigations report the presence of more than one type of Magnéli phase in the samples.[99,100] It has been established that oxygen non-stoichiometry has a significant effect on both electronic and thermal transport properties.[94,97,99,101–104] An increase in the oxygen deficiency improves electrical conductivity, but a reduction in the Seebeck coefficient occurs as a result of the increase in the carrier concentration. Interestingly, a switch from p-type to n-type behaviour was observed[99,102] when the oxygen partial pressure is changed during processing; low partial pressure leads to increasing oxygen vacancy concentration, causing n-type behaviour, while high partial pressure results in p-type

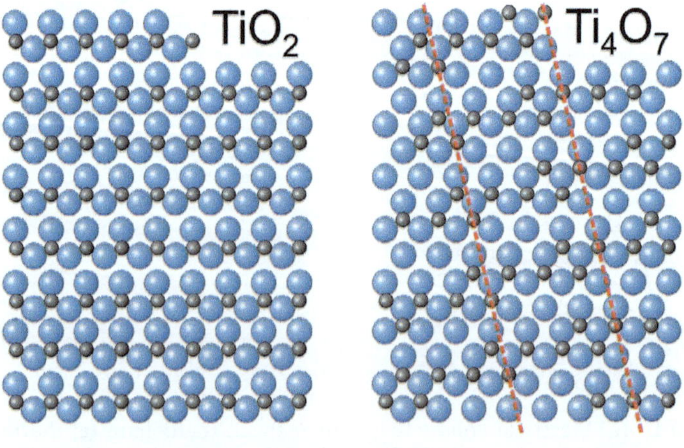

Figure 5.15 Transformation from TiO_2 to Ti_4O_7 Magnéli phase due to the formation of crystallographic shear planes as a result of increased oxygen deficiency.[98] Reproduced from ref. 98 with permission. © IOP Publishing. All rights reserved.

behaviour because Ti vacancies are formed. It was also shown by Nowotny *et al.*[103] that the Seebeck coefficients are dramatically affected by grain boundaries while they have only a minor effect on electrical conductivity. This was attributed to the oxygen activity of grain boundaries and the defects formed by the presence of oxygen vacancies. Later studies showed that depending on the oxygen non-stoichiometry there was a change from semiconductor to metallic-like electrical conductivity with increasing temperature.[3,94,97] Therefore, a maximum power factor of 1000 $\mu W\ m^{-1}\ K^{-2}$ at 973 K was attained when oxygen deficiency was optimised by the addition of 8 vol% Ti to TiO_2.[104] As well as the improvements in the power factor, the thermal conductivity could also be reduced by increasing the oxygen deficiency.[3,94,97,99] Formation of crystallographic shear planes upon the increase in oxygen vacancy concentration was responsible for the reduction in lattice thermal conductivity; falling from 5.18 $W\ m^{-1}\ K^{-1}$ for TiO_2 to 2.04 $W\ m^{-1}\ K^{-1}$ for $TiO_{1.80}$ at room temperature.[97] Additionally, the processing route has a significant effect on the thermoelectric performance of Ti-based Magnéli phases. Liu *et al.*[105] demonstrated that a $TiO_{1.76}$ Magnéli phase prepared using a high-pressure, high-temperature (HPHT) technique had superior properties (see Figure 5.16) to those prepared by conventional methods. They[105] found

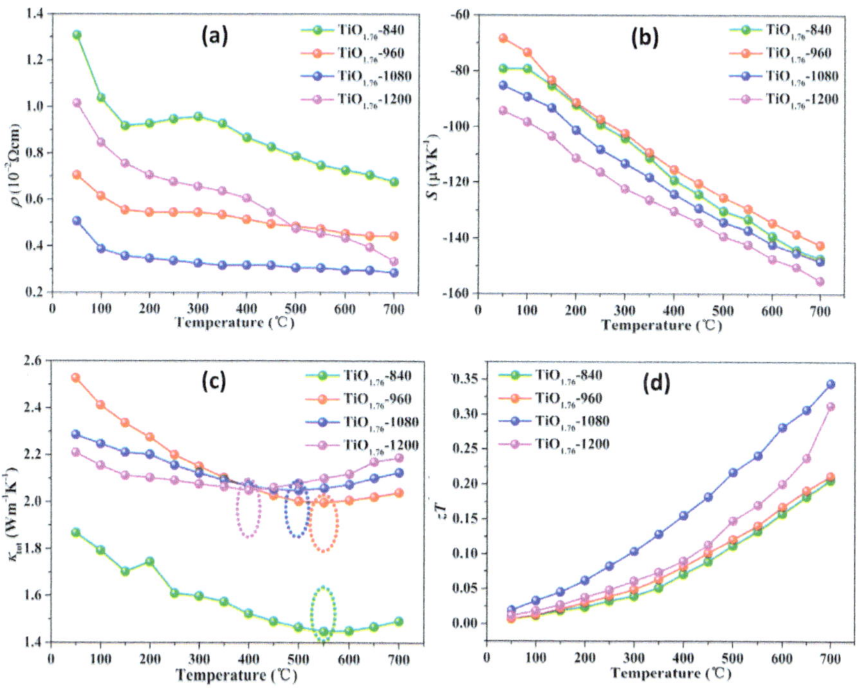

Figure 5.16 Temperature dependence of (a) electrical conductivity, (b) Seebeck coefficients, (c) thermal conductivity and (d) dimensionless figure of merit for $TiO_{1.76}$.[105] Reproduced from ref. 105 with permission from Elsevier, Copyright 2018.

that both electrical conductivity and Seebeck coefficients were increased by optimising the processing parameters, leading to a large power factor of 755 μW m^{-1} K^{-2} at 973 K, but as expected a high thermal conductivity (2.2–2.3 W m^{-1} K^{-1}). Together these gave one of the highest *ZT* values (0.35 at 973 K) reported for non-stoichiometric TiO_{2-x}. In addition to the high thermoelectric performance, the HPHT technique gave samples with high mechanical strength.

By a very different approach, Lee *et al.*[106] showed that thick film TiO_{2-x} with promising thermoelectric properties could be prepared by a spray plasma process. The resulting material mainly consisted of the rutile phase with the presence of Magnéli phases. After optimising the processing conditions a maximum power factor of 290 μW m^{-1} K^{-2} with low thermal conductivity (1.5–2 W m^{-1} K^{-1}) was achieved; giving a maximum *ZT* of 0.132 at 750 K. Their results indicated a robust method for preparing oxide materials for thermoelectric applications.

Amongst the studies to improve the properties of TiO_{2-x} phases by additives, N doping[107] and Nb/N co-doping[108] have been the most effective. Mikami and Ozaki[107] showed that by reacting TiO_2 with TiN, the resulting $Ti_6(O,N)_{11}$ exhibited the best properties among the Magnéli phases they prepared. Both the improvement in power factor and the reduction of thermal conductivity resulted in enhanced *ZT* values for N-doped TiO_{2-x}, reaching a maximum *ZT* of 0.40 at 1150 K for $Ti_8(O,N)_{15}$. Later, Liu *et al.*[108] showed that co-doping anatase TiO_2 phase by Nb and N significantly improved the power factor, reaching 987 μW m^{-1} K^{-2} at 973 K. The doping led to reduced grain size and possible reduction of Ti; the structural changes, combined with grain boundary and point defect scattering of phonons further reduced thermal conductivity, giving a maximum *ZT* of 0.35 at 973 K. For the highly reduced titania, $TiO_{1.1}$, there have been reports for *ZT* as high as 1.6, but full experimental details were not provided and the results have never been replicated.

Other Magnéli phases investigated for thermoelectric applications are the niobium and tungsten based $Nb_{3n-1}O_{8n-2}$ and W_nO_{3n-1} phases, respectively. The Nb-based compounds are attractive because of their relatively high Seebeck coefficients, but their low electrical conductivity prevents the materials achieving high power factor values. Additionally, due to the block-type structure, the material exhibits low lattice thermal conductivity.[3] Composites made of NbO_x with TiN or W additions reached *ZT* values as high as 0.23 at 1000 K, indicating potential for high-temperature thermoelectric applications. For W_nO_{3n-1} phases, a maximum *ZT* value of 0.15 at 1100 K was reported (Figure 5.17);[109] this was limited by the low σS^2 (~400 μW m^{-1} K^{-2} at 1100 K) mainly caused by the relatively low *S* (~−30 μV K^{-1} at 350 K) of these materials. The experimental results indicate that by first looking for low thermal conductivity Magnéli phases and then optimising the electronic transport properties could be a viable approach for maximising the thermoelectric properties of W-based compounds.[110]

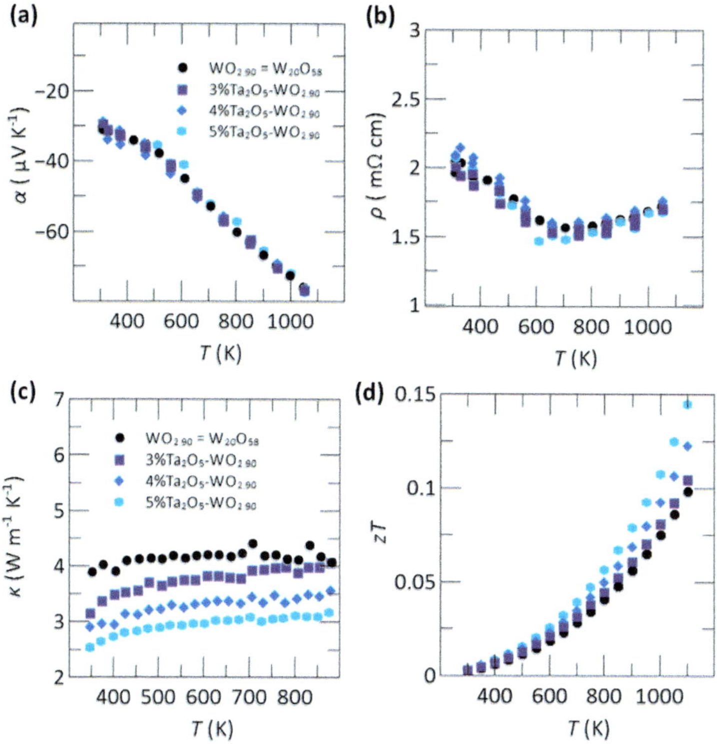

Figure 5.17 Temperature dependence of (a) Seebeck coefficient, (b) electrical resistivity, (c) thermal conductivity and (d) dimensionless figure of merit of $WO_{2.90}$ + x% Ta_2O_5 samples.[109] Reproduced from ref. 109 with permission from the Royal Society of Chemistry.

5.7 A-site Deficient Perovskites

5.7.1 Introduction

A-site deficient perovskites have emerged as very promising thermoelectric oxides. They were initially considered as electrodes for solid oxide fuel cell (SOFC) applications,[111] but attracted interest as thermoelectrics because of their complex/layered crystal structure, which is formed by alternating layers of fully and partially occupied layers of A-site cations[112–114] (Figure 5.18). This offers the possibility of enhanced scattering of phonons by the interfaces within the crystal structure and the point defects formed by cation vacancies. As a result, much lower thermal conductivity can be expected in A-site deficient perovskites, than in simple perovskites, which tend to have inherently high κ values, especially $SrTiO_3$.

Among the A-site deficient perovskites, the solid solutions of $SrTiO_3$ and $La_{2/3}TiO_3$ have been widely studied for thermoelectric applications.[4,115] They have shown three distinct regions of crystal structure based on composition; cubic for Sr-rich compositions, tetragonal for intermediate compositions

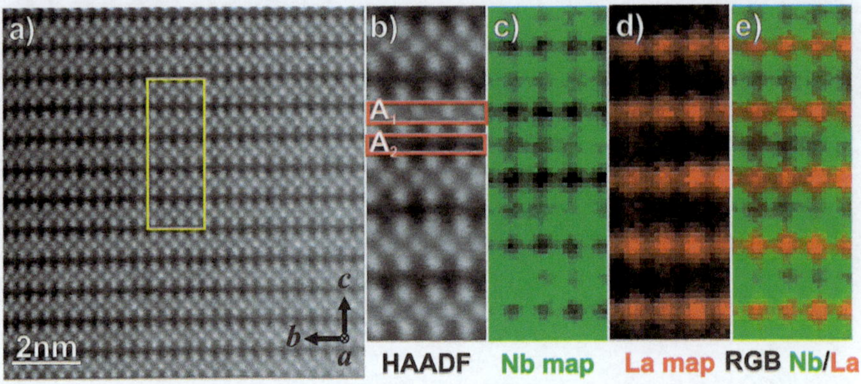

Figure 5.18 Structure of A-site deficient perovskite $La_{1/3}NbO_3$ showing alternating layers of fully and partially occupied layers of A-site cations; data based on EELS measurements along the [100] zone axis: (a) HAADF STEM image; analysis area marked by a yellow rectangle, (b) HAADF STEM signal intensity acquired simultaneously with the EELS signal, (c) integrated EELS intensity map of the Nb $L_{2,3}$ edge, (d) integrated EELS intensity map of the La $M_{4,5}$ edge and (e) RGB overlay of the Nb and La maps.[5] Reproduced from ref. 5 with permission from American Chemical Society, Copyright 2017.

and orthorhombic for La-rich compositions. The thermoelectric properties of these materials (Figure 5.19) are closely related to the crystal structure. For the simple cubic structure materials there is high electronic conductivity but also a high and temperature dependent thermal conductivity. In contrast, the La-rich compositions, with more complex crystal structures, exhibited lower electronic conductivity but also lower and temperature-stable thermal conductivity.[4]

There are also A-site deficient niobate and other titanate compositions including $La_{1/3}NbO_3$,[5] $Pr_{2/3}TiO_3$[116] and $La_{2/3}TiO_3$[115] for which the thermoelectric performance has been investigated. $La_{1/3}NbO_3$ exhibits a high Seebeck coefficient (−675 μV K^{-1}), but low electrical conductivity (5×10^{-3} S cm^{-1}) at room temperature, and low temperature-stable thermal conductivity (2 W m^{-1} K^{-1} at 500 K). However, for this niobate, only the electronic transport properties have been reported. $Pr_{2/3}TiO_3$ was shown to have a relatively high S (−120 μV K^{-1}) and reasonable σ (180 S cm^{-1}) at 400 K. $La_{2/3}TiO_3$ was reported to have a low S (−30 μV K^{-1}) and high σ (400 S cm^{-1}) at 500 K.

5.7.2 Crystal and Electronic Structure of Strontium Titanate Based Thermoelectric Materials

$SrTiO_3$ has a cubic perovskite (ABO_3) structure with space group *Pm*−3*m* at room temperature; Sr atoms (coordination number, CN = 12) occupy the A-site while the B-site is occupied by Ti (CN = 6). The lattice parameter of cubic $SrTiO_3$ is 3.905 Å.[117] Hybridisation of the O-2p states with the Ti 3d

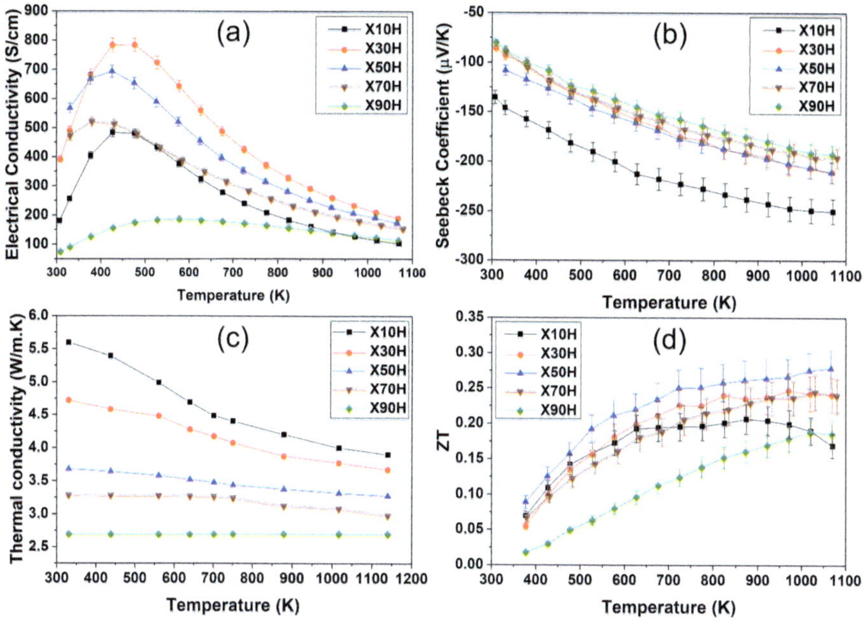

Figure 5.19 Thermoelectric properties of $Sr_{1-x}La_{2x/3}TiO_3$ (x = 0.10 to 0.90) compositions;[4] (a) electrical conductivity, (b) Seebeck coefficient, (c) thermal conductivity and (d) dimensionless figure of merit. Reproduced from ref. 4 with permission from American Chemical Society, Copyright 2017.

states results in covalent bonding, while ionic bonding occurs between Sr^{2+} and O^{2-} ions. This mixed ionic-covalent nature of the material has a significant impact on charge carrier scattering processes. The stability of perovskite-structured materials may conveniently be described by the Goldschmidt tolerance factor (τ_G),[118] given by eqn (5.3),

$$\tau_G = (R_A + R_O)/\sqrt{2}(R_B + R_O) \quad (5.3)$$

where, R_A, R_B and R_O are ionic radii of A and B-sites and oxygen atoms, respectively. For example, ideal cubic structures have $\tau_G \approx 1$, and lower τ_G values lead to lower symmetry structures such as tetragonal, rhombohedral *etc.* In this context, the τ_G for pure $SrTiO_3$ is 1.002 based on standard values for the ionic radii.[119] Thus, without any doping or the presence of vacancies, $SrTiO_3$ has an ideal cubic structure. However, incorporation of A- and B-site dopants and generation of oxygen vacancies into the structure distorts the cubic structure and modifies the properties. Investigations of $SrTiO_3$ for thermoelectric applications have focused on doping the A-sites with rare earth (RE) elements,[120–130] while B-site doping has been limited to Nb and Ta in conjunction with processing in different atmospheric environments to form oxygen vacancies for enhancing the transport properties.[131–135]

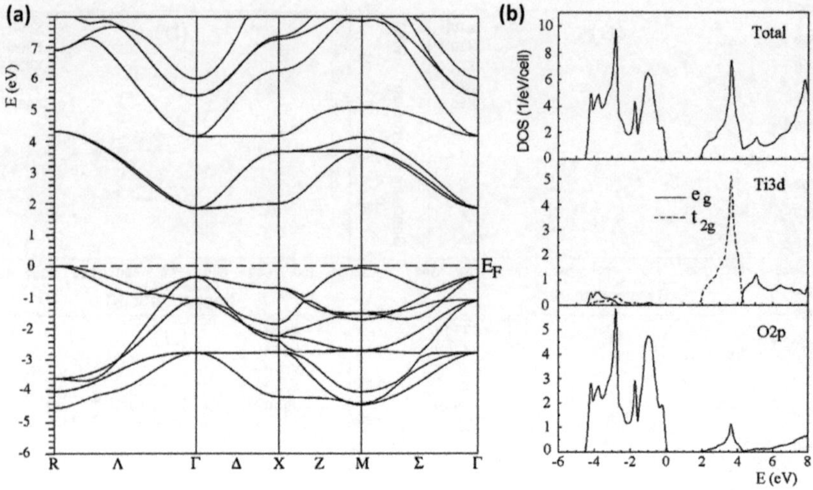

Figure 5.20 (a) Electronic band structure of cubic $SrTiO_3$; (b) DOS corresponding to the band structure illustrated in (a).[137] Reproduced from ref. 137 with permission from Elsevier, Copyright 2007.

The electronic band structure of $SrTiO_3$ comprises the valence band maxima of O-2p components of the three oxygen atoms in the structure and conduction band minima containing 3d orbitals of octahedrally-coordinated titanium atoms (Figure 5.20a).[118] The material is a wide band gap insulator in its ground state (E_g = 3.2 eV), and based on the electronic band structure and the corresponding density of states (DOS) data (Figure 5.20b), Dresselhaus *et al.*[136] correctly predicted that $SrTiO_3$ would have a high Seebeck coefficient.

5.7.3 Electronic and Thermal Transport Properties of $SrTiO_3$

There have been comparatively few studies[128,138,139] of the thermoelectric properties of undoped $SrTiO_3$, mainly because of the insulating nature of the material prepared in air. Although the carrier concentration of undoped $SrTiO_3$ (~9×10^{19} cm^{-3}) is in the desired range for thermoelectric applications and it has relatively high S values (−380 μV K^{-1} at 300 K), the low electrical conductivity limits the maximum power factor of the pristine material. Ito and Matsuda[138] prepared $SrTiO_3$ powder from a polymer complex and densified by hot pressing; they showed (Figure 5.21) that high sintering temperatures and extended processing times were desirable for improving thermoelectric properties.

The calculations of Shirai and Yamanaka[140] demonstrated that if the carrier concentration (n) of $SrTiO_3$ could be optimised, it should be possible to obtain power factors comparable with or greater than those of traditional thermoelectric materials such as Bi_2Te_3. This is explained by the non-degenerate nature of electrons, even at high concentrations, and their large

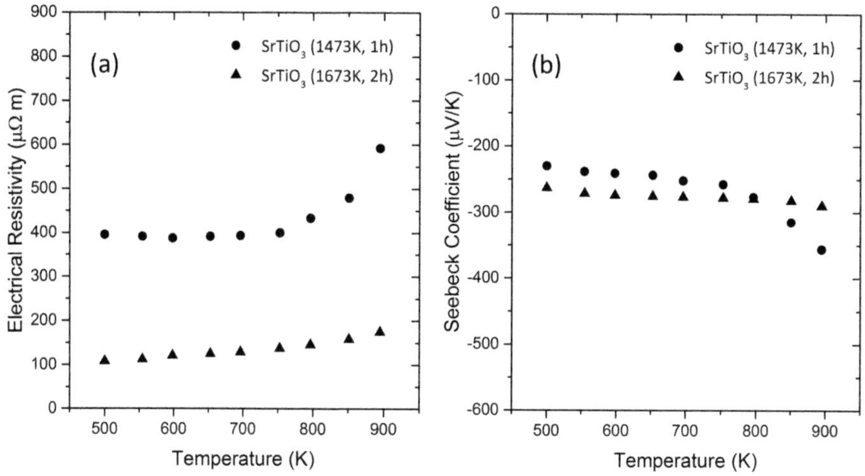

Figure 5.21 Temperature dependence of (a) electrical resistivity and (b) the Seebeck coefficients of $SrTiO_3$ ceramics prepared by hot pressing in an Ar atmosphere.[138]

effective mass in $SrTiO_3$. Additionally, there is a linear relationship between *n* and the scattering factor over a wide range of carrier concentrations, which indicates that carrier mobility is independent of impurity concentration. Nevertheless, processing conditions and the final microstructure can be very important. Noll *et al.*[141] established that grain boundaries are more resistive than the bulk grains. Therefore, reducing the concentration of grain boundaries should be beneficial for improving the electronic transport properties of $SrTiO_3$. This proposal is supported by many experimental studies showing an increase in σ with increasing average grain size.[142] This arises through improved carrier mobility with increasing average grain size, as a result of reduced scattering by grain boundaries.

Early studies of thermal transport in $SrTiO_3$ focused on single crystals[126,143] and demonstrated high thermal conductivity at room temperature (9–12 W m^{-1} K^{-1}) and a T^{-1} dependence at high temperatures, indicating Umklapp phonon scattering. In spite of the low electron component of thermal conductivity ($\kappa_{electron}$) due to the insulating nature of the material, the simple crystal structure led to a high lattice thermal conductivity ($\kappa_{lattice}$) and high overall thermal conductivity. Thermal transport investigations of polycrystalline $SrTiO_3$ were mainly in terms of the effect of average grain size[144] and the presence of defects[145] on thermal conductivity; modelling studies focused on the phonon spectrum and phonon mean free path, l_{ph}.[146,147] Wang *et al.*[144] found that the κ of $SrTiO_3$ decreased with reducing average grain size (Figure 5.22a). However, it was suggested that the mechanisms behind the reduction were not only improved phonon scattering but also increased interfacial thermal resistance. The l_{ph} also reduced with decreasing grain size. Wang *et al.*[144] also calculated the dependence of normalised cumulative

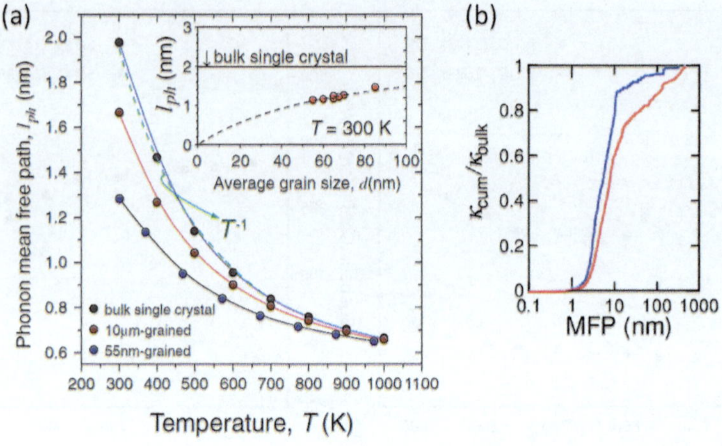

Figure 5.22 (a) Effect of microstructure on l_{ph} in $SrTiO_3$.[144] Reproduced from ref. 144 with permission from IOP Publishing, Copyright (2010) The Japan Society of Applied Physics. (b) The dependence of normalised cumulative thermal conductivity ($\kappa_{cum}/\kappa_{bulk}$) on the l_{ph} for $SrTiO_3$ at 300 K.[146] Reproduced from ref. 146 with permission from IOP Publishing, Copyright (2015) The Japan Society of Applied Physics.

thermal conductivity ($\kappa_{cum}/\kappa_{bulk}$) on the l_{ph} for $SrTiO_3$ at 300 K (Figure 5.22b), showing that heat conduction was dominated by phonons with l_{ph} smaller than 10 nm. This highlights the limited effect of nanostructuring $SrTiO_3$ as a means to reduce κ, since only features comparable in size to the l_{ph} are expected to have any significant effect on phonon transport. Recent analysis of calculated phonon dispersion curves for $SrTiO_3$ (Figure 5.23) by Feng *et al.*[146] showed that the contribution of longitudinal acoustic (LA) phonon modes to $\kappa_{lattice}$ was negligible, but the optical branch modes and the optical phonons above 2.5 THz were the main carriers of heat in cubic $SrTiO_3$; their contribution to overall lattice conductivity was greater than 60%.

At the sub-grain level, it has been shown that point defects can have a significant effect on thermal transport;[145,148] for example, increasing the concentration of oxygen vacancies leads to reduction of $\kappa_{lattice}$. The oxygen vacancies are comparable in size to l_{ph} in $SrTiO_3$, and may be randomly distributed or clustered. Similarly, cation vacancies can also reduce κ,[4,149] particularly A-site vacancies, which introduce both a mass difference and lattice strain for improved scattering of phonons.

5.7.4 Common Approaches for Tuning the Thermoelectric Properties of $SrTiO_3$

In view of the substantial body of work addressing routes to enhance, or to tune the thermoelectric performance of $SrTiO_3$-based ceramics, this section reviews the main approaches that have been employed. Whilst the focus is specifically strontium titanate materials, many of the approaches are relevant to a much wider range of thermoelectrics.

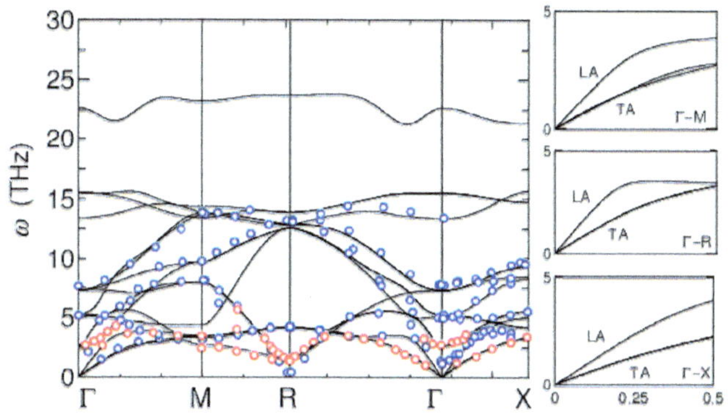

Figure 5.23 Phonon dispersion curve of cubic $SrTiO_3$.[146] The blue and red circles show the inelastic neutron scattering results at 90 and 300 K, respectively. Reproduced from ref. 146 with permission from IOP Publishing, Copyright 2015.

All the materials properties reported hereon are for $SrTiO_3$ ceramics prepared under reducing atmospheres since it has become an established and proven approach for producing high quality materials for thermoelectric applications. The options include; (i) processing route, (ii) doping on A-site, (iii) doping on B-site, (iv) combined doping on A- and B-sites, (v) A-site cation vacancies, (vi) inclusion doping and (vii) $SrTiO_3$ related Ruddlesden–Popper (RP) phases. A summary of the *ZT* values achieved using such approaches is given in Table 5.1. It can be seen that depending on the method employed, the maximum *ZT* values are quite similar, varying only from 0.20 to 0.41. Frequently, two or more of these approaches are used in conjunction with one other to achieve higher thermoelectric performance, for example, a doping strategy plus incorporation of A-site and/or oxygen vacancies.

5.7.4.1 *Effect of the Processing Route and Conditions*

Processing methods can directly affect thermoelectric properties since materials produced *via* the mixed oxide (MO) route with and without spark plasma sintering (SPS) exhibit differences in microstructures and/or oxygen deficiency level. Samples prepared by SPS frequently contain secondary phases at the grain boundaries[160] while homogenous, single phase materials can be achieved by standard MO methods with extended processing times.[120,123,124] Dehkordi *et al.*[130] showed that $SrTiO_3$ prepared by SPS contained grain boundary phases that enhanced carrier mobility (μ), which led to an increase in σ while not strongly affecting S through its strong dependency on n. This resulted in a significant improvement in the power factor σS^2. In contrast, the improvement in σS^2 for materials produced using the MO route is mainly attributed to the homogeneous microstructure with large grains, resulting in increased μ.

Table 5.1 Thermoelectric *ZT* values of $SrTiO_3$-based materials prepared using different strategies; MO, SPS and HP represent the mixed oxide route, spark plasma sintering and hot pressing, respectively.

Composition	Processing route	Maximum *ZT*	Reference
A-site doping			
$Sr_{0.9}Dy_{0.1}TiO_3$	MO	0.22 (573 K)	124
$Sr_{0.9}Nd_{0.1}TiO_3$	MO	0.28 (873 K)	150
$Sr_{0.92}La_{0.08}TiO_3$	SPS	0.37 (1045 K)	127
$Sr_{0.92}La_{0.08}TiO_3$	MO	0.28 (773 K)	151
$Sr_{0.9}Pr_{0.1}TiO_3$	MO	0.34 (1070 K)	152
$Sr_{0.9}Dy_{0.1}TiO_{3-\delta}$	MO	0.41 (1073 K)	120
$Sr_{0.85}Pr_{0.15}TiO_3$	SPS	0.35 (773 K)	122
$Sr_{0.9}Gd_{0.1}TiO_{3-\delta}$	SPS	0.37 (1006 K)	153
$Sr_{0.9}Y_{0.1}TiO_{3-\delta}$	SPS	0.30 (760 K)	154
$Sr_{0.88}Y_{0.08}TiO_{3-\delta}$	MO	0.20 (953 K)	135
B-site doping			
$SrTi_{0.80}Nb_{0.20}O_3$	HP	0.35 (1000 K)	142
$SrTi_{0.90}Ta_{0.10}O_{3-\delta}$	MO	0.30 (1000 K)	155
Combined doping			
$Sr_{0.83}La_{0.1}Dy_{0.07}TiO_3$	MO	0.36 (1045 K)	156
$Sr_{0.8}La_{0.18}Yb_{0.02}TiO_3$	MO	0.31 (1023 K)	123
$Sr_{0.8}La_{0.13}Ti_{0.95}Nb_{0.05}O_3$	MO	0.30 (1000 K)	157
$Sr_{0.97}Ti_{0.80}Nb_{0.17}W_{0.03}O_3$	MO	0.28 (1270 K)	158
A-site deficient compositions			
$Sr_{0.55}Pr_{0.30}TiO_3$	MO	0.30 (1173 K)	149
$Sr_{0.775}La_{0.15}TiO_{3-\delta}$	MO	0.41 (973 K)	115
$Sr_{0.70}La_{0.16}Ti_{0.80}Nb_{0.20}O_3$	MO	0.35 (100 K)	159

It is well established[129,161,162] that preparation of titanate materials in a reducing atmosphere leads to the formation of oxygen vacancies, as described by eqn (5.4). Assuming charge neutrality is preserved and there are no changes in the titanium oxidation state, then one oxygen vacancy introduces two additional carriers to the system (eqn (5.5)).

$$SrTiO_3(s) \xrightarrow{\text{reduction}} SrTiO_{3-\delta}(s) + \frac{\delta}{2}O_2(g) \tag{5.4}$$

$$O_o^{\times} \leftrightarrow V_o^{\cdot\cdot} + 2e' + \frac{1}{2}O_2(g) \tag{5.5}$$

However, the concentrations of carriers in $SrTiO_3$ containing high levels of oxygen vacancies ($\delta > 0.25$) were much lower than values predicted by eqn (5.5).[163] Experimental work showed that clustering occurred when high levels of oxygen vacancies were present, which then acted as sinks for the excess carriers.[164] Thus, up to a critical level, oxygen vacancies are beneficial for improving σ of $SrTiO_3$; however, a reduction in S is inevitable due to the increase in n.

In addition to the formation of oxygen vacancies, the reducing atmosphere leads to the reduction of B-site cations, which are Ti in the undoped case and usually Nb when the B-site is doped. It was shown by Hashimoto *et al.*[165] that the reduction of titanium from 4+ to 3+ was improved by increasing the temperature and reducing oxygen partial pressure (eqn (5.6)). A similar but easier reduction mechanism for Nb-doped $SrTiO_{3\pm\delta}$ samples (Nb^{5+} to Nb^{4+}) has been reported (eqn (5.7)).

$$O_o^{\times} + 2Ti_{Ti}^{\times} \xrightarrow{\text{reducing atmospehere}} 2Ti_{Ti}' + V_o^{\bullet\bullet} + \frac{1}{2}O_{2(g)} \tag{5.6}$$

$$Ti^{3+} + Nb^{5+} \xleftrightarrow{\text{equilibrium}} Ti^{4+} + Nb^{4+} \tag{5.7}$$

Since oxygen vacancies are also beneficial for the reduction of κ (due to the improved phonon scattering by vacancies, whose size is comparable to the l_{ph} in $SrTiO_3$),[145] then sample preparation of $SrTiO_3$ for thermoelectric applications generally involves first sintering in air and then annealing in a reducing atmosphere. Simultaneous reduction of κ by the presence of oxygen vacancies, and improvement in σ by the presence of additional carriers, results from the reduction of B-site cations and enhances the overall thermoelectric performance.

5.7.4.2 *Effect of A-site Doping*

Whilst the structure of $SrTiO_3$ remains cubic for many A-site dopants (for $Sr_{1-x}R_xTiO_3$, $x < 0.10$ and R is A-site dopant),[120,123] the use of dopants with lower ionic radius than Sr such as Dy, Yb, Er and Y may lead to lower symmetry structures, *e.g.* tetragonal structure with space group *I*4/*mcm*. Equally, high concentrations of dopants can distort the structure causing tetragonal symmetry,[120] and greater than 10% substitution of smaller ionic radius dopants causes secondary phase formation ($Ln_2Ti_2O_7$, Ln = lanthanide). Lattice parameter changes as a result of doping modify the electronic band structure and generate defects within the structure, thereby increasing phonon scattering and improving transport properties.[120,123,124]

Prompted by encouraging reports of a high power factor[166] and electronic transport[167] in La-doped $SrTiO_3$, (σS^2 comparable with Bi_2Te_3 [128]) many early studies investigated the effects of La doping.[124,126,129,134,165] These studies demonstrated the optimum concentration for A-site doping (10 at.%), the effects of dopants on the crystal structure and the stability of $SrTiO_3$ ceramics, and the effect of processing conditions on thermoelectric properties. Later, attention moved to the effects of rare earths, mainly lanthanide series elements, on the thermoelectric properties of $SrTiO_3$.[24,120–123,130,135,150,152,154,160] Doping with larger ionic radii lanthanides (La, Pr, *etc.*) led to a higher σS^2 while doping with smaller ionic radii lanthanides (Gd, Dy, *etc.*) was beneficial in the reduction of κ (Figure 5.24).

Doping $SrTiO_3$ with rare earth elements results in a smaller crystal structure, leading to increased localisation of electrons,[168] from which higher σ is expected. Conversely, a smaller unit cell also results in a lower effective mass (m^*)[169] and so a reduction in S is expected due to its direct dependence on m^*.

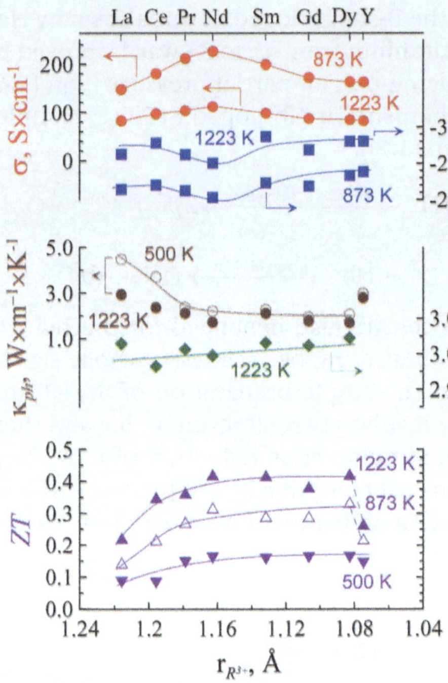

Figure 5.24 Summary of the thermoelectric properties $SrTiO_3$ ceramics doped with 10 at.% rare earth cations as a function of ionic radii of the dopant, $r_{R^{3+}}$.[120] Reproduced from ref. 120 with permission from the Royal Society of Chemistry.

Therefore, higher σS^2 values are achieved for $SrTiO_3$ doped with smaller radii rare earth dopants because of these combined effects, where the improvement in σ dominates the reduction of S. At the same time there is a simultaneous reduction in κ as a result of increased distortion in the crystal,[123] which causes increased scattering of phonons by strain field effects and by mass differences between the host and dopant cations. Thus, the highest σ was recorded for $SrTiO_3$ doped with larger cations (La, Pr, Nd and Sm) while the highest S was achieved for smaller cations (Dy and Y).[169,170] Moreover, a lower κ was also observed for $SrTiO_3$ doped with smaller cations (Figure 5.24), such as Gd, Dy, Er, Y.[120,123] Together these effects led to an improvement in the overall thermoelectric properties.

Increased σ is mainly linked to increasing n through the use of trivalent dopants on Sr^{2+} site (explained by eqn (5.7) and (5.8)). As noted earlier, the Seebeck coefficient is mainly dependent on n, m^* and also μ. Thus, an increase in n leads to a decrease in S, while increased m^* and μ are desired to achieve high S values. Therefore, increasing dopant concentration generally leads to a decrease in S. However, distortion of the crystal lattice upon doping also affects the electronic band structure and DOS around E_F; distortion caused by smaller ionic radii dopants on A-sites also improves S *via* an increase in m^* regardless of a minor reduction in μ. This ultimately leads to materials with improved σS^2 values in comparison to un-doped $SrTiO_3$.

$$\frac{x}{2}\text{Ln}_2\text{O}_3 \xrightarrow{\text{SrTiO}_{3-\delta}} (1-x)\text{Sr}_{\text{Sr}}^{\times} + x\text{Ln}_{\text{Sr}}^{\bullet} + (1-x-2\delta)\text{Ti}_{\text{Ti}}^{\times} + (x+2\delta)\text{Ti}_{\text{Ti}}' + (3-\delta)O_{\text{o}}^{\times} + \delta V_{\text{o}}^{\bullet\bullet} \tag{5.8}$$

$$n = \left[\text{Ti}_{\text{Ti}}'\right] \approx \left[\text{Ln}_{\text{Sr}}^{\bullet}\right] + 2\left[V_{\text{o}}^{\bullet\bullet}\right] \tag{5.9}$$

5.7.4.3 *Effect of B-site Doping*

Common B-site dopants for improving the thermoelectric performance of $SrTiO_3$ are Nb, Ta and W, of which Nb is the most popular. The modelling of Guo *et al.*[170] revealed that Nb doping lowers the bottom of the valence band and moves E_F into the conduction band, resulting in metallic conduction behaviour. Ohta *el al.*[171] showed that Nb doping improved S because of the increase in m^*, as a result of the decreased overlap of Ti 3d electrons. Similarly, the main defects for samples prepared in air was the formation of cation vacancies and SrO planar defects, while for samples prepared under reducing conditions it is the reduction of Ti^{4+} to Ti^{3+} (eqn (5.10) and (5.11)).[161]

$$\frac{x}{2}\text{Nb}_2\text{O}_5 \xrightarrow{\text{SrTiO}_{3-\delta}} \text{Sr}_{\text{Sr}}^{\times} + x\text{Nb}_{\text{Ti}}^{\bullet} + (1-x-2\delta)\text{Ti}_{\text{Ti}}^{\times} + (x+2\delta)\text{Ti}_{\text{Ti}}' + (3-\delta)O_{\text{o}}^{\times} + \delta V_{\text{o}}^{\bullet\bullet} \tag{5.10}$$

$$n = \left[\text{Ti}_{\text{Ti}}'\right] \approx \left[\text{Nb}_{\text{Ti}}^{\bullet}\right] + 2\left[V_{\text{o}}^{\bullet\bullet}\right] \tag{5.11}$$

Due to the comparable solubility levels, Ta doping has a similar effect to Nb in $SrTiO_3$.[155] However, the increased covalency of the bonds with Ta results in a more pronounced reduction in κ in comparison to Nb doping.[172] In contrast, the complex redox behaviour of W under reducing conditions leads to lower solubility (<6%) in $SrTiO_3$;[158] in turn this could lead to the formation of metallic inclusions in $SrTiO_3$ which could affect transport properties. Thermoelectric figure of merit ZT values as high as 0.37 have been reported for 20% Nb-doped $SrTiO_3$[171] while values of 0.30[151] and 0.13[158] were achieved at 1000 K with 10% Ta doping and 3% W doping, respectively.

5.7.4.4 *Effect of Combined A- and B-site Doping*

Combined doping on A- and B-sites is an alternative for improving the thermoelectric performance of $SrTiO_3$.[131,173,174] In this context, lanthanides are generally preferred as the A-site dopants while Nb and Ta are predominantly used for B-site doping. There are also studies utilising dual doping strategies on either A-sites alone[156,175] or B-sites alone.[158] Dual doping of A- and B-sites showed that the approach can improve σS^2 by increasing n, and in parallel reducing κ by increased phonon scattering. On the other hand, dual doping of only A-site results in increased distortion of the unit cell, and much greater mass contrast between A-site cations, dramatically reducing κ values. Combined doping on the A- and B-site using La and Nb as dopants led to a ZT value of 0.29,[173] while ZT values of 0.28 at 1270 K and 0.36 at 1045 K were obtained by dual doping only on the B-site[158] and A-site,[156] respectively. A mechanism of charge compensation for combined A- and B-site doping of $SrTiO_3$ prepared under reducing conditions can be explained in terms of the introduction of additional charge carriers (due to the dopants), the formation of oxygen vacancies and the reduction of B-site cations (eqn (5.12)). The resulting carrier concentration after dual doping is expected to be equal to the carriers introduced by A- and B-site dopants and oxygen vacancies (eqn (5.13)).

$$\begin{aligned}\frac{x}{2}\mathrm{Ln_2O_3}+\frac{y}{2}\mathrm{Nb_2O_5}\xrightarrow{\mathrm{SrTiO_{3-\delta}}}&(1-x)\mathrm{Sr_{Sr}^{\times}}+x\mathrm{Ln_{Sr}^{\bullet}}+y\mathrm{Nb_{Ti}^{\bullet}}\\&+(1-x-y-2\delta)\mathrm{Ti_{Ti}^{\times}}+(x+y+2\delta)\mathrm{Ti_{Ti}'}\\&+\delta V_{\mathrm{O}}^{\bullet\bullet}+(3-\delta)O_{\mathrm{O}}^{\times}\end{aligned} \tag{5.12}$$

$$n=\left[\mathrm{Ti_{Ti}'}\right]\approx 2\left[V_{\mathrm{O}}^{\bullet\bullet}\right]+\left[\mathrm{Ln_{Sr}^{\bullet}}\right]+\left[\mathrm{Nb_{Ti}^{\bullet}}\right] \tag{5.13}$$

5.7.4.5 *Effect of A-site Cation Vacancies*

As noted earlier, in order to effectively reduce the κ of $SrTiO_3$, atomic-scale defects are necessary to scatter the majority of phonons due to the small size of l_{ph} (1–2 nm).[144] Thus, the introduction of A-site vacancies, which have a size comparable to l_{ph}, has emerged as a new and viable route for improving the thermoelectric properties of $SrTiO_3$ by reducing κ and simultaneously tuning the electronic transport properties. The work to date has mainly focused on $SrTiO_3$–$La_{2/3}TiO_3$ solid solutions,[4,115,157] and exploits the combined effects of A-site vacancies and lanthanide doping. Popuri *et al.*[157] showed that A-site vacancies had a dramatic effect on κ values; their material exhibited glass-like, temperature-stable thermal conductivity, for samples with A-site vacancies above a critical vacancy threshold (~27%). The reduced thermal conductivity behaviour resulted from improved phonon scattering due to the significant mass difference between Sr, La and the vacant sites. Additionally, the formation of an ordered structure with a high vacancy concentration promoted interface scattering of phonons. In a similar study Lu *et al.*,[115] improved σS^2 by increasing the vacancy concentration, which resulted in one of the highest ZT values reported (0.41 at 973 K) for $SrTiO_3$ based thermoelectric materials.

Recently, Kovalevsky *et al.*[149] used Pr instead of La in the solid solution; they concentrated on the Sr-rich end of $Sr_{1-1.5x}Pr_xTiO_{3\pm\delta}$ ($x \leq 0.30$) system and confirmed the effectiveness of A-site vacancies for reducing the κ of $SrTiO_3$. However, the study raised questions about the effectiveness of A-site vacancies in comparison to the oxygen vacancies. They were answered in their later study,[176] which showed that the oxygen vacancies were more effective for low cation vacancy concentrations and cation-oxygen vacancy clustering could be utilised for further reduction of κ.

In addition to titanate solid solutions, there have been investigations of $SrTiO_3$ with niobate based A-site deficient perovskites such as $La_{1/3}NbO_3$.[5,159] Srivastava *et al.*[159] showed that the best composition was $(0.8)SrTiO_3$–$(0.2)La_{1/3}NbO_3$, which exhibited a ZT value of 0.25 at 1000 K. As all these compositions are stoichiometric, the authors also investigated the effects of additional La doping on the Sr site and Nb on the Ti site; they showed that by utilising such an approach the ZT value could be increased to 0.35 at 1000 K with only 5% excess La substitution on Sr sites. The improvement in σ arose through the extra carriers generated.

The defect chemistry of A-site deficient perovskites prepared under reducing atmospheres is complex as shown in eqn (5.14). The process involves the formation of cation vacancies on A-site, reduction of multivalent B-site cations and the formation of oxygen vacancies. As a result, n is expected to be related to the concentration of reduced B-site cations or oxygen vacancies as shown in eqn (5.15). With such a complex defect mechanism, decoupling of the thermal and electronic transport properties can be achieved.

$$x\text{Ln}_{2/3}\text{TiO}_3 \xrightarrow{\text{SrTiO}_{3-\delta}} (1-x)\text{Sr}_{\text{Sr}}^{\times} + \frac{x}{3}V_{\text{Sr}}'' + \frac{2x}{3}\text{Ln}_{\text{Sr}}^{\bullet} + (1-2\delta)\text{Ti}_{\text{Ti}}^{\times} + (2\delta)\text{Ti}_{\text{Ti}}' + \delta V_{\text{o}}^{\bullet\bullet} + (3-\delta)O_{\text{o}}^{\times} \tag{5.14}$$

$$n = \left[\text{Ti}_{\text{Ti}}'\right] + 2\left[V_{\text{Sr}}''\right] \approx \left[\text{Ln}_{\text{Sr}}^{\bullet}\right] + 2\left[V_{\text{o}}^{\bullet\bullet}\right] \tag{5.15}$$

A common observation in the investigations of A-site deficient compositions[115,149,159,176] was the formation of Ti-based Magnéli (secondary) phases, as a result of the excessive reduction of the material.[177] Above a certain oxygen vacancy threshold, in the presence of existing A-site vacancies, the perovskite lattice cannot accommodate further defects and a new perovskite is formed, which has a higher ratio of A-site cations to B-site cations; the excess Ti is accommodated as a Ti-based Magnéli phase within the material (eqn (5.16)).

$$\text{A}_{1-x}\text{TiO}_3 \xrightarrow{\text{reduction}} \text{A}_{1-x}\text{TiO}_{3-\delta} \xrightarrow{\text{further reduction}} \text{A}_{1-x}\text{Ti}_{1-y}\text{O}_{3-\delta-yn} + y\text{TiO}_n \tag{5.16}$$

5.7.4.6 *Effect of Inclusion Doping and/or Composite Approach*

Additives and/or inclusions can be employed to improve the thermoelectric performance of materials through their effect on thermal and electrical transport properties. These have been utilised in many different material systems[178–182] by either improving σS^2 and/or reducing κ. The changes in the properties were explained by their effects on electron and phonon scattering.

Wang *et al.* introduced a variety of additives into $SrTi_{0.85}Nb_{0.15}O_3$ ceramics: yttria-stabilised zirconia (YSZ);[183,184] mesoporous silica (MS);[185] potassium niobate (KTO) nanowires;[186] and TiO_2 nanotube (TNT).[132] In spite of relatively low final sample densities, the incorporation of such additives did increase density and promoted grain growth; this led to an improvement in σ without any significant change in S. This was due to the strong dependence of S on n rather than carrier mobility (μ); the changes in microstructure mainly affected μ. In addition, the inclusion additives helped to reduce thermal conductivity by interface scatterings; ZT values as high as 0.34 at 900 K were achieved by use of only 3 wt% KTO.[186] On the other hand, minor additives can also be used to tune the thermoelectric properties of $SrTiO_3$. Surprisingly, the effect of minor additives on the properties of thermoelectric oxides has not attracted much interest despite their extensive use to improve the processing and properties of other electroceramics. Additives such as B_2O_3,[187] Bi_2O_3[188] and CuO[189] are frequently used to reduce the sintering temperature *via* the formation of a liquid phase, while MnO_2,[190] Pr_6O_{11},[191] Nb_2O_5[192] and La_2O_3[193] are heavily used to decrease the sintering temperature and to modify the electrical properties of electroceramics. Among these, B_2O_3 is well-established for $SrTiO_3$-based ceramics, enabling the sintering temperature to be reduced by as much as 150 K.[187,194]

Recent studies have focused on introducing either metallic inclusions[125,182] or graphene[181,195] into the $SrTiO_3$ matrix in order to modify properties. Metallic inclusions were mainly introduced to improve σS^2 of the material because of the increased n while there is a reduction in κ from the increased scattering of phonons at the inclusion/matrix interfaces. The addition of 15% Ag to $Sr_{0.9}La_{0.1}TiO_3$ improved the ZT at 960 K from 0.08 to 0.18,[125] while addition of 2.5 wt% Cu to $Sr_{0.8}La_{0.067}Ti_{0.8}Nb_{0.2}O_{3-\delta}$ increased the ZT at 1000 K from 0.25 to 0.36.[182] The incorporation of graphene into the $SrTiO_3$ matrix appears to be another effective route for improving thermoelectric properties. Since graphene exhibits p-type thermoelectric properties[196] in comparison to the n-type nature of the matrix, there was some concern that the high σ and κ of graphene may be detrimental to the properties of the composite. However, investigations of undoped $SrTiO_3$[195] and $Sr_{0.9}La_{0.067}TiO_3$[181] revealed that by including small amounts of graphene the thermoelectric properties of the composite materials were dramatically improved. The presence of the graphene was doubly beneficial, increasing electrical conductivity and reducing thermal conductivity over a wide temperature range. The improvement in σS^2 was a result of oxygen vacancy formation at the graphene/matrix boundaries, which led to increase in both n and m^*. Thus, a significant improvement in σ was observed while maintaining S. Furthermore, κ was reduced by the inhibition of grain growth by graphene, which led to nanostructured materials, where increased phonon scattering occurred because of the higher density of grain boundaries. Therefore, ZT values as high as 0.42 at 1023 K were achieved by the addition of only 0.6 wt% graphene.[181]

5.7.4.7 $SrTiO_3$-based Ruddlesden–Popper Phases

Ruddlesden–Popper (RP) phases[197] are formed as a consequence of (SrO) planar defects due to the presence of excess Sr in $SrTiO_3$ based materials. They are represented as $SrO(SrTiO_3)_x$ (x = 1, 2, 3, ∞); the resulting layered structure comprises $SrTiO_3$ perovskite slabs separated by SrO layers. These natural superlattice structures have been extensively studied[198,199] to understand and exploit the layered nature of the material. It was shown that increasing the number of the perovskite slabs between the SrO layers ($x \rightarrow \infty$) is beneficial for improving the electronic properties, while lower x values mean smaller spacing between the SrO layers, leading to reduction of κ.[200]

In order to improve the properties of RP phase materials, lanthanide doping on Sr sites[200–202] and Nb doping on Ti site was investigated.[198,203] They showed that the electronic properties of the materials were mainly controlled by the perovskite layers; the Seebeck coefficients were significantly lower than those of $SrTiO_3$, as a result of the reduced symmetry of the TiO_6 octahedra. In addition, the SrO layers were effective electron scattering centres, having a dramatic effect on σ values. Therefore, lower ZT values (0.14 at 1000 K) in comparison to $SrTiO_3$ were reported.[198] However, later studies showed that substitution of lanthanides on the Sr site was beneficial for improving the symmetry of the TiO_6 octahedra and improving the thermoelectric performance of the materials. Using such an approach, Wang *et al.*[203] achieved a ZT value of 0.24 at 1000 K for 5% Gd-doped $SrO(SrTiO_3)_2$.

5.8 Modelling

One of the newer approaches in thermoelectrics is high throughput screening of materials databases to identify structures and materials that might be suitable for thermoelectric applications.[204] This method employs atomic-scale modelling using density functional theory (DFT) to calculate the band structures[205–207] and thermoelectric properties[208] of the materials. However, both atomistic and DFT modelling can also be employed using first principles approaches to (i) interpret experimental results and define the mechanisms operating at the atomic scale, and (ii) predict structures and compositions that should give optimum combinations of thermoelectric properties.[5,24,28,80,86,209–214] In recent years, much of this work has focused on $SrTiO_3$-based materials and layered cobaltites.

Ren *et al.*[211] used the Debye–Callaway model to simulate the effect of point defects and nano-grains on thermal conductivity in oxide-based thermoelectrics, demonstrating that the oxygen vacancy concentration needed to be controlled in order to maximise ZT. For $SrTiO_3$, Bilc *et al.*[212] showed that highly anisotropic flat-and-dispersive bands can maximise the thermoelectric power factor, and can aid electronic transport in bulk materials. From combined experimental and modelling work, Azough *et al.*[4] demonstrated that the thermal conductivity of La doped $SrTiO_3$ can be reduced and controlled through an interplay of the substituted La and A-site vacancies and the formation of a layered structure; a critical mass of A-site vacancies is needed to reduce thermal conductivity. More generally, for $SrTiO_3$ Yeandel *et al.*[209] explored the effect of tilt boundaries and predicted that efforts to further reduce thermal conductivity in nanostructured material should focus on the inter-boundary region. Charge transport has also been addressed and Kang *et al.*,[214] reviewing existing scattering mechanisms in $SrTiO_3$, suggested that the origin of the high temperature dependence was probably band conduction, with deformation potential scattering on elongated Fermi surfaces.

For the misfit-layered cobaltate $(M_2CoO_3)_{0.6}CoO_2$ (where M = Mg, Ca, Sr, Ba). Baran *et al.*[24] applied DFT and high-resolution transmission electron microscopy to investigate the relationship between chemically-induced strain and charge transfer in the CoO_2 layers of misfit-layered cobaltites by the substitution of different alkaline earths. They found that the selected different interlayer cation had little effect on the magnetic properties, but generated internal strain between the rock-salt structured M_2CoO_3 and hexagonal CoO_2 subsystems, changing the hybridisation of the cations with the environment. Their results highlighted the importance of doping $Ca_3Co_4O_{9+x}$ compounds to tune the thermoelectric properties *via* the effects of charge transfer and the strain imposed on the CoO_2 layer.

In a related investigation of the misfit-layered cobaltate $[Bi_{0.87}SrO_2]_2[CoO_2]_{1.82}$, Baran *et al.*[28] used a combination of experimental and computational approaches to investigate the crystallographic and electronic structures and the transport properties; the basic structure comprises an incommensurate stacking of hexagonal CoO_2 and double rock-salt $BiSrO_2$ layers (Figure 5.25). They developed a model which suggested a level of Bi deficiency in the rock-salt layers; this leads to opening of the band gap and increases p-type electronic conductivity due to the formation of Co^{4+} species

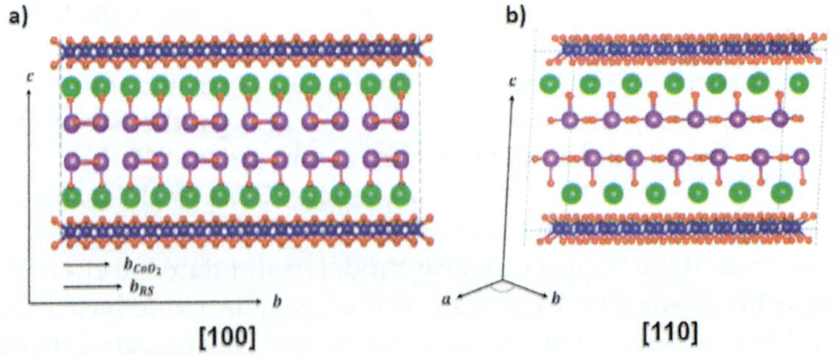

Figure 5.25 Unit cell of $[Bi_{0.87}SrO_2]_2[CoO_2]_{1.82}$ along (a) [100] and (b) [110] directions; the green, blue, red and pink spheres correspond to Sr, Co, O and Bi atoms, respectively.[28] Reproduced from ref. 28 with permission from American Chemical Society, Copyright 2016.

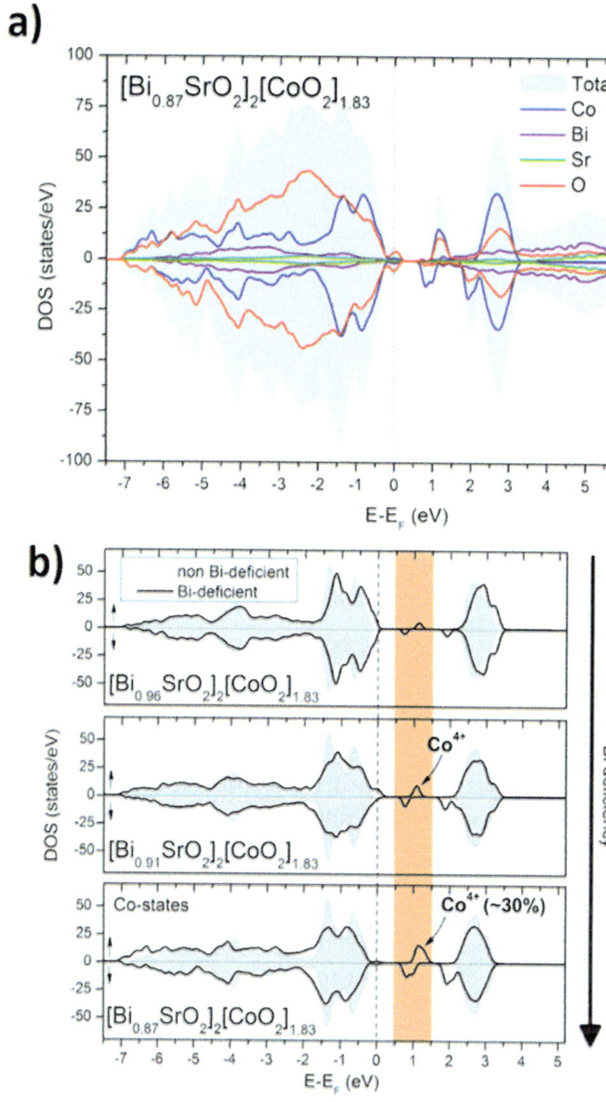

Figure 5.26 DOS modelling of $[Bi_{0.87}SrO_2]_2[CoO_2]_{1.82}$ reveals Bi deficiencies in the Bi-O layers inducing charge redistribution in the CoO_2 layers; independent EELS analysis confirms presence of some Co^{4+} states, localised to the CoO_2 layers.[28] Reproduced from ref. 28 with permission from American Chemical Society, Copyright 2016.

that serve as itinerant holes within the predominantly Co^{3+} framework of the CoO_2 layer (Figure 5.26). Calculation of Seebeck coefficients based on existing structural models gave a poor fit to experimental data (Figure 5.27); only by taking account of minor structural differences arising from Bi deficiencies (determined by atom level electron microscopy) was good agreement achieved (Figure 5.27).

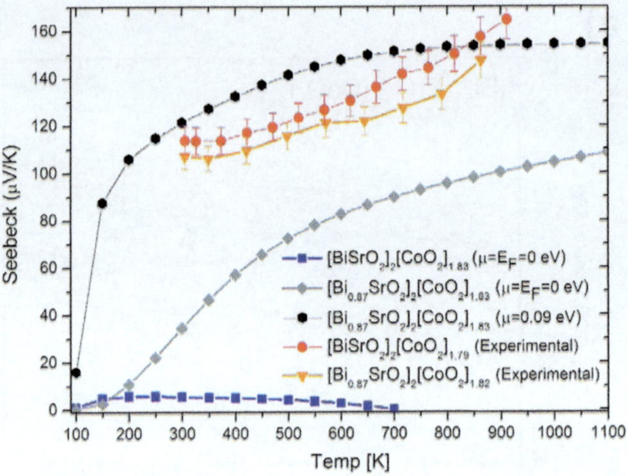

Figure 5.27 Calculated and experimental Seebeck coefficient data for $[Bi_{0.87}SrO_2]_2[CoO_2]_{1.82}$, showing good agreement between the data once the structural model took the Bi deficiencies in the layers into account.[28] Reproduced from ref. 28 with permission from American Chemical Society, Copyright 2016.

By a combination of atom level electron microscopy and modelling Kepaptsoglou *et al.*[5] explored the crystal structure and thermoelectric properties of ceramics of the A-site-deficient perovskite $La_{1/3}NbO_3$. The material exhibits promising high temperature thermoelectric properties with high Seebeck coefficients in excess of $-600\ \mu V\ K^{-1}$ and low thermal conductivity. First-principles electronic structure calculations showed that donor doping did not cause a significant reduction of Seebeck coefficients while band convergence effects at temperatures above 800 K contributed to the transport properties (see Figure 5.28a); indicating possible routes for property enhancement through control of the A-site occupancies. The modelling of thermal conductivity suggested weak anisotropy with respect to crystallographic direction with values around 2.3 W m^{-1} K^{-1} from 300–1200 K (see Figure 5.28b).

Collectively, these results show that modelling is a powerful tool to help understand the processes controlling charge and thermal transport in established and candidate thermoelectric materials, especially when combined with high level electron microscopy yielding chemical and structural information at the atomic level.

5.9 Devices and Modules

Thermoelectric modules can be used to generate power from waste heat or to pump heat for active cooling.[215] The thermoelectric conversion efficiency of the module is determined by the ratio of the electrical power generated (W)

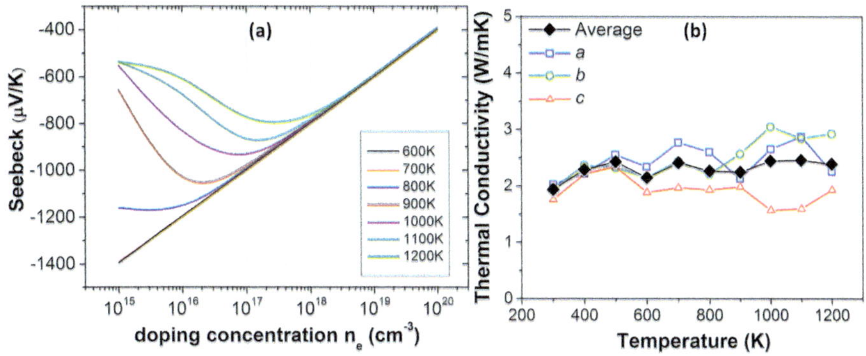

Figure 5.28 (a) Calculated Seebeck coefficient dependence of $La_{1/3}NbO_3$ on the doping concentration for various temperatures, using a rigid-band model that assumes no shape change in the band structure but only a shift in the Fermi energy due to the changing electron doping concentration. (b) Calculated thermal conductivity for different crystallographic directions.[5] Reproduced from ref. 5 with permission from American Chemical Society, Copyright 2017.

to the thermal energy input (Q_H) and the conversion efficiency is given by eqn (5.1) when the *ZT* term specifically refers to the module and not the individual materials. Other important parameters that define the performance of a thermoelectric generator are the maximum power output (P_{max}) and theoretical open circuit voltage (V_o)[216] given by

$$P_{max} = \frac{V_o^2}{4R_I} \tag{5.17}$$

$$V_o = (S_p - S_n) \times (T_H - T_C) \times N \tag{5.18}$$

where R_I is the internal resistance of the module and *N* is the number of legs. Here, R_I is the sum of the resistance of the legs and the contact resistance between the legs and the electrode ($R_I = R_{legs} + R_{contact}$). Due to such parasitic losses, V_o obtained from the actual device is lower than the estimated values, leading to lower power output from TEGs in comparison to theoretical predictions.

Most current commercial thermoelectric devices employ Bi_2Te_3 based compounds as p- and n-type legs. However, their peak operational temperature window is limited to near room temperature[68,217] so that segmented or cascade-type systems[10] involving two or more materials in each leg have been used in demanding applications where wide temperature ranges must be exploited, such as in the NASA Gallileo and Cassini missions.[218] For significantly higher temperatures (T < 600 °C), PbTe-based compounds[219] and SiGe-based alloys[220] have often been the alloys of choice. All of these materials involve expensive and/or rare (*e.g.* Te)/toxic starting materials[221] typically costing one to several hundred

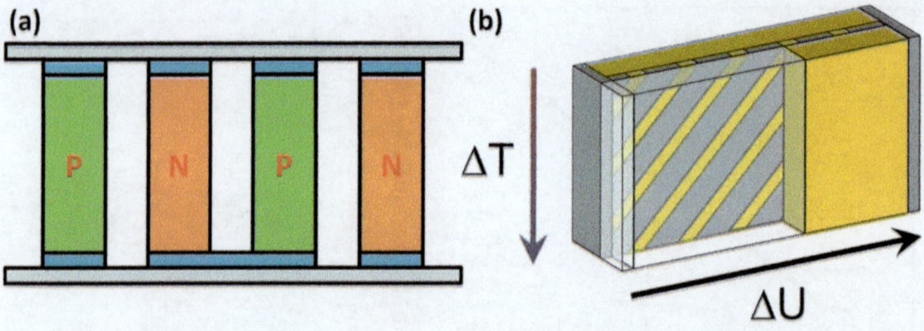

Figure 5.29 (a) Traditional construction of thermoelectric modules,[224] and (b) transverse multilayer construction (screen printed); figure adapted from Töpfer *et al.*[225] Reproduced from ref. 225 with permission from John Wiley and Sons, © 2017 The American Ceramic Society.

$ kg^{-1}.[222] Since the devices are usually subject to thermal cycling in normal operation and may also be subject to mechanical stresses, then the thermal and mechanical stability can be an important consideration for long term operations. Finally, these compounds tend to oxidise when in contact with air, which requires the use of solid/gas filler in the module to prevent oxidation. As noted earlier (Section 5.1), oxides have the potential to overcome the limitations of high temperature stability, cost and weight imposed by traditional thermoelectric materials. Indeed, the oxides (*e.g.* ZnO and $SrTiO_3$-based *etc.*) offer advantages in terms of cost, weight, high temperature stability and are predominantly based on earth-abundant materials. Furthermore, there is encouraging evidence of a significantly enhanced thermal operational window when composites are employed.[181] Nevertheless, oxide thermoelectric modules tend to deliver less power than expected because of the contact resistance at electrode/oxide junctions.[216] There are also engineering challenges to address because of differences in the thermal expansion coefficient between the oxides and electrode material, which may lead to cracking or exfoliation, and limit long-term stability of the modules. Finally, specialist electrode materials are required for the highest temperature ranges of operation.[223]

Even though the efficiency and stability of existing oxide TEGs have limited their practical applications, there are numerous examples of oxide modules in the literature. Whilst many have operated with the traditional construction (Figure 5.29a) with 4 to more than 100 pairs of p–n elements, there are growing numbers of examples of alternative and novel constructions including multilayer and transverse multilayer geometry (Figure 5.29b). Table 5.2 presents data for a selection of oxide thermoelectric modules used for power generation. These modules were prepared using different device geometry, number of p–n couples and joining techniques. However, the open circuit voltage, maximum power and power density enable the performance of the modules to be compared.

Table 5.2 Properties and performance of typical oxide thermoelectric modules.

Materials	No. of p–n couples	Joining technique	T_{Hot} (K)	ΔT (K)	V_{open} (V)	P_{max} (mW)	Dimensions (mm)	Power density (W m^{-2})	Reference
p-$Ca_{2.7}Bi_{0.3}Co_4O_9$ n-$Ca_{0.92}La_{0.08}MnO_3$	8	Pt paste	773	390	0.9	63.5	3 × 3	441	226
p-$Ca_{2.7}Bi_{0.3}Co_4O_9$ n-$La_{0.9}Bi_{0.1}NiO_3$	140	Ag paste	1072	551	4.5	150	1.3 × 1.3 × 5	317	227
p-$NaCo_2O_4$ n-$Zn_{0.98}Al_{0.02}O$	12	Diffusion welding	839	462	0.8	58	3 × 4 × 10	201	224
p-$Ca_{2.7}Bi_{0.3}Co_4O_9$ n-$CaMn_{0.98}Mo_{0.02}O_3$	8	Ag paste	897	565	1	170	5 × 5 × 4.5	425	228
p-$Ca_{2.7}Bi_{0.3}Co_4O_9$ n-$CaMn_{0.98}Mo_{0.02}O_3$	8	Ag	1273	975	0.7	340	5 × 5 × 4.5	850	229
p-$Ca_3Co_4O_9$ n-$(ZnO)_7In_2O_3$	44	Ag paste	1100	673	1.8	423	15 × 15 × 27	21	230
p-$Ca_{2.76}Cu_{0.24}Co_4O_9$ n-$Ca_{0.8}Dy_{0.2}MnO_3$	4	Ag paste	937	321	0.28	31	7 × 9 × 25	62	231
p-$Ca_3Co_4O_9$ n-$Zn_{0.98}Al_{0.02}O$	6	Ag	773	248	0.12	2.26	4 × 4 × 10	12	232
p-$Ca_3Co_4O_9$ n-$Ca_{2.9}Nd_{0.1}MnO_3$	1	Ag paste	1175	727	—	95	8.5 × 6	932	233
p-$Ca_3Co_4O_9$ n-$Zn_{0.98}Al_{0.02}O$	4	Ag	1173	700	0.67	256	4 × 4 × 8	2000	216
p-HH/$Ca_3Co_4O_9$ n-$Zn_{0.98}Al_{0.02}O$	4	Ag	1173	700	0.76	829	4 × 4 × 8	6500	216
p-$Ca_3Co_4O_9$ n-(Sr,La)TiO_3-Graphene	7	Welding	693	400	0.60	—	15 × 15 × 5	—	181
p-Li : Co_3O_4 n-TiO_2	7	Ni/spray	773	461	0.19	49	60 × 2.5 × 1.2	—	234
p-La_2CuO_4 n-Nd_2CuO_4	Trans ML[a]	AgPd/ML[a]	293	190	—	200	30 × 5 × 5	—	225
p-$Ca_3Co_{3.9}O_{9+d}$ n-$CaMnO_3$–$CaMn_2O_4$	ppn-ML	ML	900	160	0.21	570	20 × 5 × 2.5	—	235

[a]ML = multilayer structure.

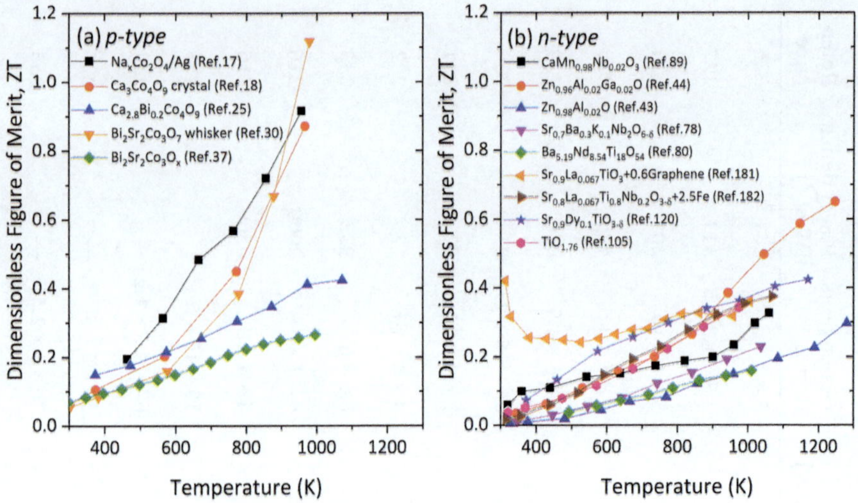

Figure 5.30 Compilation of *ZT* data for (a) p-type[17,18,25,30,37] and (b) n-type[43,44,78,80,89,105,120,181,182] oxide thermoelectrics.

5.10 Oxide Thermoelectrics: Conclusions and Prospects

Over the past 30 years a wide variety of oxide thermoelectrics have been developed and six main groups have emerged (Section 5.1); for most materials the maximum *ZT* has remained stubbornly around 0.3 to 0.4 (Figure 5.30), with just a few exceptions. Whilst the peak *ZT* values of oxides are still modest compared to many of the established metallic and newer selenide materials,[236] there are very encouraging signs in terms of increasing *average ZT*, which is much more important than peak *ZT*,[237] and significant increase in the thermal window of operation (*e.g.* for n-type (Sr, La) TiO_3-graphene composites). In addition, there is increasing power factor capability exhibited by oxides at both low and high temperatures (Figure 5.31). Combining these developments with the inherent advantages of oxides, namely high temperature stability, flexibility in material design, lighter in weight, cheaper and environmentally friendly, all suggest that oxide thermoelectrics will deliver much more if they can be used to their full capability.

To go forward to the next generation of higher performance oxide thermoelectrics we need to exploit all the available knowledge of interfaces and material design rules. For example, Gaultois *et al.*[239] used machine learning and data mining techniques to develop design principles to identify promising user-designed thermoelectrics. For oxides, high,

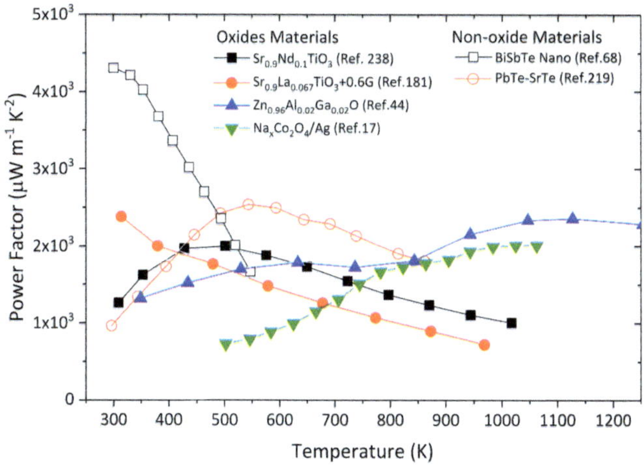

Figure 5.31 Power factor as a function of temperature for common oxide thermoelectrics[17,44,181,238] and two non-oxides (based on Bi–Sb–Te[68] and PbTe-SrTe[219]).

metallic-like electrical transport was deemed to be essential, with electrical resistivity below the Mott maximum resistivity.[240] Similarly, Terasaki[241] showed that *ZT* for oxides can be above unity even for low-mobility materials exhibiting so-called Ioffe–Regel conductivity at high temperature, and hole-doped Mott insulators can be even better than weakly correlated materials. Throughout the literature there is a recurring theme of the effects and importance of interfaces on thermal and charge transport in oxides. In $SrTiO_3$ it was shown that embedded particles could enhance electrical conductivity (by generating additional charge carriers) and also reduce thermal conductivity.[3,182] Buscaglia *et al.*[242] demonstrated that La doped $SrTiO_3$ ceramics having nanocrystalline grain sizes (~24 nm) enabled significant reduction in thermal conductivity to near the theoretical limit. Recently, Kang *et al.*[214] showed that in addition to grain size, the grain boundary resistance was an important parameter and needed to be minimised such that the electrical conductivity characteristics approached that of a single crystal (decreasing with increasing temperature). The same group[243] predicts that by minimising grain boundary resistance and controlling grain boundary structure the average *ZT* for $SrTiO_3$ should be higher than 0.5 from 300 to 1000 K, with peak *ZT* approaching 0.7 near the diffusive limit of lattice thermal conductivity (Figure 5.32). Once this can be realised, oxides will be able to compete with the more established materials; greater insight into the materials will come through the active collaboration of experimentalists with modellers and atom-level electron microscopy.

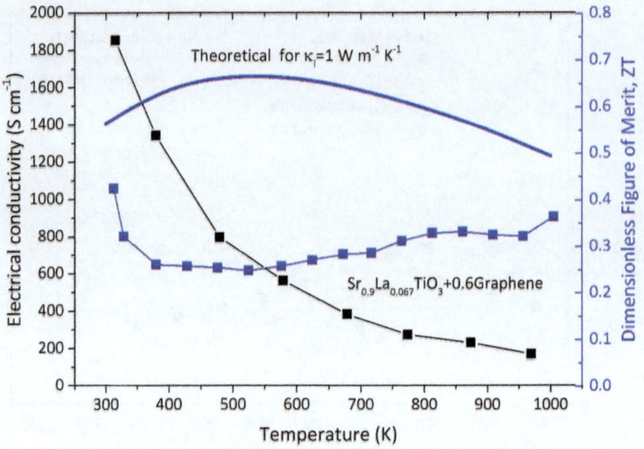

Figure 5.32 Single crystal-like electrical conductivity (left hand axis) and *ZT* as a function of temperature for $SrTiO_3$-graphene composites,[181] plus predicted *ZT* data based on minimum lattice diffusivity.[243]

Acknowledgements

The authors are grateful to the EPSRC for the provision of funding for this work (EP/H043462, EP/I036230/1, EP/L014068/1, EP/L017695/1 acknowledged by RF). All research data supporting this publication are directly available within the publication.

References

1. T. J. Seebeck, *Ann. Phys.*, 1826, **82**, 133–160.
2. J. Yang and T. Caillat, *MRS Bull.*, 2006, **31**, 224–229.
3. M. Backhaus-Ricoult, J. Rustad, L. Moore, C. Smith and J. Brown, *Appl. Phys. A: Mater. Sci. Process.*, 2014, **116**, 433–470.
4. F. Azough, S. S. Jackson, D. Ekren, R. Freer, M. Molinari, S. R. Yeandel, P. M. Panchmatia, S. C. Parker, D. H. Maldonado, D. M. Kepaptsoglou and Q. M. Ramasse, *ACS Appl. Mater. Interfaces*, 2017, **9**, 41988–42000.
5. D. Kepaptsoglou, J. D. Baran, F. Azough, D. Ekren, D. Srivastava, M. Molinari, S. C. Parker, Q. M. Ramasse and R. Freer, *Inorg. Chem.*, 2017, **57**, 45–55.
6. D. Champier, *Energy Convers. Manage.*, 2017, **140**, 167–181.
7. M. Hamid, D. Abdulameer, M. Faizul, M. Sabri, S. Binti, M. Said, M. Haji, M. Bashir and A. Bashir, *Renewable Sustainable Energy Rev.*, 2014, **30**, 337–355.
8. G. Kieslich, G. Cerretti, I. Veremchuk, R. P. Hermann, M. Panthöfer, J. Grin and W. Tremel, *Phys. Status Solidi A*, 2016, **213**, 808–823.
9. G. J. Snyder and E. S. Toberer, *Nat. Mater.*, 2008, **7**, 105–114.

10. K. Koumoto, R. Funahashi, E. Guilmeau, Y. Miyazaki, A. Weidenkaff, Y. Wang and C. Wan, *J. Am. Ceram. Soc.*, 2013, **96**, 1–23.
11. I. Terasaki, Y. Sasago and K. Uchinokura, *Phys. Rev. B*, 1997, **56**, 685–687.
12. J. W. Fergus, *J. Eur. Ceram. Soc.*, 2012, **32**, 525–540.
13. J. He, Y. Liu and R. Funahashi, *J. Mater. Res.*, 2011, **26**, 1762–1772.
14. H. Yakabe, K. Kikuchi, I. Terasaki, Y. Sasago and K. Uchinokura, in *Proc. ICT'97. 16th International Conference on Thermoelectrics* (Cat. No.97TH8291), 1997, pp. 523–527.
15. W. Koshibae, K. Tsutsui and S. Maekawa, *Phys. Rev. B: Condens. Matter Mater. Phys.*, 2000, **62**, 6869–6872.
16. Y. Miyazaki, *Solid State Ionics*, 2004, **172**, 463–467.
17. M. Ito and D. Furumoto, *J. Alloys Compd.*, 2008, **450**, 517–520.
18. M. Shikano and R. Funahashi, *Appl. Phys. Lett.*, 2003, **82**, 1851–1853.
19. G. Xu, R. Funahashi, M. Shikano, I. Matsubara and Y. Zhou, *Appl. Phys. Lett.*, 2002, **80**, 3760–3762.
20. Q. M. Lu, J. X. Zhang, Q. Y. Zhang, Y. Q. Liu and D. M. Liu, in *Proceedings International Conference on Thermoelectrics*. IEEE, 2006, pp. 66–69.
21. H. Fukutomi, Y. Konno, K. Okayasu, M. Hasegawa and H. Nakatsugawa, *Mater. Sci. Eng., A*, 2009, **527**, 61–64.
22. H. Hao, Q. He, C. Chen, H. Sun and X. Hu, *Int. J. Mod. Phys. B*, 2009, **23**, 87–95.
23. D. Moser, L. Karvonen, S. Populoh, M. Trottmann and A. Weidenkaff, *Solid State Sci.*, 2011, **13**, 2160–2164.
24. J. D. Baran, M. Molinari, N. Kulwongwit, F. Azough, R. Freer, D. Kepaptsoglou, Q. M. Ramasse and S. C. Parker, *J. Phys. Chem. C*, 2015, **119**, 21818–21827.
25. C. Boyle, P. Carvillo, Y. Chen, E. J. Barbero, D. Mcintyre and X. Song, *J. Eur. Ceram. Soc.*, 2016, **36**, 601–607.
26. R. Funahashi, I. Matsubara and S. Sodeoka, *Appl. Phys. Lett.*, 2000, **76**, 2385.
27. J. Lingner, R. Funahashi, E. Combe, M. Letz and G. Jakob, *Appl. Phys. A*, 2015, **120**, 59–66.
28. J. D. Baran, D. Kepaptsoglou, M. Molinari, N. Kulwongwit, F. Azough, R. Freer, Q. M. Ramasse and S. C. Parker, *Chem. Mater.*, 2016, **28**, 7470–7478.
29. E. Combe, R. Funahashi, T. Barbier, F. Azough and R. Freer, *J. Mater. Res.*, 2016, **31**, 1296–1305.
30. R. Funahashi and M. Shikano, *Appl. Phys. Lett.*, 2002, **81**, 1459–1461.
31. T. Fujii, I. Terasaki, T. Watanabe and A. Matsuda, *Jpn. J. Appl. Phys., Part 2*, 2002, **41**, 3–7.
32. J. C. Diez, E. Guilmeau, M. A. Madre, S. Marinel, S. Lemonnier and A. Sotelo, *Solid State Ionics*, 2009, **180**, 827–830.
33. J. C. Diez, S. Rasekh, G. Constantinescu, M. A. Madre, M. A. Torres and A. Sotelo, *Ceram. Int.*, 2012, **38**, 5419–5424.

34. A. Sotelo, M. A. Torres, G. Constantinescu, S. Rasekh, J. C. Diez and M. A. Madre, *J. Eur. Ceram. Soc.*, 2012, **32**, 3745–3751.
35. L. H. Yin, R. Ang, Y. N. Huang, H. B. Jiang, B. C. Zhao, X. B. Zhu, W. H. Song and Y. P. Sun, *Appl. Phys. Lett.*, 2012, **100**, 173503.
36. S. Wang, Z. Bai, H. Wang, Q. Lü, J. Wang and G. Fu, *J. Alloys Compd.*, 2013, **554**, 254–257.
37. E. Combe, R. Funahashi, F. Azough and R. Freer, *J. Mater. Res.*, 2014, **29**, 1376–1382.
38. H. Leligny, D. Grebille, O. Perez, A. C. Masset, M. Hervieu and B. Raveau, *Acta Crystallogr., Sect. B: Struct. Sci.*, 2000, **56**, 173–182.
39. K. Tanabe, R. Okazaki, H. Taniguchi and I. Terasaki, *J. Phys.: Condens. Matter*, 2016, **28**, 085601.
40. H. Yamakawa, S. Lee, H. Takagi and C. A. Randall, *J. Mater. Sci.*, 2011, **46**, 2064–2070.
41. M. Karppinen, H. Fjellvåg, T. Konno, Y. Morita, T. Motohashi and H. Yamauchi, *Chem. Mater.*, 2004, **16**, 2790–2793.
42. R. Furushima, S. Tanaka, Z. Kato and K. Uematsu, *J. Ceram. Soc. Jpn.*, 2010, **118**, 921–926.
43. T. Tsubota, M. Ohtaki, K. Eguchi and H. Arai, *J. Mater. Chem.*, 1997, **7**, 85–90.
44. M. Ohtaki, K. Araki and K. Yamamoto, *J. Electron. Mater.*, 2009, **38**, 1234–1238.
45. H. Cheng, X. J. Xu, H. H. Hng and J. Ma, *Ceram. Int.*, 2009, **35**, 3067–3072.
46. N. Ma, J. F. Li, B. P. Zhang, Y. H. Lin, L. R. Ren and G. F. Chen, *J. Phys. Chem. Solids*, 2010, **71**, 1344–1349.
47. D. Bérardan, C. Byl and N. Dragoe, *J. Am. Ceram. Soc.*, 2010, **93**, 2352–2358.
48. X. Qu, W. Wang, S. Lv and D. Jia, *Solid State Commun.*, 2011, **151**, 332–336.
49. P. Jood, R. J. Mehta, Y. Zhang, G. Peleckis, X. Wang, R. W. Siegel, T. Borca-Tasciuc, S. X. Dou and G. Ramanath, *Nano Lett.*, 2011, **11**, 4337–4342.
50. S. M. Choi, K. H. Lee, K. H. Jung, Y. S. Lim and W. S. Seo, *J. Electron. Mater.*, 2013, **42**, 2056–2061.
51. S. Jantrasee, P. Moontragoon and S. Pinitsoontorn, *J. Semicond.*, 2016, **37**, 1–8.
52. Y. Michiue, H. Nishijima, Y. Suzuki and T. Mori, *Solid State Sci.*, 2017, **65**, 29–32.
53. H. Ohta, W.-S. Seo and K. Koumoto, *J. Am. Ceram. Soc.*, 1996, **79**, 2193–2196.
54. H. Kaga, R. Asahi and T. Tani, *Jpn. J. Appl. Phys.*, 2004, **43**, 3540–3543.
55. M. Kazeoka, H. Hiramatsu, W. S. Seo and K. Koumoto, *J. Mater. Res.*, 1998, **13**, 523–526.
56. Y. Masuda, M. Ohta, W.-S. Seo, W. Pitschke and K. Koumoto, *J. Solid State Chem.*, 2000, **150**, 221–227.

57. X. Liang and D. R. Clarke, *Acta Mater.*, 2014, **63**, 191–201.
58. Y. Michiue, T. Mori, A. Prytuliak, Y. Matsushita, M. Tanaka and N. Kimizuka, *RSC Adv.*, 2011, **1**, 1788.
59. X. Liang, *J. Eur. Ceram. Soc.*, 2016, **36**, 1643–1650.
60. E. Guilmeau, P. Díaz-Chao, O. I. Lebedev, A. Rečnik, M. C. Schäfer, F. Delorme, F. Giovannelli, M. Košir and S. Bernik, *Inorg. Chem.*, 2017, **56**, 480–487.
61. D. T. Alvarez-Ruiz, F. Azough, D. Hernandez-Maldonado, D. M. Kepaptsoglou, Q. M. Ramasse, S. J. Day, P. Svec, P. Svec and R. Freer, *J. Alloys Compd.*, 2018, **762**, 892–900.
62. D. T. Alvarez-Ruiz, F. Azough, D. T. Hernandez-Maldonado, D. M. Kepaptsoglou, Q. M. Ramasse, P. Svec, P. Svec Sr. and R. Freer, *J. Electron. Mater.*, 2018, **48**, 1818–1826.
63. J. A. Bock, S. Trolier-Mckinstry, G. D. Mahan and C. A. Randall, *Phys. Rev. B: Condens. Matter Mater. Phys.*, 2014, **90**, 1–8.
64. A. A. Ballman and H. Brown, *J. Cryst. Growth*, 1967, **1**, 311–314.
65. P. B. Jamieson, S. C. Abrahams and J. L. Bernstein, *J. Chem. Phys.*, 1968, **48**, 5048–5057.
66. L. A. Bursill and P. J. Lin, *Philos. Mag. B*, 1986, **54**, 157–170.
67. J. Schneck, J. C. Toledano, R. Whatmore and F. W. Ainger, *Ferroelectrics*, 1981, **36**, 327–330.
68. B. Poudel, Q. Hao, Y. Ma, Y. Lan, A. Minnich, B. Yu, X. Yan, D. Wang, A. Muto, D. Vashaee, X. Chen, J. Liu, M. S. Dresselhaus, G. Chen and Z. Ren, *Science*, 2008, **320**, 634–638.
69. S. Lee, R. H. T. Wilke, S. Trolier-mckinstry, S. Zhang and C. A. Randall, *Appl. Phys. Lett.*, 2010, **96**, 031910.
70. S. Lee, J. A. Bock, S. Trolier-McKinstry and C. A. Randall, *J. Eur. Ceram. Soc.*, 2012, **32**, 3971–3988.
71. S. Lee, S. Dursun, C. Duran and C. A. Randall, *J. Mater. Res.*, 2011, **26**, 26–30.
72. C. S. Dandeneau, T. W. Bodick, R. K. Bordia and F. S. Ohuchi, *J. Am. Ceram. Soc.*, 2013, **96**, 2230–2237.
73. Y. Li, J. Liu, Z. Wang, Y. C. Zhou, C. Wang, J. Li, Y. Zhu, M. Li and L. Mei, *Phys. Scr.*, 2015, **90**, 025801.
74. J. Li, H. Wang, W. Su, C. Wang, J. Liu, Y. Zhou, Y. Zhang, Y. Li and J. Zhai, *Ceram. Int.*, 2015, **42**, 1128–1132.
75. J. H. Chan, J. A. Bock, H. Guo, S. Trolier-Mckinstry and C. A. Randall, *J. Mater. Res.*, 2017, **32**, 1160–1167.
76. R. Ottini, C. Tealdi, C. Tomasi, I. G. Tredici, A. Soffientini, U. Anselmi-Tamburini, P. Ghigna and G. Spinolo, *J. Appl. Phys.*, 2017, **121**, 085104.
77. R. Ottini, C. Tealdi, C. Tomasi, I. G. Tredici, A. Soffientini, R. Burriel, E. Palacios, M. Castro, U. Anselmi-Tamburini, P. Ghigna and G. Spinolo, *J. Solid State Chem.*, 2018, **258**, 99–107.
78. Y. Zhang, J. Li, Y. Li, C. Wang, J. Liu, Y. Chen and W. Su, *J. Mater. Sci.: Mater. Electron.*, 2018, **29**, 9137–9141.

79. Y. Li, J. Liu, Y. Zhang, Y. Chen, J. Li, W. Su, H. Wang and C. Wang, *Ceram. Int.*, 2017, **43**, 13345–13348.
80. F. Azough, R. Freer, S. R. Yeandel, J. D. Baran, M. Molinari, S. C. Parker, E. Guilmeau, D. Kepaptsoglou, Q. Ramasse, A. Knox, D. Gregory, D. Paul, M. Paul, A. Montecucco, J. Siviter, P. Mullen, W. Li, G. Han, E. A. Man, H. Baig, T. Mallick, N. Sellami, G. Min and T. Sweet, *J. Electron. Mater.*, 2016, **45**, 1894–1899.
81. T. Kobayashi, H. Takizawa, T. Endo, T. Sato, M. Shimada, H. Taguchi and M. Nagao, *J. Solid State Chem.*, 1991, **92**, 116–129.
82. D. Flahaut, T. Mihara, R. Funahashi, N. Nabeshima, K. Lee, H. Ohta and K. Koumoto, *J. Appl. Phys.*, 2006, **100**, 084911.
83. M. Ohtaki, H. Koga, T. Tokunaga, K. Eguchi and H. Arai, *J. Solid State Chem.*, 1995, **120**, 105–111.
84. R. Funahashi, A. Kosuga, N. Miyasou, E. Takeuchi, S. Urata, K. Lee, H. Ohta and K. Koumoto, *Int. Conf. Thermoelectr.*, 2007, 124–128.
85. Y. Wang, Y. Sui, H. Fan, X. Wang, Y. Su, W. Su and X. Liu, *Chem. Mater.*, 2009, **21**, 4653–4660.
86. D. Srivastava, F. Azough, R. Freer, E. Combe, R. Funahashi, D. M. Kepaptsoglou, Q. M. Ramasse, M. Molinari, S. R. Yeandel, J. D. Baran and S. C. Parker, *J. Mater. Chem. C*, 2015, **3**, 12245–12259.
87. T. Wang, P. Nan, H. Wang, W. Su, A. Sotelo, J. Zhai, X. Wang, Y. Ran, T. Chen and C. Wang, *Inorg. Chem.*, 2018, **57**, 9133–9141.
88. G. Xu, R. Funahashi, Q. Pu, B. Liu, R. Tao, G. Wang and Z. Ding, *Solid State Ionics*, 2004, **171**, 147–151.
89. L. Bocher, M. H. Aguirre, D. Logvinovich, A. Shkabko, R. Robert, M. Trottmann and A. Weidenkaff, *Inorg. Chem.*, 2008, **47**, 8077–8085.
90. L. Bocher, M. H. Aguirre, R. Robert, D. Logvinovich, S. Bakardjieva, J. Hejtmanek and A. Weidenkaff, *Acta Mater.*, 2009, **57**, 5667–5680.
91. R. Kabir, T. Zhang, D. Wang, R. Donelson, R. Tian, T. T. Tan and S. Li, *J. Mater. Sci.*, 2014, **49**, 7522–7528.
92. M. Schrade, R. Kabir, S. Li, T. Norby and T. G. Finstad, *J. Appl. Phys.*, 2014, **115**, 103705.
93. M. Molinari, D. A. Tompsett, S. C. Parker, F. Azough and R. Freer, *J. Mater. Chem. A*, 2014, **2**, 14109.
94. M. Backhaus-Ricoult, J. R. Rustad, D. Vargheese, I. Dutta and K. Work, *J. Electron. Mater.*, 2012, **41**, 1636–1647.
95. D. Bayerl and E. Kioupakis, *Phys. Rev. B*, 2015, **91**, 1–5.
96. S. Andersson, B. Collén, U. Kuylenstierna and A. Magnéli, *Acta Chem. Scand.*, 1957, **11**, 1641–1652.
97. S. Harada, K. Tanaka and H. Inui, *J. Appl. Phys.*, 2010, **108**, 083703.
98. K. Szot, M. Rogala, W. Speier, Z. Klusek, A. Besmehn and R. Waser, *Nanotechnology*, 2011, **22**, 254001.
99. Q. He, Q. Hao, G. Chen, B. Poudel, X. Wang, D. Wang and Z. Ren, *Appl. Phys. Lett.*, 2007, **91**, 5–8.

100. K. Fuda, T. Shoji, S. Kikuchi, Y. Kunihiro and S. Sugiyama, *J. Electron. Mater.*, 2013, **42**, 2209–2213.
101. Y. Lu, M. Hirohashi and K. Sato, *Mater. Trans.*, 2006, **47**, 1449–1452.
102. M. K. Nowotny, T. Bak and J. Nowotny, *J. Phys. Chem. B*, 2006, **110**, 16283.
103. J. Nowotny, T. Bak and T. Burg, *Ionics*, 2007, **13**, 155–162.
104. Y. Lu, Y. Matsuda, K. Sagara, L. Hao, T. Otomitsu and H. Yoshida, *Adv. Mater. Res.*, 2011, **415–417**, 1291–1296.
105. H. Liu, H. Ma, C. Wang, F. Wang, B. Liu, J. Chen, G. Ji, Y. Zhang and X. Jia, *Ceram. Int.*, 2018, **44**, 19859–19865.
106. H. Lee, S. J. Han, R. C. Seshadri and S. Sampath, *Sci. Rep.*, 2016, **6**, 1–11.
107. M. Mikami and K. Ozaki, *J. Phys.: Conf. Ser.*, 2012, **379**, 012006.
108. C. Liu, L. Miao, J. Zhou, R. Huang, C. A. J. Fisher and S. Tanemura, *J. Phys. Chem. C*, 2013, **117**, 11487–11497.
109. G. Kieslich, U. Burkhardt, C. S. Birkel, I. Veremchuk, J. E. Douglas, M. W. Gaultois, I. Lieberwirth, R. Seshadri, G. D. Stucky, Y. Grin and W. Tremel, *J. Mater. Chem. A*, 2014, **2**, 13492–13497.
110. G. Kieslich, I. Veremchuk, I. Antonyshyn, W. G. Zeier, C. S. Birkel, K. Weldert, C. P. Heinrich, E. Visnow, M. Panthöfer, U. Burkhardt, Y. Grin and W. Tremel, *Phys. Chem. Chem. Phys.*, 2013, **15**, 15399–15403.
111. A. Yaremchenko, S. Patrício and J. Frade, *J. Power Sources*, 2014, **245**, 557–569.
112. P. N. Iyer and A. J. Smith, *Acta Crystallogr.*, 1967, **23**, 740–746.
113. K. Yoshii, *J. Solid State Chem.*, 2000, **149**, 354–359.
114. C. J. Howard and Z. Zhang, *Acta Crystallogr., Sect. B: Struct. Sci.*, 2004, **60**, 249–251.
115. Z. Lu, H. Zhang, W. Lei, D. C. Sinclair and I. M. Reaney, *Chem. Mater.*, 2016, **28**, 925–935.
116. N. A. Kirsanov and G. V. Bazuev, *Inorg. Mater.*, 2001, **37**, 487–490.
117. T. Ishihara, *Perovskite Oxide for Solid Oxide Fuel Cells*, Springer, 2009.
118. R. J. D. Tilley, *Perovskites: Structure-property Relationships*, John Wiley & Sons, Ltd, Chichester, UK, 2016.
119. R. D. Shannon, *Acta Crystallogr.*, 1976, **32**, 751–767.
120. A. V. Kovalevsky, A. A. Yaremchenko, S. Populoh, P. Thiel, D. P. Fagg, A. Weidenkaff and J. R. Frade, *Phys. Chem. Chem. Phys.*, 2014, **16**, 26946–26954.
121. Y. Kinemuchi, K. I. Mimura, A. Towata and K. Kato, *J. Electron. Mater.*, 2014, **43**, 2011–2016.
122. A. M. Dehkordi, S. Bhattacharya, J. He, H. N. Alshareef and T. M. Tritt, *Appl. Phys. Lett.*, 2014, **104**, 193902.
123. J. Liu, C. L. Wang, Y. Li, W. B. Su, Y. H. Zhu, J. C. Li and L. M. Mei, *J. Appl. Phys.*, 2013, **114**, 223714.
124. H. Muta, K. Kurosaki and S. Yamanaka, *J. Alloys Compd.*, 2003, **350**, 292–295.

125. L. Hanbo, Z. Ganhong, D. Zhenxiang, Y. Zhenheng, W. Haiqiu and M. Yongqing, *J. Rare Earths*, 2014, **32**, 314–319.
126. H. Muta, K. Kurosaki and S. Yamanaka, *J. Alloys Compd.*, 2005, **392**, 306–309.
127. A. Kikuchi, N. Okinaka and T. Akiyama, *Scr. Mater.*, 2010, **63**, 407–410.
128. T. Okuda, K. Nakanishi, S. Miyasaka and Y. Tokura, *Phys. Rev. B*, 2001, **63**, 113104.
129. P.-P. Shang, B. Zhang, J. Li and N. Ma, *Solid State Sci.*, 2010, **12**, 1341–1346.
130. A. M. Dehkordi, S. Bhattacharya, T. Darroudi, J. W. Graff, U. Schwingenschlögl, H. N. Alshareef and T. M. Tritt, *Chem. Mater.*, 2014, **26**, 2478–2485.
131. Y. Wang, X. Zhang, L. Shen, N. Bao, C. Wan, N.-H. Park, K. Koumoto and A. Gupta, *J. Power Sources*, 2013, **241**, 255–258.
132. N. Wang, H. He, X. Li, L. Han and C. Zhang, *J. Alloys Compd.*, 2010, **506**, 293–296.
133. J. Lingner, G. Jakob and M. Letz, *AIP Conf. Proc.*, 2012, **1449**, 343–346.
134. J. Liu, C. L. Wang, W. B. Su, H. C. Wang, P. Zheng, J. C. Li, J. L. Zhang and L. M. Mei, *Appl. Phys. Lett.*, 2009, **95**, 162110.
135. C. Chen, T. Zhang, R. Donelson, T. T. Tan and S. Li, *J. Alloys Compd.*, 2015, **629**, 49–54.
136. L. D. Hicks and M. S. Dresselhaus, *Phys. Rev. B*, 1993, **47**, 8–11.
137. I. R. Shein and A. L. Ivanovskii, *Phys. Lett. A*, 2007, **371**, 155–159.
138. M. Ito and T. Matsuda, *J. Alloys Compd.*, 2009, **477**, 473–477.
139. F. Gao, S. Yang, J. Li, M. Qin, Y. Zhang and H. Sun, *Ceram. Int.*, 2014, 1–9.
140. K. Shirai and K. Yamanaka, *J. Appl. Phys.*, 2013, **113**, 053705.
141. F. Noll, W. Münch, I. Denk and J. Maier, *Solid State Ionics*, 1996, **86–88**, 711–717.
142. S. Ohta, H. Ohta and K. Koumoto, *J. Ceram. Soc. Jpn.*, 2006, **114**, 102–105.
143. S. Ohta, T. Nomura, H. Ohta and K. Koumoto, *J. Appl. Phys.*, 2005, **97**, 034106.
144. Y. Wang, K. Fujinami, R. Zhang, C. Wan, N. Wang, Y. Ba and K. Koumoto, *Appl. Phys. Express*, 2010, **3**, 031101.
145. C. Yu, M. L. Scullin, M. Huijben, R. Ramesh and A. Majumdar, *Appl. Phys. Lett.*, 2008, **92**, 1–4.
146. L. Feng, T. Shiga and J. Shiomi, *Appl. Phys. Express*, 2015, **8**, 071501.
147. R. A. Evarestov, E. Blokhin, D. Gryaznov, E. A. Kotomin and J. Maier, *Phys. Rev. B: Condens. Matter Mater. Phys.*, 2011, **83**, 1–9.
148. L. Zhang, N. Li, H.-Q. Wang, Y. Zhang, F. Ren, X.-X. Liao, Y.-P. Li, X.-D. Wang, Z. Huang, Y. Dai, H. Yan and J.-C. Zheng, *Chin. Phys. B*, 2017, **26**, 16602.

149. A. V. Kovalevsky, A. A. Yaremchenko, S. Populoh, A. Weidenkaff and J. R. Frade, *J. Phys. Chem. C*, 2014, **118**, 4596–4606.
150. J. Liu, C. L. Wang, W. B. Su, H. C. Wang, J. C. Li, J. L. Zhang and L. M. Mei, *J. Alloys Compd.*, 2010, **492**, 2009–2011.
151. P.-P. Shang, B.-P. Zhang, Y. Liu, J.-F. Li and H.-M. Zhu, *J. Electron. Mater.*, 2011, **40**, 926–931.
152. A. V. Kovalevsky, A. A. Yaremchenko, S. Populoh, A. Weidenkaff and J. R. Frade, *J. Appl. Phys.*, 2013, **113**, 053704.
153. L. Li, Y. Liu, X. Qin, D. Li, J. Zhang, C. Song and L. Wang, *J. Alloys Compd.*, 2014, **588**, 562–567.
154. S. Bhattacharya, A. M. Dehkordi, S. Tennakoon, R. Adebisi, J. R. Gladden, T. Darroudi, H. N. Alshareef and T. M. Tritt, *J. Appl. Phys.*, 2014, **115**, 223712.
155. A. A. Yaremchenko, S. Populoh, S. G. Patrício, J. Macías, P. Thiel, D. P. Fagg, A. Weidenkaff, J. R. Frade and A. V. Kovalevsky, *Chem. Mater.*, 2015, **27**, 4995–5006.
156. H. C. Wang, C. L. Wang, W. B. Su, J. Liu, Y. Zhao, H. Peng, J. L. Zhang, M. L. Zhao, J. C. Li, N. Yin and L. M. Mei, *Mater. Res. Bull.*, 2010, **45**, 809–812.
157. S. R. Popuri, A. J. M. Scott, R. A. Downie, M. A. Hall, E. Suard, R. Decourt, M. Pollet and J.-W. G. Bos, *RSC Adv.*, 2014, **4**, 33720–33723.
158. A. V. Kovalevsky, S. Populoh, S. G. Patrício, P. Thiel, M. C. Ferro, D. P. Fagg, J. R. Frade and A. Weidenkaff, *J. Phys. Chem. C*, 2015, **119**, 4466–4478.
159. D. Srivastava, C. Norman, F. Azough, M. C. Schäfer, E. Guilmeau, D. Kepaptsoglou, Q. M. Ramasse, G. Nicotra and R. Freer, *Phys. Chem. Chem. Phys.*, 2016, **18**(38), 26475–26486.
160. A. M. Dehkordi, S. Bhattacharya, T. Darroudi, H. N. Alshareef and T. M. Tritt, *J. Appl. Phys.*, 2015, **117**, 055102.
161. P. Blennow, A. Hagen, K. K. Hansen, L. R. Wallenberg and M. Mogensen, *Solid State Ionics*, 2008, **179**, 2047–2058.
162. S. Hashimoto, F. W. Poulsen and M. Mogensen, *J. Alloys Compd.*, 2007, **439**, 232–236.
163. W. Gong, H. Yun, Y. B. Ning, J. E. Greedan, W. R. Datars and C. V. Stager, *J. Solid State Chem.*, 1991, **90**, 320–330.
164. N. Shanthi, D. D. Sarma, S. Srtio and O. Specifically, *Phys. Rev. B*, 1998, **57**, 2153–2158.
165. S. Hashimoto, L. Kindermann, F. W. Poulsen and M. Mogensen, *J. Alloys Compd.*, 2005, **397**, 245–249.
166. R. Moos, A. Gnudi and K. H. Härdtl, *J. Appl. Phys.*, 1995, **78**, 5042.
167. R. Moos and K. H. Härdtl, *J. Appl. Phys.*, 1996, **80**, 393.
168. J. B. Goodenough, *J. Appl. Phys.*, 1966, **37**, 1415–1422.
169. H. Muta, K. Kurosaki and S. Yamanaka, *J. Alloys Compd.*, 2004, **368**, 22–24.

170. X. G. Guo, X. S. Chen, Y. L. Sun, L. Z. Sun, X. H. Zhou and W. Lu, *Phys. Lett. A*, 2003, **317**, 501–506.
171. S. Ohta, T. Nomura, H. Ohta, M. Hirano, H. Hosono and K. Koumoto, *Appl. Phys. Lett.*, 2005, **87**, 97–100.
172. A. Kinaci, C. Sevik and T. Çağın, *Phys. Rev. B*, 2010, **82**, 155114.
173. D. Srivastava, F. Azough, M. Molinari, S. C. Parker and R. Freer, *J. Electron. Mater.*, 2014, **44**, 1803–1808.
174. H. C. Wang, C. L. Wang, W. B. Su, J. Liu, H. Peng, Y. Sun, J. L. Zhang, M. L. Zhao, J. C. Li, N. Yin and L. M. Mei, *Ceram. Int.*, 2011, **37**, 2609–2613.
175. J. Han, Q. Sun and Y. Song, *J. Alloys Compd.*, 2017, **705**, 22–27.
176. A. V. Kovalevsky, M. H. Aguirre, S. Populoh, S. G. Patrício, N. M. Ferreira, S. M. Mikhalev, D. P. Fagg, A. Weidenkaff and J. R. Frade, *J. Mater. Chem. A*, 2017, **5**, 3909–3922.
177. D. Neagu and J. T. S. Irvine, *Chem. Mater.*, 2010, **22**, 5042–5053.
178. L. D. Chen, X. Y. Huang, M. Zhou, X. Shi and W. B. Zhang, *J. Appl. Phys.*, 2006, **99**, 064305.
179. H. Li, X. Tang, Q. Zhang and C. Uher, *Appl. Phys. Lett.*, 2009, **94**, 10–13.
180. Y. Pei, J. Lensch-Falk, E. S. Toberer, D. L. Medlin and G. J. Snyder, *Adv. Funct. Mater.*, 2011, **21**, 241–249.
181. Y. Lin, C. Norman, D. Srivastava, F. Azough, L. Wang, M. Robbins, K. Simpson, R. Freer and I. A. Kinloch, *ACS Appl. Mater. Interfaces*, 2015, **7**, 15898–15908.
182. D. Srivastava, C. Norman, F. Azough, M. C. Schäfer, E. Guilmeau and R. Freer, *J. Alloys Compd.*, 2018, **731**, 723–730.
183. N. Wang, H. Li, Y. Ba, Y. Wang, C. Wan, K. Fujinami and K. Koumoto, *J. Electron. Mater.*, 2010, **39**, 1777–1781.
184. N. Wang, H. Chen, H. He, W. Norimatsu, M. Kusunoki and K. Koumoto, *Sci. Rep.*, 2013, **3**, 3449.
185. N. Wang, L. Han, H. He, Y. Ba and K. Koumoto, *J. Alloys Compd.*, 2010, **497**, 308–311.
186. N. Wang, H. He, Y. Ba, C. Wan and K. Koumoto, *J. Ceram. Soc. Jpn.*, 2010, **118**, 1098–1101.
187. S. M. Rhim, S. Hong, H. Bak and O. K. Kim, *J. Am. Ceram. Soc.*, 2000, **83**, 1145–1148.
188. C.-L. Huang and C.-L. Pan, *Jpn. J. Appl. Phys.*, 2002, **41**, 707–711.
189. H. Y. Park, J. Y. Choi, M. K. Choi, K. H. Cho, S. Nahm, H. G. Lee and H. W. Kang, *J. Am. Ceram. Soc.*, 2008, **91**, 2374–2377.
190. H.-Y. Park, C.-H. Nam, I.-T. Seo, J.-H. Choi, S. Nahm, H.-G. Lee, K.-J. Kim and S.-M. Jeong, *J. Am. Ceram. Soc.*, 2010, **93**, 2537–2540.
191. A. B. Alles and V. L. Burdick, *J. Appl. Phys.*, 1991, **70**, 6883–6890.
192. F. Azough, R. Freer, M. Thrall, R. Cernik, F. Tuna and D. Collison, *J. Eur. Ceram. Soc.*, 2010, **30**, 727–736.

193. X. Wang, W. Lu, J. Liu, Y. Zhou and D. Zhou, *J. Eur. Ceram. Soc.*, 2006, **26**, 1981–1985.
194. W. W. Cho, K. Kakimoto and H. Ohsato, *Mater. Sci. Eng., B*, 2005, **121**, 48–53.
195. X. Feng, Y. Fan, N. Nomura, K. Kikuchi, L. Wang, W. Jiang and A. Kawasaki, *Carbon*, 2017, **112**, 169–176.
196. Y. Xu, Z. Li and W. Duan, *Small*, 2014, **10**, 2182–2199.
197. S. N. Ruddlesden and P. Popper, *Acta Crystallogr.*, 1958, **11**, 54–55.
198. K. H. Lee, S. W. Kim, H. Ohta and K. Koumoto, *J. Appl. Phys.*, 2006, **100**, 063717.
199. A. H. Reshak, *Renewable Energy*, 2015, **76**, 36–44.
200. Y. Wang, C. Wan, X. Zhang, L. Shen, K. Koumoto, A. Gupta and N. Bao, *Appl. Phys. Lett.*, 2013, **102**, 183905.
201. Y. Wang, K. H. Lee, H. Hyuga, H. Kita, H. Ohta and K. Koumoto, *J. Electroceram.*, 2010, **24**, 76–82.
202. R. R. Sun, X. Y. Qin, L. L. Li, D. Li, J. Zhang, Y. S. Zhang and C. J. Tang, *J. Phys. Chem. Solids*, 2014, **75**, 629–637.
203. Y. Wang, K. H. Lee, H. Ohta and K. Koumoto, *J. Appl. Phys.*, 2009, **105**, 103701.
204. W. Chen, J.-H. Pöhls, G. Hautier, D. Broberg, S. Bajaj, U. Aydemir, Z. M. Gibbs, H. Zhu, M. Asta, G. J. Snyder, B. Meredig, M. A. White, K. Persson and A. Jain, *J. Mater. Chem. C*, 2016, **4**, 4414–4426.
205. G. Kresse and J. Furthmüller, *Phys. Rev. B*, 1996, **54**, 11169–11186.
206. J. P. Perdew, K. Burke and M. Ernzerhof, *Phys. Rev. Lett.*, 1996, **77**, 3865–3868.
207. G. Kresse and D. Joubert, *Phys. Rev. B*, 1999, **59**, 1758–1775.
208. G. K. H. Madsen and D. J. Singh, *Comput. Phys. Commun.*, 2006, **175**, 67–71.
209. S. R. Yeandel, M. Molinari and S. C. Parker, *Nanoscale*, 2018, **10**, 15010–15022.
210. A. Rebola, R. F. Klie, P. Zapol and S. Öğüt, *Appl. Phys. Lett.*, 2014, **104**, 1–6.
211. G. K. Ren, J. Le Lan, K. J. Ventura, X. Tan, Y. H. Lin and C. W. Nan, *npj Comput. Mater.*, 2016, **2**, 1–9.
212. D. I. Bilc, C. G. Floare, L. P. Zârbo, S. Garabagiu, S. Lemal and P. Ghosez, *J. Phys. Chem. C*, 2016, **120**, 25678–25688.
213. S. R. Yeandel, M. Molinari and S. C. Parker, *RSC Adv.*, 2016, **6**, 114069–114077.
214. S. D. Kang, M. Dylla and G. J. Snyder, *Phys. Rev. B*, 2018, **97**, 1–10.
215. T. M. Tritt, *Annu. Rev. Mater. Res.*, 2011, **41**, 433–448.
216. L. T. Hung, N. Van Nong, L. Han, R. Bjørk, P. H. Ngan, T. C. Holgate, B. Balke, G. J. Snyder, S. Linderoth and N. Pryds, *Energy Technol.*, 2015, **3**, 1143–1151.

217. W. Xie, J. He, H. J. Kang, X. Tang, S. Zhu, M. Laver, S. Wang, J. R. D. Copley, C. M. Brown, Q. Zhang and T. M. Tritt, *Nano Lett.*, 2010, **10**, 3283–3289.
218. T. M. Tritt and M. A. Subramanian, *MRS Bull.*, 2006, **31**, 188–198.
219. K. Biswas, J. He, Q. Zhang, G. Wang, C. Uher, V. P. Dravid and M. G. Kanatzidis, *Nat. Chem.*, 2011, **3**, 160–166.
220. B. Yu, M. Zebarjadi, H. Wang, K. Lukas, H. Wang, D. Wang, C. Opeil, M. Dresselhaus, G. Chen and Z. Ren, *Nano Lett.*, 2012, **12**, 2077–2082.
221. M. W. Gaultois, T. D. Sparks, C. K. H. Borg, R. Seshadri, W. D. Bonificio and D. R. Clarke, *Chem. Mater.*, 2013, **25**, 2911–2920.
222. S. Leblanc, S. K. Yee, M. L. Scullin, C. Dames and K. E. Goodson, *Renewable Sustainable Energy Rev.*, 2014, **32**, 313–327.
223. K. Koumoto, Y. Wang, R. Zhang, A. Kosuga and R. Funahashi, *Annu. Rev. Mater. Res.*, 2010, **40**, 363–394.
224. T. Souma, M. Ohtaki, M. Shigeno, Y. Ohba, N. Nakamura and T. Shimozaki, *Int. Conf. Thermoelectr. ICT, Proc.*, 2006, pp. 603–606.
225. J. Töpfer, T. Reimann, T. Schulz, A. Bochmann, B. Capraro, S. Barth, A. Vogel and S. Teichert, *Int. J. Appl. Ceram. Technol.*, 2018, **15**, 716–722.
226. I. Matsubara, R. Funahashi, T. Takeuchi, S. Sodeoka, T. Shimizu and K. Ueno, *Appl. Phys. Lett.*, 2002, **78**, 3627–3629.
227. R. Funahashi, M. Mikami, T. Mihara, S. Urata and N. Ando, *J. Appl. Phys.*, 2006, **99**, 2004–2007.
228. E. S. Reddy, J. G. Noudem, S. Hebert and C. Goupil, *J. Phys. D: Appl. Phys.*, 2005, **38**, 3751–3755.
229. R. Funahashi and S. Urata, *Int. J. Appl. Ceram. Technol.*, 2007, **4**, 297–307.
230. S. M. Choi, K. H. Lee, C. H. Lim and W. S. Seo, *Energy Convers. Manage.*, 2011, **52**, 335–339.
231. K. Park and G. W. Lee, *Energy*, 2013, **60**, 87–93.
232. P. Mele, H. Kamei, H. Yasumune, K. Matsumoto and K. Miyazaki, *Met. Mater. Int.*, 2014, **20**, 389–397.
233. C.-H. Lim, S.-M. Choi, W.-S. Seo and H.-H. Park, *J. Electron. Mater.*, 2011, **41**, 1247–1255.
234. H. Lee, R. C. Seshadri, S. J. Han and S. Sampath, *Appl. Energy*, 2017, **192**, 24–32.
235. N. Kanas, M. Bittner, T. D. Desissa, S. P. Singh, T. Norby, A. Feldhoff, T. Grande, K. Wiik and M. A. Einarsrud, *ACS Omega*, 2018, **3**, 9899–9906.
236. L.-D. Zhao, S.-H. Lo, Y. Zhang, H. Sun, G. Tan, C. Uher, C. Wolverton, V. P. Dravid and M. G. Kanatzidis, *Nature*, 2014, **508**, 373–377.
237. H. S. Kim, W. Liu, G. Chen, C.-W. Chu and Z. Ren, *Proc. Natl. Acad. Sci. USA*, 2015, **112**, 8205–8210.

238. D. Ekren, F. Azough, A. Gholinia, S. J. Day, D. Hernandez-Maldonado, D. M. Kepaptsoglou, Q. M. Ramasse and R. Freer, *J. Mater. Chem. A*, 2018, **6**, 24928–24939.
239. M. W. Gaultois, A. O. Oliynyk, A. Mar, T. D. Sparks, G. J. Mulholland and B. Meredig, *APL Mater.*, 2016, **4**, 053213.
240. M. W. Gaultois, *Design principles for oxide thermoelectric materials,* PhD Thesis, University of California Santa Barbara, 2015.
241. I. Terasaki, *APL Mater.*, 2016, **4**, 2–9.
242. M. T. Buscaglia, F. Maglia, U. Anselmi-Tamburini, D. Marré, I. Pallecchi, A. Ianculescu, G. Canu, M. Viviani, M. Fabrizio and V. Buscaglia, *J. Eur. Ceram. Soc.*, 2014, **34**, 307–316.
243. M. T. Dylla, J. J. Kuo, I. Witting and G. J. Snyder, *Adv. Mater. Interfaces*, 2019, **1900222**, 1–7.

Subject Index